Umweltrechtliche Studien
Studies on Environmental Law

Herausgegeben von / edited by
Professor Dr. Sabine Schlacke
Professor Dr. Gerd Winter

Band 46
Volume 46

Ingo Unterweger

International Law on Tuna Fisheries Management

Is the Western and Central Pacific Fisheries Commission ready for the challenge?

Nomos

The doctoral thesis and the print were funded by the DFG International Research Training Group INTERCOAST.

The Deutsche Nationalbibliothek lists this publication in the Deutsche Nationalbibliografie; detailed bibliographic data are available on the Internet at http://dnb.d-nb.de

a.t.: University of Bremen & University of Waikato (Hamilton, New Zealand), Diss., 2014

ISBN   978-3-8487-2286-0 (Print)
         978-3-8452-6391-5 (ePDF)

British Library Cataloguing-in-Publication Data
A catalogue record for this book is available from the British Library.

ISBN   978-3-8487-2286-0 (Print)
         978-3-8452-6391-5 (ePDF)

Library of Congress Cataloging-in-Publication Data
Unterweger, Ingo
International Law on Tuna Fisheries Management
Is the Western and Central Pacific Fisheries Commission ready for the challenge?
Ingo Unterweger
320 p.
Includes bibliographic references.

ISBN   978-3-8487-2286-0 (Print)
         978-3-8452-6391-5 (ePDF)

1. Edition 2015
© Nomos Verlagsgesellschaft, Baden-Baden, Germany 2015. Printed and bound in Germany.

To my parents

Acknowledgements

First of all, I would like to thank my PhD advisors Dr. Till Markus at the University of Bremen, Prof. Dr. Alexander Gillespie at the University of Waikato and Prof. Dr. Sabine Schlacke formerly of the University of Bremen and now at the University of Münster for their support, encouragement and patience. In particular, my thanks must go to Dr. Markus, who always provided clear and professional guidance. Moreover his positive energy was infectious and his motivational skills were exceptional. The same applies to Prof. Dr. Gillespie. During my time in Hamilton, I could always turn to him for sometimes unconventional but always very fruitful comments regarding my work. At the University of Bremen I am also grateful to Prof. Dr. Gerd Winter for all of his valuable contributions.

Special thanks must go to Prof. Glenn Hurry, Peter Flewelling and the staff of the Secretariat of the Western and Central Pacific Fisheries Commission, as well as to the friendly people of Pohnpei. During my research stay in Micronesia, they gave me the unique chance to get practical insights into international tuna fisheries management. I am also thankful to the German Federal Agency for Nature Conservation for supporting me with an official recommendation letter prior to my research stay.

I would also like to thank the Marum and INTERCOAST graduate school organisation and management, head, secretary financial stuff and post-docs for enabling me to undertake my research with their financial and advisory support. I am also very grateful to my colleagues at the FEU-Research Centre for European Environmental Law in Bremen and at the Faculty of Law in Hamilton for their friendship and support during my whole PhD project. In particular, I want to thank to my great office mate Dr. Lisa Marquardt as well as Dr. Bevis Fedder, Dr. Gerhard Bartzke and David Gehrmeyer.

Last, but by no means least, I want to thank my family and friends who have supported me throughout the writing process with important non-scientific advice. I am particularly indebted to my kind-hearted mother and friend Monika Unterweger who taught me to always see things positively as well as to my great Dad and best fishing mate Ulrich Unterweger.

Bremen, January 2015   Ingo Unterweger

Contents

Abbreviations

B_{MSY}	Stock biomass associated with MSY
CCM	Members, Cooperating Non-members and Participating Territories
CCSBT	Commission for the Conservation of Southern Bluefin Tuna
CMM	Conservation and Management Measure
CNM	Cooperating Non-Member
CPUE	Catch Per Unit Effort
DSU	Understanding on Rules and Procedures Governing the Settlement of Disputes
DWFN	Distant Water Fisheries Nation
EEZ	Exclusive Economic Zone
E-HSP	Eastern High Seas Pocket
FAC	Finance and Administration Committee
FAD	Fish Aggregating Device
FAME	Fisheries, Aquaculture and Marine Ecosystems Division
FAO	Food and Agriculture Organization
FFA	Pacific Islands Forum Fisheries Agency
FFC	Forum Fisheries Committee
F_{MSY}	Fishing mortality associated with MSY
FSA	Fish Stocks Agreement
FSMA	Federated States of Micronesia Arrangement for Regional Fisheries Access
IA	Arrangement implementing the Nauru Agreement
IATTC	Inter American Tropical Tuna Commission
ICCAT	International Commission for the Conservation of Atlantic Tuna
IGO	Inter-Governmental Organizations
IMO	International Maritime Organization
IOTC	Indian Ocean Tuna Commission
IPOA	International Plan of Action

IRCS	International Telecommunication Union Radio Call Signs
ISC	International Scientific Committee to study the tuna and tuna-like species of the North Pacific Ocean
ITU	International Telecommunication Union
IUU	Illegal, Unreported and Unregulated
MCS	Monitoring, control and surveillance
MHLC	Multilateral High-Level Conferences
MOU	Memorandum of Understanding
MSY	Maximum Sustainable Yield
NC	Northern Committee
NGO	Non-Governmental Organization
NPOA	National Plan of Action
OFP	Oceanic Fisheries Programme
PAE	Party Allowable Effort
PICTs	Pacific Island Countries and Territories
PNA	Parties to the Nauru Agreement
REIO	Regional Economic Integration Organization
RFMO	Regional Fisheries Management Organization
RFV	Register of Fishing Vessels
ROP	Regional Observer Programme
SC	Scientific Committee
SPC	Secretariat of the Pacific Community
TAC	Total Allowable Catch
TAE	Total Allowable Effort
TCC	Technical and Compliance Committee
TVMA	Te Vaka Moana Arrangement
TVTA	Te Vaka Toa Arrangement
UNCLOS	United Nations Convention on the Law of the Sea
VDS	Vessel Day Scheme
VMS	Vessel Monitoring System
WCPFC	Western and Central Pacific Fisheries Commission

WCPO   Western and Central Pacific Ocean

WIN   WCPFC Identification Number

Figures and tables

Figures

Tables

Introduction

Tuna and tuna-like species are scombrids and billfishes that occur for the most part in all oceans.[1] Most of them are migrating over long distances and cross national and international boundaries. As a consequence, many countries are interested in these economically important resources. The fisheries targeting them involve coastal states which are often developing countries as well as distant water fishing nations which are mainly developed countries fishing far away from their coasts.[2] For the growing global population tuna and tuna-like species, mainly sold as fresh or canned fish, represent an important source of food.[3] Ecologically they are a crucial link in the food web as they include some of the biggest marine predators.[4]

Industrial fisheries targeting tuna and tuna-like species started during the 1940s and 1950s.[5] Initially the fishing activities were mainly limited to the coastal zones before expanding to the high seas. Today, the fishing vessels are active in all parts of the ocean and the catches increased from less than 0,6 million metric tonnes in 1950 to 6.6 million metric tonnes (mt), or 8.5 percent of all marine capture fisheries (77.4 million mt) in 2010.[6] The commercially most important tuna species are Atlantic bluefin (*Thunnus thynnus*), Pacific bluefin (*Thunnus orientalis*), southern bluefin (*Thunnus maccoyii*), albacore (*Thunnus alalunga*), bigeye (*Thunnus obesus*), skipjack (*Katsuwonus pelamis*) and yellowfin tuna (*Thunnus albacores*). These so called 'principle market tuna species' can be further subdivided into 23 different stocks.[7] In 2010, they contributed about 4.3 million mt, or about 65 percent, to the total global catch of all tuna and tuna-like

---

1   B. B. Collette & C. E. Nauen, *Scombrids of the world : an annotated and illustrated catalogue of tunas, mackerels, bonitos, and related species known to date*, FAO species catalogue v. 2 (Rome: FAO, 1983); I. Nakamura, *Billfishes of the world: An annotated and illustrated catalogue of marlins, sailfishes, spearfishes and swordfishes known to date*, FAO species catalogue - Vol. 5 (Rome: FAO, 1985); See also http://www.fao.org/fishery/topic/16082/en.

2   J. Majkowski, *Global fishery resources of tuna and tuna-like species*, FAO Fisheries and Aquaculture Technical Paper - Vol. 483 (Rome: FAO, 2007), pp.16-17.

3   M. Miyake, et al., *Recent developments in the tuna industry*, FAO Fisheries and Aquaculture Technical Paper - Vol. 543 (Rome: FAO, 2010), pp.61-110.

4   B. B. Collette, et al., High Value and Long Life—Double Jeopardy for Tunas and Billfishes, *Science - Vol. 333* (2011), p.291; D. Pauly, et al., Fishing down marine food webs, *Science - Vol. 279(860)* (1998), pp.860-863.

5   Majkowski, 2007, p.7.

6   FAO, *The State of World Fisheries and Aquaculture 2012* (Rome: FAO, 2012), pp.1-104.

7   Majkowski, 2007, p.26.

species. The value-at-landing of the 2010 catch was more than US$10 billion.[8] In 2013 one single pacific bluefin tuna which is one of the most expensive species was even sold for US$ 1.7 million.[9] The three main fishing gears catching tuna and tuna-like species are purse seine, longline as well as pole and line.[10] Among these gear types, in recent years the purse-seine catches have shown the largest increase, from almost nil in 1950 to 2.2 million tonnes in 2000.[11] The other two gear types remained comparatively stable during this period.

The migratory nature of tuna and tuna-like species makes it necessary that all countries that are involved in the fisheries cooperate in conservation and management. Therefore since the second half of the 20th century a comprehensive but also highly fragmented international legal framework has developed including several provisions that require regional cooperation. The most important legal instruments are the United Nations Convention on the Law of the Sea[12] and the United Nations Agreement for the Implementation of the Provisions of the United Nations Convention on the Law of the Sea of 10 December 1982 relating to the Conservation and Management of Straddling Fish Stocks and Highly Migratory Fish Stocks (Fish Stocks Agreement)[13]. These treaties explicitly require cooperation in the management of transboundary and straddling stocks as well as of highly migratory species. Other instruments with respective provisions are the Agreement to Promote Compliance with International Conservation and Management Measures by Fishing Vessels on the High Seas (Compliance Agreement)[14], the FAO Code of Conduct for Responsible Fisheries[15], the FAO International Plans of Action[16] and the Agreement on Port

---

8   FAO, *TUNA - A global perspective* (Rome: FAO, 2013).

9   In 2013 a Pacific bluefin tuna was sold in Japan for 155 million yen ($1.7million), http://www.bbc.co.uk/news/world-asia-20919306.

10   C. Nédélec & J. Prado, *Definition and classification of fishing gear categories*, FAO Fisheries Technical Paper - No. 222 (Rome: FAO, 1990).

11   M. P. Miyake, et al., *Historical trends of tuna catches in the world*, FAO Fisheries and Aquaculture Technical Paper - Vol. 467 (Rome: FAO, 2004), p.5.

12   Third United Nations Conference on the Law of the Sea, 1973-1982.

13   Agreement for the Implementation of the Provisions of the United Nations Convention on the Law of the Sea of 10 December 1982 relating to the Conservation and Management of Straddling Fish Stocks and Highly Migratory Fish Stocks, 1995.

14   Agreement to Promote Compliance with International Conservation and Management Measures by Fishing Vessels on the High Seas, 1993.

15   Code of Conduct for Responsible Fisheries, 1995.

State Measures to Prevent, Deter and Eliminate Illegal, Unreported and Unregulated Fishing (Port State Measures Agreement)[17].

Today all fisheries targeting tuna and tuna-like species are governed by Regional Fisheries Management Organizations (tuna RFMOs) which have been developed in accordance with the provisions of the international legal instruments. Overall, there are five tuna RFMOs, which together cover the whole migratory range of tuna and tuna-like species.[18] The Inter-American Tropical Tuna Commission (IATTC) in the eastern Pacific ocean; the International Commission for the Conservation of Atlantic Tunas (ICCAT) in the Atlantic ocean; the Indian Ocean Tuna Commission (IOTC) in the Indian ocean; the Western and Central Pacific Fisheries Commission (WCPFC) in the western and central Pacific ocean; and the Commission for the Conservation of Southern Bluefin Tuna (CCSBT) in the areas where southern bluefin tuna occurs. The main duties of these intergovernmental organizations include the adoption of conservation and management measures for the regulation of all fisheries related activities, scientific research on the relevant target and non-target species, and the monitoring of compliance with fishing regulations.

So far, the overall performance of the tuna RFMOs is rather limited.[19] The main problem is that several tuna and tuna-like species, particularly some of the most commercially important ones have been harvested beyond their sustainable limits for too long and that the total capacity of the current fishing fleets is too great to allow these fish stocks to recover.[20] Since 1950 tuna catches have in-

---

16   International Plan of Action for Reducing Incidental Catch of Seabirds in Longline Fisheries, 1999; International Plan of Action for the Conservation and Management of Sharks, 1999; International Plan of Action for the Management of Fishing Capacity, 1999; International Plan of Action to Prevent, Deter, and Eliminate Illegal, Unreported and Unregulated Fishing (IUU), 2001.

17   Agreement on Port State Measures to Prevent, Deter and Eliminate Illegal, Unreported and Unregulated Fishing, 2009.

18   Convention for the Conservation of Southern Bluefin Tuna, 1993; Agreement for the Establishment of the Indian Ocean Tuna Commission, 1993; Convention for the Strengthening of the Inter-American Tropical Tuna Commission established by the 1949 Convention between the United States of America and the Republic of Costa Rica "Antigua Convention", 2003; International Convention for the Conservation of Atlantic Tunas, 1966; Convention on the Conservation and Management of Highly Migratory Fish Stocks in the Western and Central Pacific Ocean, 2000.

19   A. Cox, et al., *Strengthening regional fisheries management organisations* (Paris: OECD, 2009), p.17.

20   J. Joseph, *Managing Fishing Capacity of the World Tuna Fleet*, FAO Fisheries Circular - No. 982 (Rome: FAO, 2003).

creased continuously.[21] As a consequence, in 2009, about 33 percent of the principal market tuna species were classified as overexploited and about 38 percent as fully exploited, while only 29 percent were not fully exploited.[22]

The degree of exploitation varies depending on where the catches are made. In the areas of competence of WCPFC and IOTC comparatively few stocks are fished beyond sustainable limits. In the western and central Pacific Ocean for example bigeye tuna is currently the only principal market tuna species which is overexploited: all other tuna stocks are either fully or moderately exploited.[23] In contrast, IATTC, ICCAT and CCSBT are struggling in keeping catches within sustainable limits. Both in the Atlantic Ocean and in the eastern Pacific Ocean there are at least two overexploited principal market tuna species.[24] In particular, the stocks managed by ICCAT and CCSBT are rather overfished and overfishing is still occurring. Generally, it can be noted that, at present, the latter organizations have to rebuild several of their stocks while IOTC and WCPFC are facing the different challenge of maintaining future catches at sustainable levels.

Besides the stock status there are other reasons why the WCPFC is different from the other tuna RFMOs. One reason is the legal framework the RFMO is based on. The WCPFC only has been established in 2004 and its Convention draws on the most modern international treaties for the management of tuna and tuna-like species. Another reason is the unique characteristics of the western and central Pacific tuna fisheries. The western and central Pacific Ocean is globally the most important catch area for tuna and tuna like species. In 2009, the catch of the principal market tuna species was almost 2.5 million mt, representing more than half of the global catch of these species.[25] Unique is also the role of the developing Pacific Island countries and territories which are relying heavily on the revenues from the tuna fisheries. Already prior to the establishment of the WCPFC these countries and territories had developed a powerful

---

21   Majkowski, 2007, pp.11-18.

22   FAO, 2012, pp.53-54.

23   WCPFC, *Summary Report - Seventh Regular Session of the Scientific Committee*, (2011) pp.20-36.

24   A. Aires-da-Silva & M. N. Maunder, *Status of Yellowfin Tuna in the Eastern Pacific Ocean in 2010 and Outlook for the Future* (La Jolla, USA: IATTC, 2011); A. Aires-da-Silva & M. N. Maunder, *Status of Bigeye Tuna in the Eastern Pacific Ocean in 2010 and Outlook for the Future* (La Jolla, USA: IATTC, 2011); M. N. Maunder, *Updated Indicators of Stock Status for Skipjack Tuna in the Eastern Pacific Ocean* (La Jolla: IATTC, 2011); ICCAT, *Report for Biennial Period, 2010-11 PART II (2011) - Vol. 2 (Standing Committee on Research & Statistics)*, (2012).

25   WCPFC, *Summary Report - Seventh Regular Session of the Scientific Committee*, (2011), p.3.

subregional legal framework for the management of the fisheries within their areas of national jurisdiction.

Research objectives and structure

The current legal performance of the tuna RFMOs against the backdrop of the economical and ecological importance of tuna and tuna-like species requires a detailed evaluation. The overall objective of this work is therefore to support a broader understanding of the international legal framework for the management of fisheries targeting tuna and tuna-like species and to examine whether or not this framework enables tuna RFMOs and in particular the WCPFC to manage their fisheries sustainably.

The first chapter gives an introduction into the international legal framework for the management of fisheries targeting tuna and tuna-likes species. The purpose is to reveal the scope of the relevant international legal instruments and to identify provisions on cooperative conservation and management. The second chapter addresses the two international disputes on southern bluefin tuna and south-eastern pacific swordfish which required binding dispute settlement procedures. The crucial aspects of the disputes are presented and their influence on potential future cases is analyzed. Chapter three provides a comparative analysis of the five tuna RFMOs with regard to their current legal performance. The main areas of comparison are organizational structure, membership, accession, decision making, settlement of disputes, regulation of total catch and fishing mortality, allocation criteria, application of the precautionary approach as well as compliance and enforcement. Chapter four provides an in-depth analysis of the current international legal framework for fisheries management in the WCPO. The underlying question to be answered is whether the comparatively good stock status in the WCPO has anything to do with the RFMO management. It is intended to identify crucial provisions and to highlight any remaining challenges. The book ends with a final conclusion listing the main findings and indicating some ideas with regard to future management of tuna and tuna-like species.

Chapter I

The international law on tuna fisheries management

A. Introduction

The management of fisheries targeting tuna and tuna-like species is based on a complex international legal framework that regulates most of the issues related to human fishing activities. This framework determines, among other things, who is allowed to fish, what kind of fish can be caught, where fishing operations are allowed and how fishing must be conducted. It includes customary law, legally-binding treaties and non-binding, voluntary instruments. The goal of this chapter is to analyze the scope of these instruments and to identify which provisions are crucial for the sustainable management of tuna and tuna-like species. The fundamental instrument is the United Nations Convention on the Law of the Sea (UNCLOS). It is described as the 'foundation for the modern law relating to international fisheries' and it contains several provisions that apply to the fisheries of tuna and tuna-like species.[26] Although more modern and more specific instruments have been established since UNCLOS, it remains the basis for all other related instruments. The relevant post-UNCLOS instruments include the Fish Stock Agreement, which is the most specific treaty for the management of tuna and tuna-like species, as well as other instruments such as the Compliance Agreement, the Code of Conduct, the International Plans of Action and the Port States Measures Agreement, all of which also contain important provisions.

B. Law of the Sea Convention

A growing trend since Grotius' *Freedom of the Seas* had been the erosion of the freedom of fishing in favor of increased national jurisdiction over maritime areas and their fish resources.[27] The Truman Proclamation[28] of 1949, which, among others things,[29] addressed the conservation of coastal fisheries in certain areas

---

26   P. W. Birnie, et al., *International law and the environment* (New York: Oxford University Press, 2009), p.714; W. T. Burke, *Impacts of the UN Convention of the Law of the Sea on tuna regulation*, FAO legislative study (Rome: FAO, 1982).

27   H. Grotius, *Mare Liberum, sive de jure quod Batavis competit ad Indicana commercia dissertatio* (Lugduni Batavorum: Ludovici Elzevirij, 1609).

28   Truman Proclamation on Policy of the United States With Respect to the Natural Resources of the Subsoil and Sea Bed of the Continental Shelf, 1945.

29   Other concerns were the natural resources of subsoil and seabed of the Continental Shelf.

of the high seas[30], represented one of the most important post-World War II starting points for the expansion of sovereign rights and jurisdiction over sea areas.[31] Officially, the Proclamation was not intended to move towards an extension of sovereignty, but de facto it was a first step towards the establishment of the Exclusive Economic Zone (EEZ), something which became very important for all fisheries. By 1947, the Truman Proclamation had given Chile and Peru the confidence to make, after disputes with distant water fisheries nations over the access to their waters, unilateral proclamations to extend their sovereignty up to 200 nautical miles (nm) off land.[32] The following Santiago Declaration in 1952 was then the first international instrument to proclaim a 200 nm limit.[33] Other Latin American, African and Asian states followed with similar approaches.[34] Despite these numerous unilateral approaches there was no international agreement and no custom on a legal regime for the 200 mile zone until the third Conference on the Law of the Sea.

In 1956, the first of three United Nations Conferences on the Law of the Sea was held in Geneva.[35] During this conference the participating countries made progress in the codification of international maritime law, including fisheries law. The result of the conference was four treaties: The Convention on the High Seas (in force since 1962);[36] the Convention on the Territorial Sea and Contiguous

---

30   With regard to fisheries, it was intended to protect certain national fisheries such as the salmon fishery in the North East Pacific Ocean.

31   Above all, the USA particularly recognized the worldwide need for new sources of oil and other minerals, and expressed their interest in discovering and making available new supplies of these resources.

32   Presidential Declaration Concerning Continental Shelf, 1947; Presidential Decree No. 781, 1947.

33   Declaration on the Maritime Zone (Santiago Declaration), 1952; See also FAO, *The Exclusive Economic Zone: A historical perspective*, Essays in memory of Jean Carroz (Sri Lanka: FAO, 1987).

34   The Montevideo Declaration on the Law of the Sea 1970; The Declaration of Latin American States on the Law of the Sea (The Lima Declaration) 1970; The Declaration of Santo Domingo, 1972; Report of The Thirteenth Session of the Asian-African Consultative Committee, 1972; Conclusions in the General Report of the African States Regional Seminar on the Law of the Sea (Yaoundé Conclusions), 1972; Declaration of the Organization of African Unity on the "Issues of the Law of the Sea" (Addis Ababa Declaration), 1973. See also G. Lugten, *The impact of Extra Legal Factors in the Historical Development of International Fisheries Law* (Hobart: University of Tasmania, 1996), pp.186-216.

35   United Nations Conference on the Law of the Sea, 1958.

36   Convention on the High Seas, 1958.

24

Zone (in force since 1964);[37] the Convention on the Continental Shelf (in force since 1964);[38] and the Convention on Fishing and Conservation of Living Resources of the High Seas (in force since 1966).[39] These treaties did not deal exclusively with fisheries law, but all of them contained relevant provisions. The Convention on the High Seas had greatest impact on tuna fisheries. It reiterated the freedom of fishing on the high seas but it also required reasonable regard to the interests of other states in the exercising of this freedom.[40] The Convention on Fishing and Conservation of Living Resources of the High Seas included some progressive provisions but it did not become as significant as the other treaties.[41] It came into force last and had far fewer signatories.[42] In 1960, a second Conference on the Law of the Sea was held in Geneva, but no significant progress with regard to fisheries law was made and no new agreements were reached.[43]

In the early 1970s, there was growing dissatisfaction with the existing international legal framework for fisheries management. Developing coastal states felt resentment against distant water fishing nations (DWFNs) which were fishing in comparatively short distance from their coasts.[44] The developed countries on the other hand wanted to increase their access and control over offshore fishery resources. There was also skepticism regarding the ability of Regional Fisheries Management Organizations (RFMOs) to manage fisheries in the face of increasing fishing pressure. At that time two tuna RFMOs had been established so far: The Inter American Tropical Tuna Commission (IATTC) in 1950 and the International Commission for the Conservation of Atlantic Tunas (ICCAT) in 1969. Due to the expansion of fisheries in the eastern Atlantic and the development of new offshore fishing grounds in the eastern Pacific the biggest increases in tuna catches had taken place within their areas of competence.[45]

All fishing states had in common that they wanted to end the disputes over the extent of coastal state jurisdiction. Between 1972 and 1974 territorial disputes of

---

37   Convention on the Territorial Sea and the Contiguous Zone, 1958.

38   Convention on the Continental Shelf, 1958.

39   Convention on Fishing and Conservation of the Living Resources of the High Seas, 1958.

40   Convention on the High Seas, Art.2.2.

41   R. Churchill & A. V. Lowe, *The law of the sea* (Manchester: Manchester University Press, 1999), p.287.

42   E. J. Molenaar, Non-Participation in the Fish Stocks Agreement: Status and Reasons, *The International Journal of Marine and Coastal Law - Vol. 26(2)* (2011), pp.196-197.

43   Second United Nations Conference on the Law of the Sea, 1960.

44   Churchill & Lowe, 1999, pp.287-288.

45   Majkowski, 2007, p.11.

Iceland versus the United Kingdom and Northern Ireland and Iceland versus Federal Republic of Germany even had to be settled by the International Court of Justice.[46] However, despite the increasing overall awareness of the need for new instruments regulating fisheries it has to be considered that the main driver for formal international negotiations after the second Conference on the Law of the Sea was the growing interest in seabed resources.[47]

From 1973 to 1982 the third United Nations Conference on the Law of the Sea took place over several sessions in both New York and Geneva.[48] Even though the initial focus had been on the seabed, there was also an important debate on rights and responsibilities with regard to fish resources. The participating countries made proposals defining different legal zones depending on their particular interest and, among other things, they discussed the possible legal regimes for fisheries within these zones. A type of EEZ had been proposed at an early stage of the third conference, and it was clear that the new treaty would contain provisions relevant to EEZs, including preferential fishing rights.[49]

On 10 December 1982, the United Nations Convention on the Law of the Sea (UNCLOS) was adopted, and it entered into force on 16 November 1994. The new Convention represented a comprehensive reform of maritime law, establishing rules to govern all uses of the oceans and its resources. In addition, it constituted – to a large extend – a codification of existing and emerging international customary law. It contained new provisions for environmental protection and, for the first time, provisions for an international court for maritime disputes, the International Tribunal for the Law of the Sea.[50] With respect to fisheries, coastal states gained sovereign rights and jurisdiction over most of the fish

---

46   United Kingdom of Great Britain and Northern Ireland v. Iceland, International Court of Justice, (1974); Federal Republic of Germany v. Iceland, International Court of Justice, (1974).

47   The recognition of manganese nodules as a potential ore source raised concerns about a possible parcelling out of the ocean. Questions with regard to the different claims of territorial waters were raised by the Ambassador of Malta to the United Nations, Arvid Pardo, in 1967. He pointed out that threatening conflicts concerning the oceans should be solved by an effective international regime, proposing, among others, the declaration of the seaward side of the continental shelf as common heritage of mankind.

48   Third United Nations Conference on the Law of the Sea.

49   http://www.un.org/law/diplomaticconferences/.

50   G. Eiriksson, *The International Tribunal for the Law of the Sea* (The Hague: Martinus Nijhoff Publishers, 2000); W. G. Vitzthum, *Handbuch des Seerechts* (München: C. H. Beck, 2006), p.41.

resources in their EEZs, and the principles of sustainable and shared use were implemented on a global scale for the first time.[51]

For fisheries targeting tuna and tuna-like species particularly the zonal regulations and the provisions for the conservation and management of highly migratory species were important. Already during the negotiations a need to cooperate at an international level for the conservation and management of tuna and tuna-like species had been identified.[52] The USA proposed that, due to their highly migratory nature, stock-specific management should be introduced, while other countries, such as Canada, India, Kenya and Sri Lanka, favored regional regulations for exploration, exploitation, conservation and development.[53] In the end, UNCLOS is reflecting the different positions. Annex I provides an explicit list of highly migratory species including the most important tuna and tuna like species and Article 64 requires cooperation in ensuring conservation and promoting the objective of optimum utilization of these species both within and beyond the EEZ.[54]

UNCLOS is currently ratified by 162 countries.[55] It is remarkable that six of the 20 most important (in terms of total catch weight) tuna fishery nations have not yet ratified the Convention. Those countries are: Taiwan (2nd most important tuna fishery nation), Ecuador (9th), Iran (12th), the USA (13th), Venezuela (15th) and Colombia (17th).[56] However, the lack of ratification does not necessarily

---

51   Further information on the establishment of the EEZ: A. L. Hollick, The Origins of 200-Mile Offshore Zones, *The American Journal of International Law - Vol. 71(3)* (1977); M. Dahmani, *The fisheries regime of the exclusive economic zone*, Publications on ocean development - No. 11 (Dordrecht: Martinus Nijhoff, 1987), pp.38-42; D. J. Attard, *The exclusive economic zone in international law* (New York: Oxford University Press, 1987), pp.1-31; F. Orrego Vicuña, *The exclusive economic zone: regime and legal nature under international law*, Cambridge studies in international and comparative law. New series (Cambridge: Cambridge University Press, 1989), pp.3-15.

52   R. L. Allen, *International management of tuna fisheries - Arrangements, challenges and a way forward*, FAO Fisheries and Aquaculture Technical Paper - Volume 536 (Rome: FAO, 2010), p.1.

53   U.N. Document A/AC.138/SC.II/L.9, section III SBC Report 1972, at 175, 176 (USA), 1972; U.N. Document A/AC.138/SC.II/L.38, article 10, reproduced in III SBC Report 1973, at 82,84 (Canada, India, Kenya and Sri Lanka), 1973.

54   M.H. Nordquist, *United Nations Convention on the Law of the Sea, 1982: A Commentary Volume II Article 1 to 85 Annexes I & II Final Act, Annex II* (Dordrecht: Martinus Nijhoff, 1993), p.649.

55   See the website of the UN Division for Ocean Affairs and the Law of the Sea.

56   J. Majkowski, *Global fishery resources of tuna and tuna-like species* (Rome: Food and Agriculture Organization of the United Nations, 2007), p.16.

mean that the respective states are not at all bound by the rules and principles of UNCLOS. Many of these rules and principles which are relevant for the management of fisheries targeting tuna and tuna-like species express or indicate existing obligations within international customary law or they are restated in other treaties to which the respective countries are parties.

1.   Zonal regulations

Tuna and tuna-like species migrate over long distances with no regard to national boundaries. Depending on the species, age and season, they can be found in all of the legal zones defined by UNCLOS. Because of this constant migration through EEZs and between the EEZs and the high seas, the most relevant provisions for tuna and tuna-like species are laid down in Part V and Part VII, Section 2, of UNCLOS. Part V deals with the rights and obligations of states within the EEZs and the areas beyond and adjacent to it. Part VII, Section 2, addresses the conservation and management of the living resources on the high seas. Further relevant provisions can be found in Part IV, which deals with archipelagic waters, which are important as nursery grounds for tuna and tuna-like species, as well as in Part II, which covers internal waters and territorial seas, although these are less important because of the lower levels of tuna and tuna-like species present. In order to understand the whole context of zonal regulation this sub-chapter will outline the most important fisheries-related provisions that apply within each zone.

a)   Internal waters and territorial sea

Internal waters are the waters on the landward side of the baseline of the territorial sea.[57] The baseline is the outer limit of the internal waters, marked by the low-water line along the coast.[58] The territorial sea extends up to a limit not exceeding 12 nautical miles, measured from the baselines.[59]

Internal waters and the territorial sea are in comparison to the Exclusive Economic Zone (EEZ) and the high seas not crucial for tuna fisheries management as only a minor part of the catch of the principal market tuna species is made in this area.[60] However, especially juvenile tunas are also present in an area which is comprised by the territorial sea. A study for the Philippine waters has shown

---

57   United Nations Convention on the Law of the Sea, 1982, Art.8.

58   Further definitions of other baselines can be found in Arts.6-14.

59   United Nations Convention on the Law of the Sea, Art.3.

60   Statement is based on a personal communication of the author with John Hampton, the Manager of the Oceanic Fisheries Programme of the Secretariat of the Pacific Community.

that smaller yellowfin tuna mainly occur between 0 and 19 nautical miles from the coast.[61]

Article 2 of the UNCLOS provides that the coastal state exercises full sovereignty over the airspace, seabed, subsoil and water column of the territorial sea, including exclusive rights for fishing. However, this sovereignty has to be exercised subject to UNCLOS and to other rules of international law.[62] That also includes conservation conventions to which the state is party and which apply in the territorial sea.[63] Coastal states may further adopt laws and regulations to conserve marine living resources, to prevent non-compliance and to preserve the environment.[64] Unlike the provisions for the EEZ, the coastal state is not obliged to provide access to surplus fish stocks within these areas.[65] All the fishing activities of foreign flagged vessels within the internal waters and territorial sea require the permission of the coastal state.[66] Despite this sovereignty, all states enjoy the right of 'innocent passage' through territorial seas.[67] Fishing activities are explicitly excluded from the definition of innocent passage.[68]

b)   Archipelagic states

Archipelagic states are coastal states which are constituted by one or more archipelagos,[69] possibly including other islands.[70] As with internal waters, the archipelagic states' sovereignty is extending to the air space over the archipelagic waters, the seabed and subsoil, and the resources therein.[71] Generally it can be recognized that the archipelagic States are controlling fishing in vast are-

---

61   M. Yesaki, *Observations on the biology of yellowfin (Thunnus albacares) and skipjack (Katsuwonus pelamis) tunas in Philippine waters*, FAO./UN D P. Indo-Pac. Tuna Dev. Mgt.Programme (Rome: FAO, 1983), p.12.

62   United Nations Convention on the Law of the Sea, Art.2.3.

63   Birnie, et al., 2009, p.716.

64   United Nations Convention on the Law of the Sea Arts.21 d, e and f.

65   Ibid. Art.62(2).

66   W. T. Burke, *The new international law of fisheries: UNCLOS 1982 and beyond* (Oxford: Clarendon Press, 1994), p.39.

67   United Nations Convention on the Law of the Sea Art.17.

68   Ibid. Art.19(2)(i).

69   An archipelago defines under UNCLOS to "a group of islands, including parts of islands, interconnecting waters and other natural features which are so closely interrelated that such islands, waters and other natural features form an intrinsic geographical, economic and political entity, or which historically have been regarded as such"; ibid. Art.46(a).

70   Ibid. Art.46(a).

71   Ibid. Art.49(2) and Art.2.

as of the sea.[72] Archipelagic waters are, like internal waters, on the landward side of the baseline.[73] The definition of baselines is of major importance with regard to archipelagic waters. The breadth of territorial sea and the EEZ on the seaward side is measured from the archipelagic baseline. The maximum length of this baseline - which is drawn between the outermost points of the archipelago above sea level - cannot exceed 100 nautical miles or, in exceptional cases, 125 nautical miles.[74]

The distribution of catches of yellowfin and bigeye tuna in the Western and Central Pacific Ocean, highlights the importance of archipelagic states.[75] Particularly the vast archipelagic waters of Indonesia and the Philippines are playing a crucial role for breeding and during the early stages of life of yellowfin and bigeye tuna.[76] Due to the migratory nature of these species fishing activities in archipelagic waters have the potential to affect fish stock status even in regions which are far beyond the archipelagic waters. Therefore, if an archipelagic state does not meet the conservation needs, the exercise of its sovereign rights can become a problem for other states which are interested in the same fish stocks.[77] A potential limitation of exclusive control is provided by Article 51 which states that an archipelagic State, without prejudice to its sovereignty over the resources and the area concerned "shall respect existing agreements with other States and shall recognize traditional fishing rights and other legitimate activities of the immediately adjacent neighbouring States in certain areas falling within archipelagic waters".[78] However, the impact of this recognition is rather limited as it excludes all DWFN and those neighbouring states which are not immediately adjacent.[79]

---

72   Birnie, et al., 2009, p.716.

73   United Nations Convention on the Law of the Sea Art.48.

74   Ibid. Art.47(2); Up to three percent of the total number of baselines enclosing any archipelago may exceed the 100 nautical miles up to a maximum length of 125 nautical miles.

75   A. Langley, et al., *Stock assessment of yellowfin tuna in the western and central Pacific Ocean (Rev.1 - 03 August 2011)* (Pohnpei: WCPFC, 2011), Figure 6; N. Davies, et al., *Stock assessment of bigeye tuna in the western and central Pacific Ocean* (Pohnpei: WCPFC, 2011), Figure 4.

76   M. Bailey, et al., Towards better management of Coral Triangle tuna, *Ocean & Coastal Management - Vol. 63* (2012), p.31.

77   Q. Hanich, Distributing the bigeye conservation burden in the western and central pacific fisheries, *Marine Policy - Vol. 36(2)* (2012), p.328.

78   United Nations Convention on the Law of the Sea, Art.51.

79   Burke, 1982, p.17.

c)   Exclusive Economic Zone

The Exclusive Economic Zone (EEZ) is located beyond and adjacent to the internal waters and extends 200 nautical miles from the baseline.[80] This zone represents the most important catch area for tuna and tuna-like species. About 60 percent of commercially-exploited tuna species are caught within EEZs.[81] The EEZs in the WCPO are particularly important for the global tuna production. In 2007, more than one million metric tonnes of the tuna species were caught within EEZs of Pacific Island countries.[82]

Coastal states have sovereign rights and jurisdiction within their own EEZs. This includes the right to explore and exploit, conserve and manage any fish resources,[83] a right which is, however, subject to several responsibilities. States which are exercising their sovereign rights shall have due regard to the rights and duties of other States and shall act in a manner compatible with the provisions of UNCLOS.[84] Coastal states also have to promote the objective of optimum utilization of the fish resources within their EEZ.[85] As part of this, they are required to determine a total allowable catch (TAC), and to allocate it between their own nationals and those of third countries which are entitled to fish in their EEZ.[86]

With regard to the management of highly migratory species like the tuna and tuna-like species Article 64 of UNCLOS requires international cooperation including the contribution and exchange of scientific information, catch and fishing effort statistics, and other data relevant to the conservation of highly migratory species.[87] The coastal state has the power and jurisdiction to determine the TAC as long as a stock is not endangered by overexploitation and as long as the obligations under Article 64 are respected.[88]

Unlike in the internal waters and territorial sea, there are provisions for the EEZ which, under certain circumstances, require coastal states to grant access to the

---

80   United Nations Convention on the Law of the Sea Arts.55 and 57.

81   R. L. Allen, et al., *Conservation and management of transnational tuna fisheries* (Ames, Iowa: Blackwell, 2010), p.3.

82   R. Gillett, *Marine fishery resources of the Pacific Islands*, FAO Fisheries and Aquaculture Technical Paper - Vol. 537 (Rome: FAO, 2010), p.38.

83   Art.56(1)(a).

84   United Nations Convention on the Law of the Sea, Art.56(2).

85   Ibid. Art.62(1).

86   Ibid. Art.61(1).

87   United Nations Convention on the Law of the Sea, Art.61(5).

88   Ibid. Art.61(2).

surplus of the total allowable catch (TAC). A coastal state with insufficient capacity to harvest the entire TAC shall give other states, through agreements or arrangements, access to any surplus.[89] Today most fish species, including tuna and tuna-like species, are at least fully utilized and the existing fishing capacities of coastal states are often greater than would be necessary to harvest up to the TAC. As a result, the provisions on the mandatory granting of access are rarely applied in practice.[90] Instead, particularly in the WCPO, voluntary agreements providing access to states other than the coastal state are very common. In this area nationals of states given access to the fishing resources must pay for the access. While fishing they have to comply with all requirements established in the laws and regulations of the coastal state.[91] Access is also granted via agreements between coastal states and private foreign entities.[92] In the WCPO a commonly applied approach is charter arrangements. Usually, local companies from small island developing states charter vessels from DWFNs for fishing operations in order to develop their tuna fisheries.[93] A critical aspect of all bilateral arrangements is that the conditions of most of them are often not transparent and accessible for the public.[94]

Within the EEZ the coastal State has to "promote the objective of optimum utilization of the living resources" without prejudice to the conservation requirements.[95] Conservation and management measures have to be designed in order to achieve the maximum sustainable yield (MSY).[96] This is defined as the largest amount of fish that can be taken continuously from a self-regenerating stock while still maintaining the average size of the stock.[97] The MSY is achieved when both fishing mortality and recruitment to the stock are maximized at the same time.[98] Despite the fact that the MSY is not defined by UNCLOS, it can be

---

89   Ibid. Art.62(2).

90   J. Joseph, et al., 'Addressing the Problem of Excess Fishing Capacity in Tuna Fisheries', *Conservation and Management of Transnational Tuna Fisheries* (Iowa: Wiley-Blackwell, 2010), p.19.

91   United Nations Convention on the Law of the Sea, Art.62(4).

92   OECD, *Fishing for Coherence - Proceedings of the Workshop on Policy Coherence for Development in Fisheries* (Paris: OECD Publishing, 2006), pp.73-102.

93   Ibid. p.224.

94   E. Havice, The structure of tuna access agreements in the Western and Central Pacific Ocean: Lessons for Vessel Day Scheme planning, *Marine Policy - Vol. 34(5)* (2010).

95   United Nations Convention on the Law of the Sea Art.61(2).

96   Ibid. Art.61(3).

97   Birnie, et al., 2009, pp.590-593.

98   H. Scott Gordon, The Economic Theory of a Common-Property Resource: The Fishery, *Journal of Political Economy - Vol. 62(2)* (1954), pp.124-142.

described as the fundamental concept underpinning all important management decisions with regard to fisheries in the EEZs.[99] In principle, no catch limit beyond the MSY is in line with UNCLOS.[100] However, fishing beyond the MSY could be permitted due to specific social, environmental or economic reasons.[101] In this work it is assumed that coastal states are legitimately allowed to exceed MSY levels within the EEZ, but only as long as the overexploitation does not endanger the fish stocks.[102] However, it has also to be recognized that some authors have decided to assume that the exploitation limit provided under UNLCOS is not only violated where stocks are endangered by overexploitation but already where states continuously fish beyond MSY.[103]

In practice it can be seen that there is a significant number of tuna and tuna-like species for which the MSY levels have been exceeded despite the provisions of UNCLOS. In the past this was especially the case for the stocks managed by ICCAT and CCSBT but also for some of the stocks managed by IATTC and the more recently established IOTC and WCPFC.[104]

d)   High seas

The high seas comprise all parts of the ocean apart from EEZs, territorial seas, internal waters and archipelagic waters.[105] Not included by this term is the "Area" which includes the seabed and ocean floor and subsoil thereof, beyond the limits of national jurisdiction.[106] The high seas play an important role for the sustainable management of tuna and tuna-like species. About 40 percent of

---

99   Burke, 1994, p.44.

100   S. B. Kaye, *International fisheries management* (The Hague: Kluwer Law International, 2001), pp.100-101; J. F. Caddy & R. Mahon, *Reference points for fisheries management* FAO Fisheries Technical Paper - Volume 347 (Rome: FAO, 1995).

101   United Nations Convention on the Law of the Sea, Art.61(3).

102   Churchill & Lowe, 1999, p.289; J.A. de Yturriaga, *The International Regime of Fisheries: from UNCLOS 1982 to the Presidential Sea* (The Hague: Kluwer Law International, 1997), pp.118-119; Burke, 1994, p.53; R. Barnes, 'The Convention on the Law of the Sea: An Effective Framework for Domestic Fisheries Conservation?', *The Law of the Sea Progress and Prospects* (New York: Oxford University Press, 2006), pp.242-243.

103   See, for example, T. Markus, *European Fisheries Law - From Promotion to Management* (Groningen: Europa Law Publishing, 2009), p.277.

104   For detailed information on the stock status in the tuna RFMOs see chapter 3.

105   United Nations Convention on the Law of the Sea, Art.86.

106   Ibid., Art.1.1(1)

commercially-exploited tuna species are caught in these waters. [107] In principle, all states enjoy the freedom of fishing on the high seas.[108] However, there are limitations: the freedom of fishing has to be exercised with due regard for the interests of other states[109] and in accordance with the relevant provisions of UNCLOS and related treaties.[110] In addition, it is subject to the rights, duties and interests for coastal states regarding straddling stocks, highly migratory species and other species groups.[111] The provisions with particular requirements for conservation and management of the tuna and tuna-like species will be discussed in the next section.

Another requirement that limits the freedom of fishing is the duty to cooperate. For the conservation and management of high seas living resources States are required to cooperate by establishing Regional Fisheries Management Organizations (RFMOs).[112] The fishing of tuna and tuna-like species in particular requires joint measures to ensure their conservation on the high seas.[113] For the high seas there is a provision in place, similar to that in the EEZ, which requires the determination of a TAC for the fish stocks and the establishment of conservation and management measures.[114] The overall aim of the measures for the high seas is to maintain or restore populations of harvested species at levels which can produce the MSY.[115]

A difficulty is to limit excess fishing capacity which comes along with new entrants to the fishery. According to the provisions of UNCLOS it is impossible to close the fishery to new entrants if they want to join the RFMO.[116] Another major issue with regard to participation is, if, how, to what extent, and by what means states can be excluded from fishing on the high seas if they are involved in Illegal Unreported and Undocumented (IUU) fishing. UNCLOS did not enable RFMOs to substantially and actively intervene in such IUU fishing activities

---

107   R. L. Allen, et al., *Conservation and Management of Transnational Tuna Fisheries* (Ames: Wiley-Blackwell, 2010), p.3.

108   United Nations Convention on the Law of the Sea, Art.87(1)(e).

109   Ibid. Art.87(2).

110   Ibid. Art.116(a).

111   Ibid. Art.116(b).

112   United Nations Convention on the Law of the Sea Art.118.

113   Ibid. Art.117.

114   Ibid. Art.119.

115   Ibid. Art.119(1).

116   Joseph, et al., 2010, p.19.

because flag states had exclusive jurisdiction over the vessels flying their flag on the high seas.[117]

2. Regulations for specific stocks/species

Most of the tuna and tuna-like species are divided into separate stocks.[118] The stock concept "defines semi-discrete groups of fish with some definable attributes of interest to managers."[119] These groups of fish migrate between two or more EEZs and between the EEZs and the high seas. In order to take their migration patterns into account, specific regimes have been established within UNCLOS. However, the biological differences between the resulting legal categories of transboundary stocks, straddling stocks and highly migratory species are not always clear.[120] In particular there is no real biological or economic justification to distinguish between the two categories of straddling stocks and highly migratory species.[121] The separation goes back to the negotiations of UNLOS where some states requested an explicit list for highly migratory species.[122] In fact, because both 'straddle', they could be classified as one category.[123]

Despite this lack of clarity, the provisions in Articles 63(1), 63(2) and 64 give further guidance for the conservation and management of tuna and tuna-like species by the coastal states and other states whose nationals fish for the species concerned.[124] These provisions should be read in conjunction with the articles

---

117   United Nations Convention on the Law of the Sea, Art.94; E. Meltzer & S. Fuller, *The quest for sustainable international fisheries : regional efforts to implement the 1995 United Nations Fish Stocks Agreement : an overview for the May 2006 review conference* (Ottawa: NRC Research Press, 2009), p.63.

118   The seven principal market tuna species are divided into 23 stocks; Majkowski, 2007, p.26.

119   Gavin A. Begg, et al., Stock identification and its role in stock assessment and fisheries management: an overview, *Fisheries Research - Vol. 43(1-3)* (1999), p.3.

120   S. M. Garcia, et al., *World review of highly migratory species and straddling stocks*, FAO Fisheries Technical Paper - Vol. 337 (Rome: FAO, 1994).

121   G. R. Munro, Internationally shared fish stocks, the high seas, and property rights in fisheries, *Marine Resource Economics - Vol. 22(4)* (2007), p.427.

122   For further information see the description of the development of UNCLOS earlier in this chapter.

123   A. Tahindro, Conservation and management of transboundary fish stocks: Comments in light of the adoption of the 1995 agreement for the conservation and management of straddling fish stocks and highly migratory fish stocks, *Ocean Development and International Law - Vol. 28(1)* (1997), p.2.

124   Birnie, et al., 2009, p.722.

which deal with the EEZ and the high seas. This subchapter is intended to show how conservation and management requirements differ depending on whether a tuna stock is classified as transboundary, straddling or highly migratory.

a)   Transboundary stocks

Transboundary stocks are described in Article 63(1) as a "stock or stocks of associated species [that] occur within the Exclusive Economic Zones of two or more coastal states".[125] The term 'transboundary stocks' is not used in the Article but it is widely known and describes the transboundary migration patterns of the species. Unlike for highly migratory species, UNCLOS does not provide a list that clearly indicates which stocks are included by this term, but, based on their known migration patterns, it can be stated that tuna and tuna-like species could be qualified as transboundary stocks.

The Article describes how coastal states with adjacent EEZs have to 'seek' to agree upon the measures necessary to coordinate and ensure the conservation and development of any such transboundary stocks, but the wording is very weak.[126] The Article does not provide principles of catch allocation, nor does it elaborate on management and conservation objectives.[127] However, such principles exist for the stocks in the EEZ and apply therefore to transboundary stocks as well.[128]

States sharing transboundary stocks are required to cooperate, either directly or through RFMOs, in order to adopt any necessary conservation and management measures.[129] However, the implicit duty to cooperate is weak because the states only need to 'seek' to agree.[130] No requirement to reach an agreement is

---

125   United Nations Convention on the Law of the Sea, Art.63(1).

126   Kaye, 2001, pp.157-158.

127   G. R. Munro, et al., *The conservation and management of shared fish stocks- legal and economic aspects*, FAO Fisheries Technical Paper - Vol. 465 (Rome: FAO, 2004), p. 9.

128   United Nations Convention on the Law of the Sea, Art.61.

129   Ibid. Art.63.

130   R. Churchill & D. Owen, *The EC common fisheries policy* (New York: Oxford University Press, 2010), p.95.

imposed on the states.[131] As a result, the real value of this article might be questioned.[132]

b)   *Straddling stocks*

Again, the term 'straddling stocks' is not used in UNCLOS, but it is generally recognized and refers to what Article 63(2) describes as, "the same stock or stocks of associated species occur[ing] both within the [EEZ] and in an area beyond and adjacent to the zone".[133] It was first formally recognized in the Fish Stocks Agreement, where 'straddling stocks' is used in the full title.[134] As with transboundary stocks, UNCLOS does not provide a list of the stocks concerned, but the migration patterns of tuna and tuna-like species show that these species could also be categorized as straddling stocks.[135]

Coastal states and any other states fishing for such 'straddling stocks' have to seek to agree upon the measures necessary for the conservation of the stocks in the adjacent high seas area.[136] Again, the choice of the term 'seek' exposes the weak nature of this provision. As with transboundary stocks, there are no guidelines given on how a TAC should be established and allocated between the fishing states involved or how conservation and management measures should be designed.[137] In addition, the measures are only to be adopted for the high seas.[138] The provisions on straddling stocks should be read in combination with Article 116. Here, it is stated that the right to fish on the high seas is subject to the rights, duties and interests of coastal states. The sovereign rights of a coastal state with regard to the stocks occurring in its EEZ have to be respected by distant water fishing states. This is based on the assumption that, although both the coastal state and the distant water fishing state are subject to the weak-

---

131   Munro, et al., 2004, p. 9; W. T. Burke, *Annex 1 - 1982 Convention on the Law of the Sea provisions on conditions of access to fisheries subject to national jurisdiction*, Report of the expert consultation on the conditions of access to the fish resources of the exclusive economic zones (Rome: FAO, 1983).

132   Kaye, 2001, p.110.

133   Churchill & Owen, 2010, p.95.

134   de Yturriaga, 1997, p.125.

135   Burke, 1994, pp.136-137. See also the straddling stocks selected by the FAO: Garcia, et al., 1994

136   United Nations Convention on the Law of the Sea, Art.63(2).

137   Dahmani, 1987, pp.114-115.

138   Churchill & Owen, 2010, p.95.

ly worded *pactum de negotiando* principle, the distant water fishing state is also subject to the rights, duties and the interests of the coastal state.[139]

The coastal state and any other states fishing for straddling stocks must, either directly or through RFMOs, seek to agree upon the measures necessary for the conservation of these stocks in the adjacent high seas area. The states only need to 'seek' to agree. The provisions for straddling stocks contain no guidelines on how such an agreement should be reached.

c)   *Highly migratory species*

The provisions for highly migratory species are more detailed and more specific than those for either transboundary or straddling stocks. Article 64 deals exclusively with highly migratory species and the provisions cover their migratory range both on the high seas and in the EEZ. [140] Although UNCLOS does not provide a definition of the term 'highly migratory species', it does list several species in Annex I.[141] That list which is referred to in Article 64(1) contains 17 categories comprising, *inter alia*, both tuna and billfish species. The largest group is composed of several tuna species, including the principal market tuna species:

- albacore tuna (*Thunnus alalunga*);
- bluefin tuna (*Thunnus thynnus*);
- bigeye tuna (*Thunnus obesus*);
- yellowfin tuna (*Thunnus albacares*);
- southern bluefin tuna (*Thunnus maccoyii*); and
- skipjack tuna (*Katsuwonus pelamis*);

more neritic tuna species:

- blackfin tuna (Thunnus atlanticus);
- little tuna (Euthynnus alleteratus, or little tunny, and E. affinis, or kawakawa); and

---

139   Kaye, 2001, pp.159-160.

140   E. Hey, 'The Fisheries Provisions of the LOS Convention', *Developments in international fisheries law* (The Hague: Kluwer Law International, 1999), p.25.

141   Kaye, 2001, p.124.

- frigate mackerel (Auxis thazard, referred to as frigate tuna, and A. rochei, called bullet tuna);[142]

and billfish species:

- marlins, represented by six species belonging to the genus Tetrapturus (Tetrapturus angustirostris, T. belone, T. pfluegeri, T. albidus, T. audax, T. georgei) and three species of the genus Makaira (Makaira indica, M. mazara and M. nigricans, although the latter two have recently been merged taxonomically under a single species, Makaira nigricans);[143]

- sailfishes, with two species (Istiophorus platypterus and I. albicans); and

- swordfish (Xiphias gladius).

There are two principal issues regarding this apparently comprehensive list which have to be mentioned. One issue is that in Annex I there are some species which might not be highly migratory and other potential highly migratory species are not included in the list.[144] Five of the species listed in Annex I are among those regarded by the FAO as neritic or not truly highly migratory,[145] while species with recognized migration patterns[146] are missing.[147] However, this issue which goes back to the negotiation history UNCLOS[148] has no significant impact on conservation and management as tuna RFMOs are seeking long term conservation and sustainable use of tuna and tuna-like species without distinguishing between highly migratory and neritic tuna species. In addition,

---

142 J-J. Maguire, *The state of world highly migratory, straddling and other high seas fishery resources and associated species*, FAO Fisheries and Aquaculture Technical Paper - Vol. 495 (Rome: FAO, 2006), p.9.

143 V. P. Buonaccorsi, et al., Geographic distribution of molecular variance within the blue marlin (Makaira nigricans): A hierarchical analysis of allozyme, single-copy nuclear DNA, and mitochondrial DNA markers, *Evolution - Vol. 53(2)* (1999).

144 Birnie, et al., 2009, p.722.

145 Blackfin tuna (Thunnus atlanticus), little tuna (Euthynnus alletteratus; Euthynnus affinis), frigate mackerel (Auxis thazard; Auxis rochei).

146 Among others: longtail tuna (Thunnus tongol), black skipjack (Euthynnus lineatus).

147 Garcia, et al., 1994.

148 A. Serdy, One fin, two fins, red fins, bluefins: some problems of nomenclature and taxonomy affecting legal instruments governing tuna and other highly migratory species, *Marine Policy - Vol. 28(3)* (2004), p.236; Burke, 1994, p.200.

the WCPFC is the only tuna RFMO which in its constituent agreement is referring explicitly to Annex I.[149]

Another issue is the discrepancies between the nomenclature in Annex I of UNCLOS and the names used by the FAO, RFMOs or the fishing countries themselves.[150] This might lead to problems with regard to the exact identification of the species. For example, the UNCLOS term bluefin tuna actually corresponds to the FAO term Northern bluefin tuna.[151] There are differences in definitions between different languages (e.g. English, French and Spanish), but also within the same language spoken in different countries. In Australia, New Zealand and the United Kingdom, Northern bluefin tuna is called Pacific bluefin tuna, while in Australia, the term Northern bluefin tuna is used for longtail tuna (*Thunnus tonggol*). It is therefore recommended that the 'one species, one name' principle of the FAO should gain wider use.[152] A practical next step could be the rectification of the English, French and Spanish wordings of Annex I, under Article 79 of the Vienna Convention on the Law of Treaties.[153] However, history shows that it is not that easy to make changes to Annex I.[154] A more promising approach was chosen by the IOTC which has developed an own list in Annex B of its Agreement including the FAO names in English, French and Spanish as well as the scientific name.[155]

More important than the nomenclature are the conservation and management requirements for highly migratory species which apply in addition to other provisions of Part V.[156] Article 64(1) requires the conservation but also, in contrast to the provisions for transboundary and straddling stocks, the optimum utilization of these species. The use of the term 'optimum' makes clear that within the EEZs `full utilization' is not mandatory and that the coastal state is not obliged to set the allowable catch at any specific level of utilization.[157] This provision is part of the generally accepted requirement that the management of the highly migratory tuna and tuna-like species like the management of all oth-

---

149　Convention on the Conservation and Management of Highly Migratory Fish Stocks in the Western and Central Pacific Ocean, Art.1(f).

150　Nordquist, 1993, pp.995-999.

151　Serdy, 2004, pp.237-238.

152　Ibid. p.245.

153　Vienna Convention on the Law of Treaties with final act of the conference, declarations and resolutions, 1969.

154　Birnie, et al., 2009, p.722.

155　Agreement for the Establishment of the Indian Ocean Tuna Commission, Annex B.

156　United Nations Convention on the Law of the Sea, Art.64(2).

157　Burke, 1994, p.215.

er all living and non-living resources within the EEZ is subject to the sovereign rights of the coastal state.[158]

However, the responsibility for the achievement of the objective of a coordinated and coherent management of highly migratory species by meeting the conservation requirements is to be shared by several cooperating states and not to be borne by the coastal state alone.[159] With two clear obligations for cooperation, Article 64(1) takes into account the fact that cooperation is essential for the sustainable use of highly migratory species. The first obligation requires that "coastal State and other States whose nationals fish in the region for the highly migratory species listed in Annex I shall cooperate directly or through appropriate international organizations."[160] The wording 'shall cooperate' shows a clear duty to cooperate in the management of highly migratory species.[161] The second addresses regions where no appropriate international organization exists. In such regions, the coastal state and other states whose nationals harvest these species in the region should cooperate to establish an organization and participate in its ongoing work.[162]

3. *Compliance and enforcement*

Sustainable fisheries management relies on effective compliance and enforcement of fisheries rules by coastal states, flag states and port states. In particular it is important to prevent Illegal Unreported and Unregulated (IUU) fishing which is undermining all conservation and management efforts.[163] The overall proportion of illegal tuna catches, in comparison to other species, is relatively low.[164] However, for some of the species, especially the valuable bluefin tunas,

---

158 United Nations Convention on the Law of the Sea, Art.56(1)(a). Burke, 1994, pp.214-215; B. M. Tsamenyi, Treaty on Fisheries between the Governments of Certain Pacific Island States and the Government of the United States of America: The Final Chapter in United States Tuna Policy, *The Brooklyn Journal of International Law - Vol. 15(2)* (1989), p.217; Attard, 1987, p.185; de Yturriaga, 1997, pp.128-129.

159 Kaye, 2001, pp.126-127.

160 United Nations Convention on the Law of the Sea, Art.64(1), sent.1.

161 de Yturriaga, 1997, p.128; Churchill & Owen, 2010, p.95.

162 United Nations Convention on the Law of the Sea, Art.64(1), sent.2.

163 For further information on IUU fishing see the sections on the International Plan of Action to Prevent, Deter, and Eliminate Illegal, Unreported and Unregulated Fishing (IPOA-IUU) as well as on the Agreement on Port State Measures to Prevent, Deter and Eliminate Illegal, Unreported and Unregulated Fishing in this chapter.

164 D.J. Agnew, et al., Estimating the Worldwide Extent of Illegal Fishing, *PLoS ONE 4(2): e4570* (2009), pp.3-4.

there is a significant proportion unreported catches.[165] It has to be recognized that, in general, the highly migratory nature of tuna and tuna-like species is making it difficult to prevent IUU fishing.

A legal difficulty is the different provisions for the EEZs and the high seas. Tuna migrate between the legal zones and neither the coastal state nor the states fishing on the high seas can ensure sustainable management on their own. The sovereign rights of the coastal state are basically limited to the EEZ where it is allowed to enforce laws and regulations in order to ensure compliance.[166] The enforcement measures which can be used include the boarding of vessels, the inspection of vessels, and arrest and judicial proceedings. When necessary, the coastal state is also allowed to conduct a 'hot pursuit'.[167] The right of a hot pursuit is defined as "the right of a [coastal] state to chase and arrest a vessel which has committed an offense within its waters."[168]

The provisions on the high seas differ from those within the EEZ. A vessel fishing on the high seas and flying the flag of a certain state is subject to the exclusive jurisdiction of this state.[169] This means that only the flag state has the right and duty to exercise jurisdiction and control in administrative, technical and social matters.[170] The boarding and inspection of a vessel on the high seas or the arresting of crew members by a third state is only possible if it happens in accordance with another treaty, for example, one established by RFMO member states. No provisions are provided by UNCOLS for a hot pursuit if a violation is occurring on the high seas.

Tuna RFMOs play a crucial role in ensuring compliance and enforcement. They have developed comprehensive conservation and management measures as well as specific tools and programmes.[171] Traditional tools to strengthen flag State´s responsibility are vessel register, vessel monitoring systems and regional observer programmes. These tools are used by all tuna RFMOs. In contrast, the

---

165   The International Consortium of Investigative Journalists, *Looting the Seas* (Washington: The Center for Public Integrity, 2012), pp.8-49.

166   United Nations Convention on the Law of the Sea, Arts. 62(4)(k) and 73(1).

167   Ibid. Arts.111(1) and 111(2).

168   M. J. Gamboa, *A Dictionary of International Law and Diplomacy* (Quezon City: Phoenix Press, 1973), pp.139-140.

169   United Nations Convention on the Law of the Sea Art.92(1).

170   Ibid. Art.94.

171   A detailed analysis of these measures will be provided in the third chapter.

more recently developed port states measures and trade measures have been established only in some of the tuna RFMOs.[172]

4. International obligations for the protection of the environment

UNCLOS introduced a number of fundamental changes in international environmental law, putting increased responsibility on the contracting parties.[173] These changes apply in all waters and the relevant provisions take account of the unintended effects of multispecies fisheries like some of those targeting tuna and tuna-like species. Part XII of UNCLOS is stressing the obligation to protect and preserve the environment.[174] The general provisions which make no explicit reference to fisheries include the obligation that states must protect and preserve the marine environment in all waters.[175] The states have the sovereign right to exploit their natural resources[176] but they are not released from the obligation to protect and preserve the marine environment.[177] Another important provision is the duty "to protect and preserve rare or fragile ecosystems as well as the habitat of depleted, threatened or endangered species."[178] This provision has been criticized because according to the wording it seems that measures have to be taken only when damage already has been done.[179] All states are fur-

---

172   M. A. Palma, et al., *Promoting Sustainable Fisheries: The International Legal and Policy Framework to Combat Illegal, Unreported and Unregulated Fishing* (Leiden: Martinus Nijhoff Publishers, 2010), pp. 63, 193. A. Fabra, et al., *Closing the gap: Comparing tuna RFMO port State measures with the FAO Agreement on Port State Measures* (Philadelphia: Pew Environment Group 2011), G. Schneider, *Current Debate Over Potential Use of Trade Measures* (Washington: U.S. Department of Commerce - National Marine Fisheries Service, 2000), M. Lack, *Catching On? Trade-related Measures as a Fisheries Management Tool* (Cambridge: TRAFFIC International, 2007).

173   J. M. Van Dyke, 'Allocating Fish Across Jurisdictions', *Conservation and Management of Transnational Tuna Fisheries* (Iowa: Wiley-Blackwell, 2010), p.164.

174   United Nations Convention on the Law of the Sea Arts.192-237.

175   Ibid. Art.192.

176   Ibid. Art.193.

177   R. Wolfrum, 'The Protection of the Marine Environment after the Rio Conference: Progress or Stalemate?', *Recht zwischen Umbruch und Bewahrung: Völkerrecht, Europarecht, Staatsrecht (Beiträge zum ausländischen öffentlichen Recht und Völkerrecht)* (Berlin: Springer Verlag, 1995), p.1009.

178   United Nations Convention on the Law of the Sea Art.194(5).

179   Wolfrum, 1995, p.1009.

ther required to cooperate globally and regionally to achieve that same goal, namely to protect and preserve the marine environment.[180]

The progressive approach of UNCLOS with regard to the protection of the environment can be seen in particular in the emphasis on the linkage between target species and associated or dependent species. For sustainable management of tuna and tuna-like species this is important as it has to be considered that the fisheries targeting these species may also have a negative impact on other species.[181] Possible effects for both target and non-target species are a reduction of abundance and spawning potential, changes in age and size structure, sex ratio, genetics and species composition.[182] In Articles 61 and 119, among other points, it is laid down that conservation and management measures have to be qualified by relevant environmental factors and must take into account the interdependence of stocks. Both coastal states and states fishing on the high seas have to take measures designed to maintain or restore populations of target species at levels which can produce the maximum sustainable yield while taking into consideration the effects on associated or dependent species maintaining or restoring populations of these species above levels at which their reproduction may become seriously threatened.[183] These provisions for associated or dependent species are somewhat weakly worded, with non-target species having to be considered only 'with a view' to maintain or restore populations at levels above those at which their reproduction may become seriously threatened.[184] There is no clear obligation to formulate a coherent policy concerning associated or dependent species.[185]

---

180   United Nations Convention on the Law of the Sea Art.197.

181   Pauly, et al., 1998, pp.860-863; D. A. O'Connell, Tuna, Dolphins, and Purse Seine Fishing in the Eastern Tropical Pacific: The Controversy Continues, *UCLA Journal of Environmental Law and Policy - Vol. 23(1)* (2005), pp.77-100; E. L. Gilman, Bycatch governance and best practice mitigation technology in global tuna fisheries, *Marine Policy - Vol. 35(5)* (2011), pp.590-609.

182   S.M. Garcia, et al., *The ecosystem approach to fisheries. Issues, terminology, principles, institutional foundations, implementation and outlook*, FAO Fisheries Technical Paper- No. 443 (Rome: FAO, 2003), p.10.

183   United Nations Convention on the Law of the Sea, Arts. 61(3), 61(4), 119(1)(a) and 119(1)(b).

184   Ibid. Arts.64 (4) and 119(1)(b).

185   Wolfrum, 1995, p.1010.

5. Scientific research

In all fisheries sound scientific information is crucial for sustainable conservation and management. In particular, scientific data is the basis for all stock assessments and management decisions of coastal states and RFMOs. It has to be recognized that despite the economic importance of tuna and tuna like species there are still significant uncertainties regarding their migration patterns, stock sizes or stock statuses.[186] The main reason for the limited information is, again, that tuna and tuna-like species are migrating over long distances and that their migration routes and abundance are influenced by a variety of factors. These are human influences like fishing and a number of environmental factors such as changes in water temperature or in the availability of food.

For the sustainable management of these species it is crucial that all states cooperate in marine scientific research at an international level. International organizations have the right to conduct but also the obligation to promote and facilitate the development and conduct of marine scientific research.[187] Relevant states and competent international organizations are, required to promote such cooperation in order to both prevent and limit damage to the marine environment.[188] Bilateral and multilateral agreements have to be established in order to create favourable research conditions.[189] In order to make any research accessible to the maximum number of stakeholders, it is also important that states and competent international organizations are required to publicize and to disseminate their objectives and other relevant information.[190] Today scientific research on tuna and tuna like species is conducted by the tuna RFMOs in cooperation with coastal states or subregional organizations.[191]

---

186   J. Zhu, et al., Implications of uncertainty in the spawner–recruitment relationship for fisheries management: An illustration using bigeye tuna (Thunnus obesus) in the eastern Pacific Ocean, *Fisheries Research - Vol. 119-120* (2012), pp.89-93; D. Kolody, et al., Salvaged pearls: lessons learned from a floundering attempt to develop a management procedure for Southern Bluefin Tuna, *Fisheries Research - Vol. 94(3)* (2008), pp.339-341; J. Sibert & J. Hampton, Mobility of tropical tunas and the implications for fisheries management, *Marine Policy - Vol. 27(1)* (2003), pp.87-95; S. Losada, et al., The Status of Atlantic Bluefin Tuna, *Science - Vol. 328* (2010), pp.1353-1354.

187   United Nations Convention on the Law of the Sea Art.239.

188   Ibid. Art.242.

189   Ibid. Art.243.

190   Ibid. Art.244(1).

191   For detailed information on scientific research within tuna RFMOs see chapter III.

6. Dispute settlement

UNCLOS introduced progressive and far-reaching provisions on compulsory dispute settlement.[192] These provisions were intended to be a counterweight to the increased jurisdictional competence of coastal states.[193] Disputes concerning the interpretation or application of UNCLOS have to be settled by peaceful means and in accordance with the relevant provisions in the Charter of the United Nations.[194] The states concerned are free to settle a dispute by any peaceful means of their own choice,[195] but if a peaceful settlement cannot be reached by them, the dispute must be submitted to the International Tribunal for the Law of the Sea, the International Court of Justice, or an Arbitral Tribunal.[196] The choice of tribunal depends on the parties involved.[197] In general, whichever court or tribunal is chosen, it has jurisdiction over any dispute "concerning the interpretation or application of an international agreement related to the purposes of th[e] Convention, which is submitted to it in accordance with the agreement."[198] However, there are some exceptions when coastal state's sovereign rights over the conservation and management of living resources and scientific research within their EEZ are affected.[199] This relates to issues like the determination of the total allowable catch, the harvesting capacity, the allocation of surpluses to other states, and the terms and conditions established in a coastal state's own conservation and management laws and regulations.[200] That such disputes within the EEZ are not subject to compulsory jurisdiction is open to criticism, but it is a great success that there are provisions which include the principle of compulsory adjudication for the settlement of disputes concerning

---

192   United Nations Convention on the Law of the Sea Arts.279-299. R. Churchill & A. V. Lowe, *The law of the sea* (Manchester: Manchester University Press, 1998), pp.453-459; M. Gaertner, The Dispute Settlement Provisions of the Convention on the Law of the Sea: Critique and Alternatives to the International Tribunal for the Law of the Sea, *San Diego Law Review - Vol. 19* (1982), p.577; T. Treves, Dispute-Settlement Clauses in the law of the Sea Convention and their Impact on the Protection of the Marine Environment: Some Observations, *Review of European Community & International Environmental Law - Vol. 8(1)* (1999), p.6.

193   R. Rayfuse, The Future of Compulsory Dispute Settlement Under the Law of the Sea Convention, *Victoria University of Wellington Law Review - Vol. 36* (2005), p.683.

194   United Nations Convention on the Law of the Sea, Art.279.

195   Ibid. Art.280.

196   Ibid. Art.287.1.

197   Ibid. Arts.286 and 287.

198   Ibid. Arts.288.1 and 288.2.

199   Ibid. Art.297.

200   Ibid. Art.297(3)(a).

fisheries on the high seas.[201] To date, compulsory dispute settlement procedures have been applied in two fisheries cases, all of which were related to tuna and tuna-like species.[202] These cases on southern bluefin tuna and swordfish in the south eastern Pacific Ocean will be explicitly addressed in chapter II.

C. Development after the adoption of the Convention on the Law of the Sea

The codification in UNCLOS of previously-existing laws and policies regarding the marine environment represented a great step forward for international fisheries management. However, some continuing gaps and unclear provisions have weakened its effect. In practice, UNCLOS has not been able to prevent the decline of many valuable fish stocks, including tuna and tuna-like species.[203] The overexploited fish stocks in the North Pacific, the Bering Sea, the Antarctic, the North Atlantic and the North Sea in particular demonstrate that the provisions of UNCLOS could not on their own guarantee sustainable fisheries.[204] What happened to the Southern bluefin tuna shows how insufficient the provisions are. It was heavily overfished up to the 1990s and stocks have not yet recovered.[205]

During the 1992 Rio Conference, a series of problems were identified which showed that the management of high seas fisheries in particular had been deficient.[206] These were the overutilization of stocks and the inadequate adoption, monitoring and enforcement of effective conservation measures, as well as issues of unregulated fishing, overcapitalization, excessive fleet size, vessel reflagging, insufficiently selective gear, unreliable databases and a lack of suffi-

---

201 de Yturriaga, 1997, p.150.

202 Southern Bluefin Tuna Case - Australia and New Zealand v. Japan, Arbitral Award of August 4, 2000, First Arbitral Tribunal constituted under Part XV ("Settlement of Disputes"), Annex VII ("Arbitration") of the United Nations Convention of the Law of the Sea (UNCLOS), (2000); Case concerning the Conservation and Sustainable Exploitation of Swordfish Stocks in the South-Eastern Pacific Ocean (Chile / European Union) - Order 2000/3, ITLOS, (2000).

203 FAO, *FAO yearbook*, Fishery and Aquaculture Statistics 2007 (Rome: FAO, 2007), pp.30-35; D. A. Balton, Strengthening the Law of the Sea: The new agreement on straddling fish stocks and highly migratory fish stocks, *Ocean Development and International Law - Vol. 27(1-2)* (1996), p.125; Pauly, et al., 1998.

204 Birnie, et al., 2009, p.731.

205 Miyake, et al., 2004, p. 71; For details see also
http://www.fao.org/fishery/statistics/tuna-atlas/query/en.

206 Agenda 21: Earth Summit - The United Nations Programme of Action from Rio, 1992.

cient cooperation between states.[207] In order to strengthen the relevant provisions within UNCLOS, several agreements were established to deal specifically with these problems. These included a UN conference on straddling and highly migratory stocks which aimed to enhance effective cooperation through RFMOs.[208]

In 1995, this conference led to the Agreement for the Implementation of the Provisions of the United Nations Convention on the Law of the Sea of 10 December 1982 relating to the Conservation and Management of Straddling Fish Stocks and Highly Migratory Fish Stocks. Since then, this Agreement has become the most important treaty for the management of tuna and tuna-like species. Other important developments initiated by the Rio Conference led to the 1993 Agreement to Promote Compliance with International Conservation and Management Measures by Fishing Vessels on the High Seas,[209] the 1995 Code of Conduct for Responsible Fisheries,[210] the International Plans of Action[211] as well as the more recent 2009 Agreement on Port State Measures to Prevent, Deter and Eliminate Illegal, Unreported and Unregulated Fishing[212].

1. United Nations Fish Stocks Agreement

The Agreement for the Implementation of the Provisions of the United Nations Convention on the Law of the Sea of 10 December 1982 relating to the Conservation and Management of Straddling Fish Stocks and Highly Migratory Fish Stocks[213], commonly known as the Fish Stocks Agreement, was adopted on the 4th of August 1995 and entered into force on the 11th of December 2001. As an

---

207 Ibid. Chapter 17.45.

208 Ibid. Chapter 17.50.

209 Agreement to Promote Compliance with International Conservation and Management Measures by Fishing Vessels on the High Seas.

210 Code of Conduct for Responsible Fisheries.

211 International Plan of Action for Reducing Incidental Catch of Seabirds in Longline Fisheries; International Plan of Action for the Conservation and Management of Sharks; International Plan of Action for the Management of Fishing Capacity; International Plan of Action to Prevent, Deter, and Eliminate Illegal, Unreported and Unregulated Fishing (IUU).

212 Agreement on Port State Measures to Prevent, Deter and Eliminate Illegal, Unreported and Unregulated Fishing.

213 Agreement for the Implementation of the Provisions of the United Nations Convention on the Law of the Sea of 10 December 1982 relating to the Conservation and Management of Straddling Fish Stocks and Highly Migratory Fish Stocks.

elaboration of Articles 63(2) and 64 as well as of section 2 of Part VII,[214] it addresses the key uncertainties within UNCLOS that had resulted in unsustainable fishery for straddling and highly migratory fish stocks.[215] As a result, the objective of the Fish Stocks Agreement is to "ensure the long-term conservation and sustainable use of straddling fish stocks and highly migratory fish stocks through effective implementation of the relevant provisions of UNCLOS".[216] Its key provisions address conservation, sustainable use and the precautionary approach, compatibility of the provisions for the EEZs and the high seas, international cooperation, flag state obligations, compliance and enforcement, as well as dispute settlement.

The Fish Stocks Agreement deals explicitly with straddling stocks and highly migratory species. However, it gives no additional information or clarification about which species are included by the terms 'straddling stocks' or 'highly migratory species'.[217] The only guidance is therefore still the list of highly migratory species provided by Annex I of UNCLOS. As the name of the new treaty already indicates, transboundary stocks and stocks which are exclusive to one EEZ are not covered by the provisions of the Agreement.[218]

Only the countries that have ratified the Fish Stocks Agreement are bound by its provisions. It has been ratified by 80 countries.[219] Unfortunately, 9 of the 20

214 T. Henriksen, et al., *Law and politics in ocean governance: the UN Fish Stocks Agreement and regional fisheries management regimes*, Publications on ocean development (Leiden: Martinus Nijhoff Publishers, 2006), p.1 Churchill & Owen, 2010, p.99.

215 See: C. Warbrick, et al., The Straddling Stocks Agreement of 1995 - an Initial Assessment, *International & Comparative Law Quarterly - Vol. 45(2)* (1996), p.463; Balton, 1996, p.125; D. Freestone & Z. Makuch, 'The New International Environmental Law of Fisheries: The 1995 UN Straddling Stocks Agreement', *Yearbook of International Environmental Law - Vol. 7* (Oxford: Clarendon Press 1998), p.3; P. Davies & C. Redgewell, 'The International Legal Regulation of Straddling Fish Stocks', *British Yearbook of International Law - Vol.67* (Oxford: Oxford University Press, 1996), p.199; M. Hayashi, 'The Straddling and Highly Migratory Fish Stocks Agreement', *Developments in international fisheries law* (The Hague: Kluwer Law International, 1999), p.55.

216 Agreement for the Implementation of the Provisions of the United Nations Convention on the Law of the Sea of 10 December 1982 relating to the Conservation and Management of Straddling Fish Stocks and Highly Migratory Fish Stocks, Art.2.

217 C. Hedley, et al., *The Implications of the UN Fish Stocks Agreement (New York, 1995) for Regional Fisheries Organisations and International Fisheries Management*, Working paper prepared for the European Parlament (Luxembourg: Directorate-General for Research, 2003), p.5.

218 P. W. Birnie & A. E. Boyle, *International law and the environment* (New York: Oxford University Press, 1999), p.673.

219 See the website of the UN Division for Ocean Affairs and the Law of the Sea.

most important (in terms of total catch weight of the principal market tuna species) tuna fishing countries have not yet ratified the Fish Stocks Agreement.[220] These are: Taiwan[221] (2nd most important); the Philippines (4th); Ecuador (9th); Mexico (10th); Iran (12th); Venezuela (15th); Colombia (17th); China (18th); and Vanuatu (19th). The main reasons for non-ratification are a lack of capacity, technical, juridical or policy differences, and political differences.[222] Obviously, for the effectiveness of the Agreement, it is important for it to be ratified by the largest number of countries possible. Only those states which have expressly consented can be bound by the Agreement.[223] This is due to the *pacta tertiis rule* which lays down that a treaty does not "create either obligations or rights for a third State without its consent".[224] Ideally, the large number of non-parties should be reduced in the future. Given what we know of the history of ratification of other international treaties but also due to the efforts made during the review conferences of the Fish Stocks Agreement, it is still possible that the number of ratifying parties will increase. Ratification or accession by the Philippines, for example, could take place in the near future.[225] Even without further participation, the Fish Stocks Agreement already has an impact on non-

---

220   Majkowski, 2007, p.16.

221   Referring to Taiwan in Article 1(3) the Fish Stocks Agreement can also apply *mutatis mutandis* 'to other fishing entities whose vessels fish on the high seas'. This provision thus allows Taiwan to accept the Agreement despite the uncertainty with regard to its international status. This possible inclusion of Taiwan, which has the second largest national tuna catch, is important for the sustainable management of tuna and tuna-like species as it provides the opportunity to close any loopholes which allow unregulated trade and fishing. See also Birnie, et al., 2009, p.734; P. S. C. Ho, The impact of the UN Fish Stocks Agreement on Taiwan's participation in international fisheries fora, *Ocean Development and International Law - Vol. 37(2)* (2006).

222   ICSP7, *Report of the Seventh Round of Informal Consultations of States Parties to the Agreement for the Implementation of the Provisions of the United Nations Convention on the Law of the Sea of 10 December 1982 relating to the Conservation and Management of Straddling Fish Stocks and Highly Migratory Fish Stocks*, (2008), p.14.

223   E. Franckx, *Pacta Tertiis and the Agreement for the Implementation of the Provisions of the United Nations Convention on the Law of the Sea of 10 December 1982 relating to the Conservation and Management of Straddling Fish Stocks and Highly Migratory Fish Stocks*, FAO Legal Papers - Vol. 8 (Rome: FAO, 2000).

224   Vienna Convention on the Law of Treaties with final act of the conference, declarations and resolutions, Art.34.

225   Molenaar, 2011, p.199.

parties.[226] Some of the non-parties are bound implicitly to certain provisions because of their membership in newer RFMOS, such as WCPFC, IATTC or SEAFO[227], which implemented provisions of the Fish Stocks Agreement in their constituent instruments.[228]

The Fish Stocks Agreement has been reviewed twice. In 2006, the first Review Conference was held in New York, according to the provisions of Article 36 of the Agreement and pursuant to paragraph 16 of the General Assembly resolution 59/25.[229] It was attended by delegations from parties and non-parties, as well as representatives from both inter-governmental organizations (IGOs) and non-governmental organizations (NGOs). The aim of the conference was to assess the effectiveness of the Agreement in achieving the conservation and management of straddling and highly migratory fish stocks and to propose means of strengthening its substance and methods of implementation.[230] Among others, the Conference came up with recommendations for future actions. In 2010, a Resumed Review Conference took place in New York in line with resolutions 63/112 and 64/72 of the General Assembly.[231] This conference, like the previous one, was instructed to analyse the effectiveness of the Fish Stocks Agreement. The difference was an added emphasis on assessing the implementation of the recommendations made by the first Review Conference. In general, many delegations recognized the positive developments which had been made following the adoption of those recommendations.[232] However, concerns about various issues, such as the status of bigeye tuna or yellowfin tuna in the Pacific Ocean, were also expressed.[233] The Conference underlined the continuing importance

---

226 United Nations General Assembly, *Report of the Resumed Review Conference on the Agreement for the Implementation of the Provisions of the United Nations Convention on the Law of the Sea of 10 December 1982 relating to the Conservation and Management of Straddling Fish Stocks and Highly Migratory Fish Stocks*, (2010) Para.111; Henriksen, et al., 2006, p.13; Hayashi, 1999, p.82.

227 South East Atlantic Fisheries Organisation.

228 Molenaar, 2011, p.229.

229 United Nations General Assembly, *Report of the Review Conference on the Agreement for the Implementation of the Provisions of the United Nations Convention on the Law of the Sea of 10 December 1982 relating to the Conservation and Management of Straddling Fish Stocks and Highly Migratory Fish Stocks*, (2006), Para.1.

230 Ibid. Para.2.

231 United Nations General Assembly, *Report of the Resumed Review Conference on the Agreement for the Implementation of the Provisions of the United Nations Convention on the Law of the Sea of 10 December 1982 relating to the Conservation and Management of Straddling Fish Stocks and Highly Migratory Fish Stocks*, (2010), Para.3.

232 Ibid. Paras.24 and 39.

233 Ibid. Para.27.

of both the full implementation of and compliance with conservation and management measures, in accordance with international law, the precautionary approach and the best scientific evidence available.[234]

This section will analyze the basic requirements of the Fish Stocks Agreement and it will show the most important developments in the context of deficits of the UNCLOS. Findings of the Review Conferences and recommendations with respect to the fisheries of tuna and tuna-like species will be related with the requirements where possible. With regard to the 2010 Conference the section will look at the core results of the assessment as well as the main recommendations in the context of previous recommendations.

a)   Conservation and management

The Fish Stocks Agreement sets out general conservation and management obligations that require parties to ensure long term sustainability and to promote optimum utilization based on both the best scientific evidence available and the ability to maintain or restore stocks at levels which can produce the MSY.[235] Some of the respective obligations follow directly from Articles 61, 62 and 119 of UNCLOS, while others address the causes of the crisis which had subsequently evolved in marine fisheries.[236]

An indicator for the environmental perspective of the Fish Stocks Agreement is the requirement to apply the precautionary approach.[237] Though implied, UNCLOS did not contain explicit provisions for its application.[238] In contrast, the precautionary approach of the Fish Stocks Agreement is described as the most detailed one in a global treaty on fisheries management, having the potential to significantly improve the management of straddling and highly migratory fish stocks.[239] The precautionary approach requires states to take measures

---

234   Ibid. Annex, Preamble.

235   Agreement for the Implementation of the Provisions of the United Nations Convention on the Law of the Sea of 10 December 1982 relating to the Conservation and Management of Straddling Fish Stocks and Highly Migratory Fish Stocks, Arts.5(a) and (b).

236   Balton, 1996, p.136.

237   Birnie, et al., 2009, p.737.

238   D. Freestone, 'International Fisheries Law Since Rio: The Continued Rise of the Precautionary Principle', *International Law and Sustainable Development - Past Achievements and Future Challenges* (New York: Oxford University Press, 1999), p.141; L. Cordonnery, A note on the 2000 Convention for the Conservation and Management of Tuna in the Western and Central Pacific Ocean, *Ocean Development and International Law - Vol. 33(1)* (2002), pp.6-7.

239   Churchill & Owen, 2010, p.100; Balton, 1996, pp.135-136.

against overfishing, even when the available scientific information is uncertain, unreliable or inadequate.[240] Precautionary conservation and management measures have to be adopted for new or exploratory fisheries[241] or in extreme situations where fishing activities are a serious threat to the sustainability of stocks.[242] The requirement to assess the impacts of fishing on those species which are associated with or belonged to the same ecosystem as the target stocks[243] is also new, as is the requirement to take measures to prevent or eliminate overfishing and excess fishing capacity.[244]

The introduction of the precautionary approach was a paradigm shift towards the application of precautionary limit and target reference points.[245] The reason for this shift was the fact that MSY as the exclusive management technique had disregarded relevant variables in allocation decisions, which had then led to the overexploitation of many stocks.[246] With the new approach all catches are intended to be kept below MSY and, if a reference point is reached, automatic management measures are to be initiated in order to rebuild the stock.[247] All reference points have to be stock-specific, and states are required to take action if reference points are breached.[248] Limit or conservation reference points indicate "a state of a fishery and/or a resource which is considered to be undesirable and which management action should avoid" while target or management reference points indicate "to a state of a fishing and/or resource which is considered to be desirable and at which management action, whether during development or stock rebuilding, should aim."[249] Annex II to the Agreement gives guidelines for the application of limit and target reference points. Limit reference points are intended to "set boundaries which are intended to constrain harvesting within safe biological limits within which the stocks can produce

240 Agreement for the Implementation of the Provisions of the United Nations Convention on the Law of the Sea of 10 December 1982 relating to the Conservation and Management of Straddling Fish Stocks and Highly Migratory Fish Stocks, Art.6.2.

241 Ibid. Art.6.6.

242 Ibid. Art.6.7.

243 Ibid. Art.5(d).

244 Ibid. Art.5(h).

245 Birnie, et al., 2009, p.737.

246 J. M. McDonald, Appreciating the precautionary principles as an ethical evolution in ocean management, *Ocean Development and International Law - Vol. 26(3)* (1995), pp.274.

247 Ibid. pp.275.

248 Agreement for the Implementation of the Provisions of the United Nations Convention on the Law of the Sea of 10 December 1982 relating to the Conservation and Management of Straddling Fish Stocks and Highly Migratory Fish Stocks Art.6.3(b).

249 Caddy & Mahon, 1995, p. 8; Kaye, 2001, p.236.

maximum sustainable yield."[250] In contrast the target reference points are intended "to meet management objectives."[251]

During the Resumed Review Conference in 2010 it was noted that, since 2006, conservation and management measures had not always been adopted in line with the best available scientific information and the precautionary approach.[252] About 30 percent of tuna and tuna-like species were still either overexploited or depleted.[253] A related problem was ongoing overcapacity and the high level of subsidies. Their further reduction to sustainable levels, as well as a determination of target levels and assessments of fishing activity was seen as necessary in order to give full effect to the recommendations made in 2006.[254] Therefore, the Conference recommended that both states and RFMOs should urgently commit themselves to improving the status of all straddling and highly migratory fish stocks which were overexploited or depleted.[255] In this regard, the time-area closures for the regulation of certain fisheries which had been implemented by ICCAT and IATTC were seen as examples of good practice. For some stocks, in particular swordfish, bigeye tuna and bluefin tuna fisheries, this measure had been identified to be more successful than the allocation of quotas alone.[256] It was further recommended that, for fisheries with poor information, the guidelines for precautionary reference points be applied in order to determine such reference points, and that action be taken when certain levels were exceeded.[257] In addition, the Conference suggested that the ecosystem approach should be fully implemented as soon as possible.[258] The core issues in this area where a need for action was identified included non-target species, associated or dependent species, unregulated fisheries, the fishing of juveniles, and the selectivity of fishing gear.[259] Therefore, the Conference recommended the strengthen-

---

250   Agreement for the Implementation of the Provisions of the United Nations Convention on the Law of the Sea of 10 December 1982 relating to the Conservation and Management of Straddling Fish Stocks and Highly Migratory Fish Stocks Annex II, Para.2, sent.2.

251   Ibid. Annex II, Para.2, sent.3.

252   United Nations General Assembly, *Report of the Resumed Review Conference on the Agreement for the Implementation of the Provisions of the United Nations Convention on the Law of the Sea of 10 December 1982 relating to the Conservation and Management of Straddling Fish Stocks and Highly Migratory Fish Stocks*, (2010) Para.47.

253   Ibid. Para.8.

254   Ibid. Para.65 and Annex, Paras.I(e) and (m).

255   Ibid. Annex, Paras.I(a) and (j).

256   Ibid. Para.64.

257   Ibid. Annex, Para.I(i).

258   Ibid. Annex, Para.I(f).

259   Ibid. Para.54.

ing of efforts to study and address environmental factors affecting marine ecosystems,[260] and to assess the impacts of fishing, other human activities and environmental factors on target stocks and any associated and dependent species.[261]

b)   *Compatibility of EEZ and the high seas measures*

Ensuring the compatibility of measures throughout the whole migration range
of tuna and tuna-like species represents one of the biggest challenges for fishing
states and RFMOs. In order to manage the stocks sustainably, it is necessary
that the measures for the EEZ, mainly adopted by coastal states, and the
measures for the high seas, mainly adopted by RFMOs, are compatible. Within
UNCLOS, the provisions on compatibility are neither very specific nor comprehensive. The respective provisions in the Fish Stocks Agreement were therefore
an attempt to overcome the prior lack of co-operation and compatibility that
had characterized state relations in tuna RFMOs.[262]

One of the major achievements of the Fish Stocks Agreement with regard to the
integrated management of tuna and tuna-like species are certain conservation
requirements which apply in both the EEZs and on the high seas.[263] The relevant provisions of Articles 6 and 7 apply explicitly in both of these legally defined zones, in contrast to all other parts of the Agreement.[264] Article 6 contains
provisions for the application of compatible precautionary measures, and Article 7 deals explicitly with the compatibility of all conservation and management
measures.[265] The rules laid down in Article 7 are similar to those in Articles
63(2) and 64 of UNCLOS, but much more detailed. Multilateral measures for
straddling fish stocks apply, as in UNCLOS, to the high seas only, while those

---

260   Ibid. Annex, Para.I(n).

261   Ibid. Annex, Para.I(o).

262   P. Örebech, et al., The 1995 United Nations Straddling and Highly Migratory Fish Stocks
       Agreement: Management, Enforcement and Dispute Settlement, *The International Journal
       of Marine and Coastal Law - Vol. 13* (1998), p.120.

263   C. Safina, 'Tuna Conservation', *Tuna: Physiology, ecology, and evolution* (San Diego:
       Academic Press, 2001), p.423.

264   Agreement for the Implementation of the Provisions of the United Nations Convention
       on the Law of the Sea of 10 December 1982 relating to the Conservation and Management
       of Straddling Fish Stocks and Highly Migratory Fish Stocks, Art.3.1.

265   Agreement for the Implementation of the Provisions of the United Nations Convention
       on the Law of the Sea of 10 December 1982 relating to the Conservation and Management
       of Straddling Fish Stocks and Highly Migratory Fish Stocks, Art.7.

for highly migratory species apply both within and beyond the EEZ.[266] The central obligation to achieve compatibility is laid down in Article 7(2):

"Conservation and management measures established for the high seas and those adopted for areas under national jurisdiction shall be compatible in order to ensure conservation and management of the straddling fish stocks and highly migratory fish stocks in their entirety. To this end, coastal States and States fishing on the high seas have a duty to cooperate for the purpose of achieving compatible measures in respect of such stocks."[267]

These obligations were new and they incorporated several specific duties which had to be fulfilled by the party states. These include the requirement for states to take into account previously existing measures, the biological characteristics of the stocks, relationships between the stocks, as well as the dependence of the different fishing states on the stocks.[268] States are required to make every effort to agree on compatible conservation and management measures within a reasonable period of time.[269] The detailed provisions also address dispute settlement in cases where no agreement can be reached,[270] as well as the development of provisional agreements[271]. During the negotiation of the Fish Stocks Agreement, the issue of the compatibility of EEZ and high seas measures was the subject of several controversial debates. Most of the states opposed mandatory compatible measures, arguing that they would restrict the freedom of high seas fishing through creeping jurisdiction.[272] In view of these controversies, the relevant provisions in Article 7 can be seen as a great achievement for the integrated management of tuna and tuna-like species.

The level of cooperation between coastal states and states fishing on the high seas was noted as continuing problem during the 2010 Resumed Review Conference. Enhanced cooperation was recommended in order to ensure the compatibility of measures for the high seas and for areas under national jurisdic-

---

266   Ibid. Arts.7.1(a) and (b) See also Balton, 1996, p.137.

267   Agreement for the Implementation of the Provisions of the United Nations Convention on the Law of the Sea of 10 December 1982 relating to the Conservation and Management of Straddling Fish Stocks and Highly Migratory Fish Stocks, Art.7.2.

268   Ibid. Arts.7.2 (a) - (e).

269   Ibid. Art.7.3.

270   Ibid. Art.7.4.

271   Ibid. Arts.7.5 and 7.6.

272   Hayashi, 1999, p.63.

tion.[273] Already in 2006 the Conference had criticized the fact that the provisions of the Agreement with respect to compatibility had not been fully applied for some fisheries due to a lack of cooperation in conservation and management between coastal states and DWFNs.[274]

c) International cooperation

International cooperation is crucial for the sustainable management of tuna and tuna-like species. The relevant provisions in the Fish Stocks Agreement are based on the key elements of Articles 63(2), 64 and 116-119 of UNCLOS. These articles encouraged states to agree, directly or through RFMOs or other arrangements, upon measures for straddling stocks,[275] and they also required them to cooperate, again, directly or through RFMOs or other arrangements, in the conservation and management of highly migratory species.[276] States fishing on the high seas were further explicitly required to cooperate by establishing RFMOs.[277] However, within UNCLOS, the duty to cooperate was rather limited, because no detailed information was given on how to establish RFMOs or how they should operate.[278] By providing more detail on these issues, the Fish Stocks Agreement represented an attempt to strengthen the legal basis for existing RFMOs by improving their effectiveness.[279] The states concerned were required, through RFMOs, to agree on and to comply with conservation and management measures, to agree on allocations of allowable catch or levels of fishing effort,[280] and to review the status of target and non-target species.[281]

---

273 United Nations General Assembly, *Report of the Resumed Review Conference on the Agreement for the Implementation of the Provisions of the United Nations Convention on the Law of the Sea of 10 December 1982 relating to the Conservation and Management of Straddling Fish Stocks and Highly Migratory Fish Stocks*, (2010) Annex, Para.I(b).

274 United Nations General Assembly, *Report of the Review Conference on the Agreement for the Implementation of the Provisions of the United Nations Convention on the Law of the Sea of 10 December 1982 relating to the Conservation and Management of Straddling Fish Stocks and Highly Migratory Fish Stocks*, (2006) Annex, Para.17.

275 United Nations Convention on the Law of the Sea, Art.63.2.

276 Ibid. Art.64.

277 Ibid. Arts.116–119.

278 de Yturriaga, 1997, pp.158-161.

279 Agreement for the Implementation of the Provisions of the United Nations Convention on the Law of the Sea of 10 December 1982 relating to the Conservation and Management of Straddling Fish Stocks and Highly Migratory Fish Stocks, Art.13.

280 Ibid. Art.10 (b).

281 Ibid. Art.10 (d).

The mandatory cooperation requirement of the Fish Stocks Agreement applies to both highly migratory fish stocks and straddling fish stocks.[282] RFMOs have to be open in a non-discriminatory manner to potential member states with a real interest in fishing for these stocks in order to enable all relevant states to cooperate.[283] Despite this general openness, access to the fishery resources can only be granted to state parties which are members of the RFMO, or non-members which agreed to apply the conservation and management measures established by the RFMO.[284] Any state party whose vessels want to fish in the area of competence of an RFMO is thus obliged to join the RFMO or to apply the conservation and management measures of the RFMO.[285] Although this requirement applies only to parties to the Fish Stocks Agreement it has been a remarkable step forward as it substantiates the duty to cooperate of UNCLOS and gives RFMOs a key role in the management and conservation of straddling and highly migratory fish stocks.[286] Particularly the inclusion of non-members is a progress as it allows their active participation in the management framework if they apply the general rules to their vessels.[287] The expanded duty to cooperate is also designed to facilitate the exclusion of 'free-riding' parties who have not taken responsibility for conservation or management. If such parties do not prevent their vessels from fishing in the region of an RFMO, then the members of the RFMO and any other parties to the Fish Stocks Agreement are able, and indeed are bound, to take action consistent with the Fish Stocks Agreement and international law in order to deter any activities of vessels which might undermine the effectiveness of conservation and management measures.[288] Such action includes trade measures or port states measures which are increasingly discussed at an international level and which are applied by

---

282   Ibid. Art.8.1.

283   Ibid. Art.8.3.

284   Ibid. Art.8.4.

285   Ibid. Art.8.4.

286   Hedley, et al., 2003, p.36

287   Agreement for the Implementation of the Provisions of the United Nations Convention on the Law of the Sea of 10 December 1982 relating to the Conservation and Management of Straddling Fish Stocks and Highly Migratory Fish Stocks Art.17.1.

288   Ibid. Arts.17.4, 20.7 and 33.2.

some of the tuna RFMOs.[289] Prima facie, such an obligation may conflict with the *pacta tertiis rule* in international law, since such actions would indirectly encroach on the freedom of fishing on the high seas.[290] It has to be considered that even if such measures would help to improve compliance with widely accepted conservation purposes/standards under international law, they still would restrict the freedom of fishing and at worst constitute exclusive claims over high seas resources.

The Fish Stocks Agreement provided further new requirements for the establishment of RFMOs.[291] In areas where RFMOs do not exist, the fishing states must cooperate to establish such an organization and must also participate in its work.[292] The states are required to agree on a range of issues, including the stocks to which any conservation and management measures will apply, on the area of application, on the relationship with other organizations, and on mechanisms for obtaining scientific data.[293] Like several other parts of the Fish Stocks Agreement, these requirements were crucial during the negotiations for the Convention of the WCPFC.[294] Now they are less relevant because in the meantime all areas where tuna and tuna-like species occur are covered by the areas of competence of the five already existing tuna RFMOs.

Mechanisms for international cooperation and the inclusion of non-members in RFMOs were important issues for the 2010 Review Conference. Among others the Conference recommended the strengthening of efforts to agree on the participatory rights of members, new members and cooperating non-members of

---

289 Lack, 2007; B. Le Gallic, The use of trade measures against illicit fishing: Economic and legal considerations, *Ecological Economics - Vol. 64(4)* (2008); C.A. Roheim & J. Sutinen, *Trade and Marketplace Measures to Promote Sustainable Fishing Practices*, ICTSD *Natural Resources, International Centre for Trade and Sustainable Development*, Series Issue Paper No. 3 (Geneva: 2006); M. W. Lodge, et al., *Recommended Best Practices for Regional Fisheries Management Organizations* (London: Chatham House, 2007), pp.54-60, Fabra, et al., 2011; A detailed analysis of these measures and their application in tuna RFMOs is provided in chapter 3.

290 Franckx, 2000, pp.13-20.

291 Agreement for the Implementation of the Provisions of the United Nations Convention on the Law of the Sea of 10 December 1982 relating to the Conservation and Management of Straddling Fish Stocks and Highly Migratory Fish Stocks Art.9.

292 Ibid. Art.8.5.

293 Ibid. Arts.9(a)-(d).

294 Convention on the Conservation and Management of Highly Migratory Fish Stocks in the Western and Central Pacific Ocean.

RFMOs.[295] The delegations stated that, particularly to help address the problem of illegal, unreported and unregulated fishing, all states with a real interest in the fisheries should be able to become full members of RFMOs, and that the status of a cooperating non-member should not be indefinite, but rather serve as a stepping stone towards full membership.[296] Many delegations emphasized the need to improve the effectiveness of the management measures adopted by RFMOs, and to increase transparency in decision-making processes and compliance with agreed measures.[297] More widely, the Conference recommended a modernization of all RFMOs in order to reflect the modern approaches of more recent international instruments.[298]

Unfortunately, no recommendations were made with regard to the allocation of fishing quotas via RFMOs, despite the fact that some delegations noted that Article 10 of the Agreement had not established suitable criteria and that transparency and fairness in the development of such criteria was necessary.[299] The Conference did not, therefore, take the opportunity to give further guidance on this important issue. This also applies to guidance on the participation of intergovernmental organizations and non-governmental organizations. No recommendations were adopted, despite the fact that delegations noted the lack of opportunities for participation.[300]

d)   *Compliance and enforcement*

The provisions of UNCLOS on compliance and enforcement did not effectively prevent overfishing. Prior to the Fish Stocks Agreement, illegal and unregulated fishing in particular, including in the fisheries of tuna and tuna-like species, presented a serious problem. Fishing vessel owners often circumvented the provisions on compliance and enforcement by reflagging their vessels to so-called flag of convenience states which did not had, nor enforced adequate fisheries legislation at all. The Fish Stocks Agreement, as one of several instruments adopted after the 1992 Rio Conference, addressed these issues with provisions that went beyond those of UNCLOS. The most important provisions of the

---

295   United Nations General Assembly, *Report of the Resumed Review Conference on the Agreement for the Implementation of the Provisions of the United Nations Convention on the Law of the Sea of 10 December 1982 relating to the Conservation and Management of Straddling Fish Stocks and Highly Migratory Fish Stocks*, (2010) Annex, Para.II(h).

296   Ibid. Paras.82-84.

297   Ibid. Para.74.

298   Ibid. Annex, Para.II(a).

299   Ibid. Para.85.

300   Ibid. Paras.86-87.

Agreement cover flag state duties, regional cooperation and port state measures.[301]

According to the Fish Stocks Agreement, a flag state has to make sure that any vessels flying its flag and fishing on the high seas comply with the conservation and management measures of the relevant RFMOs.[302] UNCLOS did not indicate what specific measures had to be adopted by the flag state for managing these vessels. In contrast, the Fish Stocks Agreement contains numerous provisions dealing with the duties of the flag state. The measures to be taken by the flag state include: the control of vessels on the high seas; the establishment of regulations and a national record of authorized vessels; the marking of fishing vessels; the recording and reporting of positions, catches and effort; the verification of the catch; monitoring, control and surveillance; the regulation of transshipment; and the ensuring of compliance with sub-regional, regional and global measures.[303] The flag state itself is required to enforce these measures and to investigate potential violations immediately and fully.[304]

Another important issue regarding compliance and enforcement is the requirement for the states parties to strengthen regional cooperation.[305] Therefore, the Fish Stocks Agreement includes, amongst others, provisions for mutual boarding and inspection procedures on the high seas.[306] These provisions, which went beyond those of UNCLOS are intended to solve the problem of 'free-riding'. They lay down that parties to the Fish Stocks Agreement which are also members of an RFMO may board and inspect the vessels of other states party to the Agreement found in the high seas area covered by that RFMO.[307] It is not necessary that the boarded vessel´s state is also a member of the respective RFMO, as long as both states are party to the Fish Stock Agreement. Vessels

---

301 Agreement for the Implementation of the Provisions of the United Nations Convention on the Law of the Sea of 10 December 1982 relating to the Conservation and Management of Straddling Fish Stocks and Highly Migratory Fish Stocks, Arts.18-23.

302 Ibid. Art.18.1.

303 Ibid. Art.18.3.

304 Ibid. Arts.19.1(a) and (b).

305 Ibid. Art.20.1.

306 Ibid. Art.21.

307 Ibid. Art.21.1.

which are neither parties to the Fish Stocks Agreement or to the respective RFMO cannot be boarded due to the *pacta tertiis rule*.[308]

The Fish Stocks Agreement contains further progressive provisions with regard to the adoption of port state measures.[309] These measures are another way to ensure compliance of fishing vessels. Port states have both the right and the obligation to take measures, in accordance with international law, to promote the effectiveness of conservation and management measures adopted by RFMOs, as long as they do not discriminate against the vessels of any particular state.[310] Among other measures, port states are allowed to inspect documents, fishing gear and any catch on board fishing vessels, but only when such vessels are voluntarily in its ports or at its offshore terminals.[311]

During the 2010 Resumed Review Conference, some progress was noted on monitoring, control and surveillance, and compliance and enforcement, but the delegations also identified several areas for improvement. Generally, it was recognized that flag states performed poorly with regard to complying with the requirements of the international legal instruments.[312] The Conference recommended the full implementation of flag state responsibilities, and the development of criteria to assess flag state performance and to address persistent failure.[313] It was highlighted that a special focus on the implementation of the detailed provisions of the Agreement for both national waters and the high seas was required.[314] The Conference recommended an annual assessment of compliance with the measures of RFMOs and, where appropriate, the cooperation by non-members with those measures.[315] Particular concerns were raised with regard to IUU fishing. While some delegations supported the development of a

---

308 E. Franckx, *Pacta Tertiis and the Agreement for the Implementation of the Provisions of the United Nations Convention on the Law of the Sea of 10 December 1982 relating to the Conservation and Management of Straddling Fish Stocks and Highly Migratory Fish Stocks* FAO Legal Papers - Vol. 8 (Rome: FAO, 2000), p.19.

309 Agreement for the Implementation of the Provisions of the United Nations Convention on the Law of the Sea of 10 December 1982 relating to the Conservation and Management of Straddling Fish Stocks and Highly Migratory Fish Stocks, Art.23.1.

310 Ibid. Art.23.1.

311 Ibid. Art.23.2.

312 United Nations General Assembly, *Report of the Resumed Review Conference on the Agreement for the Implementation of the Provisions of the United Nations Convention on the Law of the Sea of 10 December 1982 relating to the Conservation and Management of Straddling Fish Stocks and Highly Migratory Fish Stocks*, (2010) Para.93.

313 Ibid. Annex, Para.III(d).

314 Ibid. Paras.88-89.

315 Ibid. Annex, Para.III(a).

new instrument, others called for coastal states to be allowed to intervene where flag states were unwilling or unable to take action against vessels flying their flag.[316] The adoption of the FAO Agreement on Port State Measures to Prevent, Deter and Eliminate Illegal, Unreported and Unregulated Fishing (Port States Measures Agreement) was seen by many delegations as crucial to combating IUU fishing.[317]

The effort made by the RFMOs to strengthen compliance and enforcement schemes was recognized by the delegations. However, a lack of compliance had been identified in RFMO performance reviews, and several delegations urged the RFMOs to make efforts to further strengthen their schemes through independent observer coverage on board, international boarding and inspection, and harmonized catch documentation.[318] The efforts of ICCAT and WCPFC, which had led to improvements in both their schemes, were positively received. As a result, IUU vessels in the ICCAT Convention area had decreased from 500 a year to fewer than 10, and the WCPFC, following its adoption of a boarding and inspection scheme in accordance with Articles 21 and 22 of the Agreement, had conducted 28 high-seas boardings and inspections within 12 months.[319]

Further topics were transshipment and market-related measures, both of which were also seen as critical tools for combating IUU fishing. It was recommended that states and RFMOs should prevent illegally harvested fish or fish products from entering the market, develop catch documentation schemes and other market-related measures, strengthen law enforcement cooperation and facilitate the commerce of sustainably-caught fish.[320] A number of delegations expressed concern over the transshipment of catches on the high seas and acknowledged the difficulty of monitoring fishing activities in that area.[321] They proposed that transshipment should only occur in designated ports in order to allow more reliable monitoring of catches. Several delegations emphasized the need for RFMOs to extend control measures throughout the whole market chain, including through the adoption of catch documentation schemes.[322] The Conference

---

316   Ibid. Paras.94-96.

317   The Port States Measures Agreement will be discussed in another sub-chapter.

318   United Nations General Assembly, *Report of the Resumed Review Conference on the Agreement for the Implementation of the Provisions of the United Nations Convention on the Law of the Sea of 10 December 1982 relating to the Conservation and Management of Straddling Fish Stocks and Highly Migratory Fish Stocks*, (2010) Paras.100-103.

319   Ibid. Para.104.

320   Ibid. Annex, Para.III(c).

321   Ibid. Para.106.

322   Ibid. Para.107.

recommended strengthening measures to monitor and regulate transshipment activity and to support the monitoring of vessels fishing on the high seas.[323]

e)   *Dispute settlement*

The Fish Stocks Agreement provides a compulsory and binding dispute settlement mechanism to resolve conflicts in a peaceful manner.[324] This mechanism had been described as one of the three pillars of the whole Agreement.[325] All of the dispute settlement provisions of UNCLOS[326] have been incorporated in the Fish Stocks Agreement and they apply, mutatis mutandis, "to any dispute between parties to this Agreement concerning the interpretation or application of this Agreement",[327] as well as "to any dispute between States Parties to this Agreement concerning the interpretation or application of a subregional, regional or global fisheries agreement relating to straddling fish stocks or highly migratory fish stocks to which they are parties, including any dispute concerning the conservation and management of such stocks".[328] That means that the UNCLOS provisions not only concern the interpretation or application of disputes in respect of the Agreement, but also regarding the interpretation or application of the constitutive instruments of the tuna RFMOs.[329] Particularly interesting is the extended application of the UNCLOS provisions.[330] Now they

---

323   Ibid. Annex, Para. III(g).

324   Agreement for the Implementation of the Provisions of the United Nations Convention on the Law of the Sea of 10 December 1982 relating to the Conservation and Management of Straddling Fish Stocks and Highly Migratory Fish Stocks Arts.27-32.

325   S. N. Nandan, *Statement of the chairman, ambassador Satya N. Nandan, on 4 August 1995, upon the adoption of the agreement for the implementation of the provisions of the United Nations Convention on the Law of the Sea of 10 December 1982 relating to the conservation and management of straddling fish stocks and highly migratory fish stocks*, Sixth session of the United Nations Conference on straddling fish stocks and highly migratory fish stocks (New York, USA: United Nations, 1995) The two other identified pillars are the conservation and management measures based on the precautionary approach and the provisions for compliance and enforcement.

326   United Nations Convention on the Law of the Sea, Part XV.

327   Agreement for the Implementation of the Provisions of the United Nations Convention on the Law of the Sea of 10 December 1982 relating to the Conservation and Management of Straddling Fish Stocks and Highly Migratory Fish Stocks, Art.30.1.

328   Ibid. Art.30.2.

329   Hedley, et al., 2003, p.ix.

330   L. Juda, 'The United Nations Fish Stocks Agreement', *Yearbook of International Co-operation on Environment and Development* (London: Earthscan Publications, 2001), p.55.

apply to all parties to the Agreement, "whether or not they are also parties to UNCLOS."[331]

A new provision in the Fish Stocks Agreement is an additional step prior to binding dispute settlement. Disputing states can refer their dispute to an ad hoc expert panel which has been set up by the disputing states.[332] This panel will then confer with the states concerned in order to resolve the dispute, ideally without recourse to binding procedures. Overall, the dispute settlement regime within the Fish Stocks Agreement marked a real step forward. For disputing parties, it provided a further option for resolving and managing disputes, and its binding nature allowed for negotiations during disputes to be improved or at least speed up.[333] So far there has been no recourse to this dispute settlement regime.

2. FAO Compliance Agreement

The Agreement to Promote Compliance with International Conservation and Management Measures by Fishing Vessels on the High Seas (Compliance Agreement) was adopted on 24 November 1993 by the FAO Conference through Resolution 15/93, and it entered into force on 24 April 2003. The adoption had followed the FAO Technical Consultation Council on High Seas Fishing[334] and the 102nd session of the FAO Council in 1992 where it was "agreed that the issue of reflagging of fishing vessels into flags of convenience to avoid compliance with agreed conservation and management measures, [...] should be addressed immediately by FAO, with a view to finding a solution which could be implemented in the near future."[335] The main reason for the negotiations towards the Compliance Agreement was the fact that some of the vessels fishing, particularly those on the high seas, were flying the flags of states which were not able or willing to control their activities.[336] Vessel owners often changed their flags to "flags of convenience" in order to circumvent the compli-

---

331   Agreement for the Implementation of the Provisions of the United Nations Convention on the Law of the Sea of 10 December 1982 relating to the Conservation and Management of Straddling Fish Stocks and Highly Migratory Fish Stocks, Arts.30.1 and 30.2.

332   Ibid. Art.29.

333   Örebech, et al., 1998, p.133.

334   FAO, *Report of the Technical Consultation on High Seas Fishing, FAO Fisheries Report - No. 484*, (1992), Para.45.

335   FAO, *Report of the FAO Council - 102nd Session*, (1992), Para.58.

336   Kaye, 2001, pp.213-214.

ance measures to which their states of origin were obliged to adhere.[337] The resulting lack of compliance affected sustainable management of fisheries in general but in particular those fisheries targeting high seas stocks like tuna and tuna-like species.

The Compliance Agreement reiterates the provisions of UNCLOS regarding the effective control of vessels fishing on the high seas and it specifies the requirements for flag states regarding the compliance with international conservation and management measures on the high seas.[338] It complements the Fish Stocks Agreement and some provisions like those on the record of fishing vessel have also been imported in the Fish Stocks Agreement.[339] With currently 39 parties the Compliance Agreement has received, in comparison to the Fish Stocks Agreement,[340] significantly less support by the fishing states.[341] Today, even with the overall increased number of parties there are still few parties which are recognized flag of convenience states.[342] Both the low number of parties in general and the low number of parties which are known as flag of convenience states show the limitations of the Compliance Agreement. In order to work effectively the Agreement depends on the willingness of flag states to ratify and implement it.[343]

The overall objective of the Compliance Agreement is "to deal with the problem of fishing vessels re-flagging into flags of convenience to avoid compliance with agreed conservation and management measures."[344] Its provisions apply to fishing vessels that are used or intended for fishing on the high seas.[345] Only

---

337 D. A. Balton, 'The Compliance Agreement', *Developments in international fisheries law* (London: Kluwer Law International, 1999), p.34.

338 Palma, et al., 2010, p.60.

339 Agreement for the Implementation of the Provisions of the United Nations Convention on the Law of the Sea of 10 December 1982 relating to the Conservation and Management of Straddling Fish Stocks and Highly Migratory Fish Stocks, Annex I, Art.4; Agreement to Promote Compliance with International Conservation and Management Measures by Fishing Vessels on the High Seas Art.VI; Meltzer & Fuller, 2009, p.19.

340 The Fish Stocks Agreement has currently 80 parties.

341 http://www.fao.org/fishery/topic/14766/en.

342 Belize, Georgia or Mauritius. For all flags of convenience countries see: http://www.itfglobal. org/ flags-convenience/flags-convenien-183.cfm

343 Meltzer & Fuller, 2009, p.19.

344 J. Swan, *Fishing Vessels operating under open registers and the exercise of flag State responsibilities - Information and options*, FAO Fisheries Circular - No. 980 (Rome, Italy: FAO, 2002), p.11.

345 Agreement to Promote Compliance with International Conservation and Management Measures by Fishing Vessels on the High Seas Art.II(1).

vessels of less than 24 metres are exempted, as long as this exception does not undermine the objective and purpose of the Agreement.[346] With regard to these smaller vessels, the flag state is required to take effective measures should the actions of any such fishing vessel undermine the effectiveness of international conservation and management measures.[347] Article III sets out the responsibilities of the flag state and contains some of the most important provisions of the Agreement.[348] Put simply, any flag state which is party to the Agreement must take the necessary action to ensure that its fishing vessels do not engage in any activity that undermines the effectiveness of international conservation and management measures.[349] This provision applies to all measures, irrespective of whether a state party is member of the RFMO which has adopted the measure.[350] In order to underline the responsibilities of the flag state, it is laid down that no party should allow a vessel to fly its flag while fishing on the high seas unless it is authorized by the appropriate authorities of that party.[351] Further clauses specify the details of this authorization and contain the requirements for the granting or cancellation of such an authorization.[352] The marking of vessels provides information on their operations and facilitates enforcement measures. The Compliance Agreement therefore obliges parties to ensure that any vessels flying their flag are marked in such a way that they can be readily identified in accordance with generally accepted standards.[353] In addition, each party must maintain a record of its fishing vessels and all vessels flying its flag have to be listed in that record.[354] Other provisions of the Compliance Agreement aim on improved international cooperation concerning the exchange of information.[355] To this end, RFMOs are required to promote the objectives of the Agreement.[356]

---

346   Ibid. Art.II(2).

347   Ibid. Art.III(1)(b).

348   W. R. Edeson, et al., *Legislating for sustainable fisheries: a guide to implementing the 1993 FAO Compliance Agreement and 1995 UN Fish Stocks Agreement* (Washington: World Bank, 2001).

349   Agreement to Promote Compliance with International Conservation and Management Measures by Fishing Vessels on the High Seas Art.III(1)(a).

350   Balton, 1999, p.49.

351   Agreement to Promote Compliance with International Conservation and Management Measures by Fishing Vessels on the High Seas Art.III(2).

352   Ibid. Arts.III(4) and (5)(a).

353   Ibid. Arts.III(6-8).

354   Ibid. Art.IV.

355   Ibid. Arts.V and VI.

356   Ibid. Art.V(3).

3. FAO Code of Conduct for Responsible Fisheries

In 1995, the FAO Code of Conduct for Responsible Fisheries (Code of Conduct, or simply Code) was unanimously adopted during the 28th Session of the FAO Conference as "the most complete and up-to-date expression of the principles of sustainable fisheries management".[357] The general need for a code of conduct in international fisheries had already been recognized in 1991 when the FAO Committee on Fisheries called for the development of "guidelines or a code of practice for responsible fishing which would take into account all the technical, socio-economic and environmental factors involved."[358] By that time it was realized "that unregulated fisheries on the high seas, in some cases involving straddling and highly migratory fish species, which occur within and outside EEZs, were becoming a matter of increasing concern."[359] The official negotiation process towards the Code of Conduct commenced with the 1992 Declaration of Cancun which further developed the concept of 'responsible fishing' to encompass "the sustainable utilization of fisheries resources in harmony with the environment."[360] Then, in the course of the negotiations towards the Fish Stocks Agreement and the Compliance Agreement the FAO Governing Bodies made the recommendation to formulate a non-binding instrument consistent with these instruments and establishing principles and standards that cover conservation, management and development of all fisheries.[361]

Due to its non-binding nature it does not, in itself, establish any legal rights or obligations.[362] Nevertheless, it provides a broad framework for national and international efforts to ensure the effective conservation, management and development of living aquatic resources, with due respect for the ecosystem and biodiversity.[363] Moreover, the non-binding character of the Code can even be interpreted as an advantage as it provides the possibility of setting out much more norms and principles than legally binding instruments.[364] Despite its non-

---

357   G. Moore, 'The Code of Conduct for Responsible Fisheries', *Developments in international fisheries law* (The Hague: Kluwer Law International, 1999), p.85.

358   FAO, *Report of the Nineteenth Session of the Committee on Fisheries, FAO Fisheries Report - No. 459*, (1991), Para 82.

359   Code of Conduct for Responsible Fisheries, Preface.

360   Declaration of Cancun, 1992.

361   Code of Conduct for Responsible Fisheries, Preface.

362   Moore, 1999, p.89.

363   Code of Conduct for Responsible Fisheries, Introduction.

364   D. M. Sodik, Non-Legally Binding International Fisheries Instruments and Measures to Combat Illegal, Unreported and Unregulated Fishing, *Australian International Law Journal - Vol. 15(1)* (2008), p.130.

binding character it created a strong urge for implementation. It was noted that especially those provisions dealing with the relationship of the Code to other international agreements were drafted with the precision of a treaty regime.[365] Certain provisions are based on relevant rules of international law, including those reflected in the UNCLOS while other provisions may be or have already been given binding effect by means of other binding instruments like the Compliance Agreement or the Fish Stocks Agreement.[366] It is explicitly stated that the Code is to be interpreted and applied "in a manner consistent with the relevant provisions" of the Fish Stocks Agreement.[367] Consistency with the Fish Stocks Agreement can be particularly identified in the provisions on general principles (Article 6), fisheries management (Article 7) and fishing operations (Article 8).[368]

The Code provides a wide range of objectives[369] which can be summarized in the establishment of "principles and criteria for national and international legal and institutional arrangements and to provide standards of conduct for persons involved in the fishery sector."[370] Similar to the other instruments that have been developed after UNCLOS it is recognized particularly in Articles 7 and 6.12 of the Code that RFMOs are playing the central role with regard to the long term conservation and sustainable use of international fisheries resources.[371] The most important provisions with regard to fisheries management of the Code are laid down in Article 6.[372] Among the inclusive 19 principles which are outlined in this article there are several which address fisheries targeting tuna and tuna-like species. Among others, states and RFMOs should apply the ecosystem[373] and precautionary approach,[374] prevent overfishing and excess fishing capacity,[375] develop further and apply selective and environmentally safe fishing gear and practices,[376] protect nursery and spawning areas[377] and ensure

---

365   W. R. Edeson, The Code of Conduct for Responsible Fisheries: An Introduction, *The International Journal of Marine and Coastal Law- Vol. 11(2)* (1996), p.235.

366   Code of Conduct for Responsible Fisheries, Art.1.1.

367   Ibid. Art.1.1.

368   Meltzer & Fuller, 2009, p.20.

369   Code of Conduct for Responsible Fisheries, Art.2.

370   Swan, 2002, p.15.

371   Meltzer & Fuller, 2009, p.20.

372   Kaye, 2001, p.223.

373   Code of Conduct for Responsible Fisheries, Arts.6.1 and 6.2.

374   Ibid. Art.6.5.

375   Ibid. Art.6.3.

376   Ibid. Art.6.6.

transparent and timely decision making.[378] The Code also includes detailed provisions on the implementation of the precautionary approach.[379] Among others it gives guidance with regard to the development of reference points.[380] States and RFMOs are requested, on the best scientific evidence available, to determine stock specific target and limit reference points, and, at the same time, the action to be taken if they are exceeded.[381] It is further stated that "when a limit reference point is approached, measures should be taken to ensure that it will not be exceeded."[382]

The progress of the implementation of the Code was evaluated in 2009.[383] The evaluation also included with CCSBT, IATTC and WCPFC three tuna RFMOs. It was found that, by that time, only the CCSBT had established stock-specific target reference points,[384] while IATTC and WCPFC had partially applied the precautionary approach,[385] but all of them had taken measures to ensure the compliance of fishing operations with management measures.[386] In the meantime there is a wider application of the precautionary approach and the IOTC has adopted recently precautionary target and limit reference points for some of the main target species.[387] Despite this progress it has to be recognized that the provisions of the precautionary approach laid down in the Code of Conduct and in the Fish Stocks Agreement are still only partially applied.[388]

During the evaluation non-governmental organizations addressed the relevancy of the Code for sustainable fisheries as well. They criticized its poor implementation and detected a lack of awareness, political will, transparency, human

---

377   Ibid. Art.6.8.

378   Ibid. Art.6.13.

379   Ibid. Art.7.5.

380   Ibid. Art.7.5.3.

381   Ibid. Art.7.5.3.a and b(sent.1).

382   Ibid. Art.7.5.3.b(sent.2).

383   68 countries, 14 Regional Fisheries Management Organizations (RFMOs) including the three tuna RFMOs CCSBT, IATTC, and WCPFC as well as six Non Governmental Organizations (NGOs) responded to a questionnaire.

384   FAO, *Progress in the Implementation of the Code of Conduct for Responsible Fisheries, Related International Plans of Action and Strategy, Twenty-eighth Session of the Committee on Fisheries,* (2009), Para.44.

385   Ibid. Para.45.

386   Ibid. Para.46.

387   IOTC, *Recommendation 12/14 On Interim Target And Limit Reference Points,* (2012).

388   A detailed analysis of the application of the precautionary approach through tuna RFMOs is made in chapter III.

and financial resources, scientific information and effective trade control as some of the key problems underlying this.[389] Similar conclusions were reached by a group of scientists that evaluated 53 countries with regard to their compliance with the Code.[390] The results showed the very poor performance of these countries and the authors suggested the establishment of other binding instruments on a national or international level.

4. FAO International Plans of Action

In 1999 and 2001, during two intergovernmental meetings four International Plans of Action (IPOAs) have been adopted in order to improve levels of compliance with the Code of Conduct as envisaged by Article 2(d).[391] All IPOAs were addressing specific fisheries related issues also affecting those fisheries targeting tuna and tuna-like species. Three of these non-binding IPOAs were adopted in 1999,[392] and the fourth one in 2001.[393] Despite the fact that IPOAs have no direct or binding effect at a national level, it can be argued that precisely because they are voluntary instruments they have the potential to achieve wider use than binding instruments which are only implemented by those parties to them.[394]

a) IPOA-Seabirds

The International Plan of Action for Reducing Incidental Catch of Seabirds in Longline Fisheries (IPOA-Seabirds) is non-binding instrument and it applies "to States in the waters of which longline fisheries are being conducted by their own or foreign vessels and to States that conduct longline fisheries on the high seas and in the EEZs of other States."[395] The reason for the development of the

---

389 FAO, *Progress in the Implementation of the Code of Conduct for Responsible Fisheries, Related International Plans of Action and Strategy, Twenty-eighth Session of the Committee on Fisheries,* (2009) Para.57.

390 T. Pitcher, et al., Not honouring the code, *Nature - Vol. 457(5)* (2009).

391 FAO, *Committee on Fisheries, 23rd session* (1999); FAO, *Committee on Fisheries, 24rd Session* (2001).

392 International Plan of Action for Reducing Incidental Catch of Seabirds in Longline Fisheries; International Plan of Action for the Conservation and Management of Sharks; International Plan of Action for the Management of Fishing Capacity.

393 International Plan of Action to Prevent, Deter, and Eliminate Illegal, Unreported and Unregulated Fishing (IUU).

394 Palma, et al., 2010, pp.19-20.

395 International Plan of Action for Reducing Incidental Catch of Seabirds in Longline Fisheries Paras.8 and 9.

IPOA-Seabirds was the fact that seabirds, such as albatrosses, petrels, fulmars and gulls, were often being caught accidentally by longline vessels like those targeting tuna and tuna like-species.[396] Therefore, the overall objective of this IPOA which is referring to Articles 7.6.9 and 8.5 of the Code of Conduct was to "reduce the incidental catch of seabirds in longline fisheries where this occurs."[397]

All states with longline fisheries were requested to conduct an assessment of these fisheries in order to determine if a problem exists with respect to incidental catch of seabirds[398] and to adopt National Plans of Action (NPOAs) to reduce unintentional catches.[399] In particular, the plans should provide optional technical and operational mitigation measures.[400] These measures are basically intended to prevent access by seabirds to baited hooks. The technical measures include weighting the longline gear,[401] setting the line below the water,[402] or bird scaring curtains.[403] The operational measures include night setting,[404] area and seasonal closures[405] or the release of live birds. [406] With regard to the implementation of the IPOA all states were requested to cooperate through RFMOs.[407]

In 2009, the FAO Committee of Fisheries evaluated the in the implementation of the IPOAs. By that time a total of 38 percent of the evaluated countries had assessed their longline fisheries with regard to seabird bycatches and the overall implementation rate of NPOA-Seabirds was 78 percent.[408] The three tuna

---

396   Ibid. Paras.1 and 2.

397   Ibid. Para.10.

398   Ibid. Para.12.

399   Ibid. Para.12.

400   Ibid. Para.16.

401   International Plan of Action for Reducing Incidental Catch of Seabirds in Longline Fisheries, Technical note on some optional technical and operational measures for reducing the incidental catch of seabirds in longline fisheries, 1999, Para II.1(a).

402   Ibid. Para II.5.

403   Ibid. Para II.2.

404   Ibid. Para.III.1.

405   Ibid. Para.III.3.

406   Ibid. Para.III.5.

407   International Plan of Action for Reducing Incidental Catch of Seabirds in Longline Fisheries Paras.19 and 20.

408   FAO, *Progress in the Implementation of the Code of Conduct for Responsible Fisheries, Related International Plans of Action and Strategy, Twenty-eighth Session of the Committee on Fisheries,* (2009), Para.36.

RFMOs CCSBT, IATTC and WCPFC which were evaluated had made efforts towards the implementation of the IPOA-Seabirds including the adoption of special conservation measures and the introduction of devices to minimize seabird interaction.[409]

b) IPOA-Sharks

The International Plan of Action for the Conservation and Management of Sharks (IPOA-Sharks) is non-binding but all concerned States are encouraged to implement it.[410] The main reason for this IPOA was "concern over the increase of shark catches and the consequences which this has for the populations of some shark species in several areas of the world's oceans."[411] Particular concerns were related to the limited knowledge of sharks and the practices employed in fisheries catching them.[412] In tuna fisheries, sharks were often being caught as bycatch species.[413] The IPOA considers the multiple pressures on sharks, it applies to all shark species both as target and non-target catches[414] and it provides a framework for their conservation and management.[415]

The states concerned were requested to adopt NPOAs for the conservation and management of shark stocks (Shark plan).[416] These plans were supposed to ensure the sustainability of shark catches, to assess threats to shark populations, to contribute to the protection of biodiversity and ecosystem as well as to minimize waste and discards from shark catches.[417] The Shark plans should contain a description of the prevailing state of: shark stocks and populations, associated fisheries, management framework and its enforcement as well as objectives and strategies for achieving these objectives.[418] Possible strategies were, among oth-

---

409 Ibid. Para.52.

410 International Plan of Action for the Conservation and Management of Sharks, Para.10.

411 Ibid. Para.2.

412 Ibid. Para.3.

413 Mainly longline, purse seine, trolling.

414 International Plan of Action for the Conservation and Management of Sharks Paras.11 and 12.

415 E. J. Techera, Good Environmental Governance: Overcoming Fragmentation in International Law for Shark Conservation and Management, *American Society of International Law Proceedings - Vol. 105* (2011),p.105; H. Edwards, When Predators Become Prey: The Need for International Shark Conservation *Ocean and Coastal Law Journal - Vol. 12(2)* (2007), p.308.

416 International Plan of Action for the Conservation and Management of Sharks Para.18.

417 Ibid. Para.22.

418 Ibid. Appendix A, Para.II.

ers, a decrease in fishing effort, better data collection and monitoring of shark fisheries or to obtain utilization and trade data on shark species.[419] States were to cooperate through RFMOs to ensure the effective conservation and management of the stocks.[420]

By 2009, only 50 percent of the evaluated countries had assessed the need for an NPOA to implement the IPOA-Shark. However, all tuna RFMOs[421] and several other bodies had initiated efforts to collect information about sharks, and to assess their stock status.[422] At least IATTC and WCPFC had implemented conservation measures specifically for sharks and were promoting research on alternative gear types to minimize shark bycatch.[423]

c)   IPOA-Capacity

The International Plan of Action for the Management of Fishing Capacity (IPOA-Capacity) is non-binding but States and RFMOs "should apply [it] consistently with international law and within the framework of the respective competencies of the organizations concerned."[424] By the time of the adoption of the IPOA overcapacity was recognized as one of the main reasons for overfishing and unsustainable fisheries.[425] Tuna fisheries in particular were facing overcapacity with increasing overall capacity rates.[426] The objective of the IPOA-Capacity was therefore to achieve - and by 2005 at the latest - the efficient, equitable and transparent management of fishing capacity on a global level.[427] Where overcapacity existed, capacity was to be limited and progressively reduced, while growth in capacity was to be avoided in other areas.

States were requested to assess their capacity and to identify national fisheries and fleets where urgent measures were required.[428] NPOAs for the manage-

---

419   Ibid. Appendix A, Para.II.C.

420   Ibid. Paras.25 and 26.

421   WCPFC did not exist yet when the plan was adopted.

422   FAO, *Progress in the Implementation of the Code of Conduct for Responsible Fisheries, Related International Plans of Action and Strategy, Twenty-eighth Session of the Committee on Fisheries,* (2009), Para.6.

423   Ibid. Para.51.

424   International Plan of Action for the Management of Fishing Capacity Paras.4 and 5.

425   Ibid. Para.1.

426   Joseph, 2003.

427   International Plan of Action for the Management of Fishing Capacity, Para.7.

428   Ibid. Paras.13 and 14.

ment of fishing capacity sholud have been adopted by the end of 2002.[429] In addition, states should cooperate through RFMOs[430] and immediate steps were urged with regard to overfished transboundary, straddling, highly migratory and high seas stocks.[431] Individual, bilateral and multilateral action was requested in order "to reduce substantially the fleet capacity applied to these resources as part of management strategies to restore overfished stocks to sustainable levels."[432]

In practice, the progress of the implementation of the IPOA-Capacity was quite slow. In 2006, during the Review Conference of the Fish Stocks Agreement, it was concluded that the implementation of the IPOA-Capacity was far from complete.[433] By 2009, only 40 percent of the evaluated countries had finished the preliminary assessment.[434] At least IATTC and WCPFC had made efforts towards the implementation of the IPOA-Capacity, which included plans of action, limiting effort and catch by introducing new measures, and attempting to control the number of vessels based on a record of fishing vessels authorize to fish.[435]

d)   IPOA-IUU

The International Plan of Action to Prevent, Deter, and Eliminate Illegal, Unreported and Unregulated Fishing (IPOA-IUU) is non-binding[436] but it was described as one of the most important global instrument concerning fisheries since the Fish Stocks Agreement and FAO Code of Conduct for Responsible

---

429   Ibid. Paras.21 and 22.

430   Ibid. Para.27.

431   Ibid. Para.39.

432   Ibid. Para.40.

433   United Nations General Assembly, *Report of the Review Conference on the Agreement for the Implementation of the Provisions of the United Nations Convention on the Law of the Sea of 10 December 1982 relating to the Conservation and Management of Straddling Fish Stocks and Highly Migratory Fish Stocks*, (2006), Annex, Para.11.

434   FAO, *Progress in the Implementation of the Code of Conduct for Responsible Fisheries, Related International Plans of Action and Strategy, Twenty-eighth Session of the Committee on Fisheries*, (2009), Para.34.

435   Ibid. Para.50.

436   International Plan of Action to Prevent, Deter, and Eliminate Illegal, Unreported and Unregulated Fishing (IUU), Para.4.

Fisheries in 1995.[437] Some experts concluded that, due to the prominence of the issue of IUU fishing, the IPOA-IUU would become the most accepted voluntary instrument since the Code of Conduct.[438] It was a response to the ineffectiveness of previous international instruments addressing IUU fishing like the UNCLOS and the Compliance Agreement.[439] The main objective of this IPOA was to prevent, deter and eliminate IUU fishing by providing all states with comprehensive, effective and transparent measures.[440]

States were requested to give full effect to the relevant norms of international law. Particularly Articles 1.1, 1.2, 3.1, and 3.2 of the Code of Conduct apply to the interpretation and application of the IPOA-IUU and its relationship with other international instruments.[441] The IPOA has to be interpreted and applied in a manner consistent with the UNCLOS,[442] the Compliance Agreement and the Fish Stocks Agreement.[443] Although it was mainly designed for the high seas, it applies also within the EEZ.[444] States should address all aspects of IUU fishing through their national legislation,[445] to take measures or cooperate to ensure that nationals subject to their jurisdiction did not support or engage in

---

437 W. R. Edeson, International Plan of Action on Illegal Unreported and Unregulated Fishing: The Legal Context of a Non-Legally Binding Instrument, *The International Journal of Marine and Coastal Law* - *Vol. 16(4)* (2001), p.603.

438 Palma, et al., 2010, pp.16-20.

439 The term IUU fishing refers according to Paragraph 3 of the IPOA-IUU to several illegal and undesired activities like fishing in national waters without permission or in contravention of the laws and regulations of the coastal State; flying the flag of RFMO contracting parties without complying with the respective conservation and management measures or other relevant provisions of the applicable international law; fishing that has not been reported, or misreported, to the relevant national authority; fishing in the area of competence of a relevant RFMO which have not been reported or have been misreported, in contravention of the reporting procedures of that organization; fishing of vessels without nationality in an area of a relevant RFMO or by a fishing entity, in a manner not consistent with or contravening the respective conservation and management measures; fishing in areas or for fish stocks without applicable conservation or management measures and in a manner inconsistent with State responsibilities for the conservation of living marine resources under international law.

440 International Plan of Action to Prevent, Deter, and Eliminate Illegal, Unreported and Unregulated Fishing (IUU), Para.8.

441 Ibid. Para.5.

442 Ibid. Paras.10.

443 Ibid. Paras.11.

444 Ibid. Para.51.

445 Ibid. Paras.16-17.

IUU fishing[446], to ensure that sanctions were of sufficient severity,[447] to take measures against non-cooperating states,[448] to undertake comprehensive and effective monitoring, control and surveillance,[449] and to cooperate through the relevant RFMOs[450].

However, by 2009, less than 30 percent of states had established NPOAs to combat IUU fishing.[451] In contrast, all evaluated tuna RFMOs[452] were able to detail their efforts towards the implementation of the IPOA-IUU. These efforts included the introduction of strengthened MCS measures such as port state measures, the implementation of trade monitoring and control measures and the listing of authorized fishing vessels and IUU fishing vessels.[453]

5.   FAO Port State Measures Agreement

The Agreement on Port State Measures to Prevent, Deter and Eliminate Illegal, Unreported and Unregulated Fishing (Port State Measures Agreement) was approved by the FAO Governing Conference in November 2009.[454] One year later 17 FAO members had signed it already.[455] To date an additional six members have signed the Agreement[456] and two members have acceded to it[457]. Only

---

446   Ibid. Paras.18-19.

447   Ibid. Para.21.

448   Ibid. Para.22.

449   Ibid. Para.24.

450   Ibid. Para.28.

451   FAO, *Progress in the Implementation of the Code of Conduct for Responsible Fisheries, Related International Plans of Action and Strategy, Twenty-eighth Session of the Committee on Fisheries*, (2009), Para.33.

452   CCSBT, IATTC and WCPFC.

453   FAO, *Progress in the Implementation of the Code of Conduct for Responsible Fisheries, Related International Plans of Action and Strategy, Twenty-eighth Session of the Committee on Fisheries*, (2009), Para.53.

454   FAO, *Report of the Twenty-eighth Session of the Committee on Fisheries*, (2009), Para.65.

455   Angola, Australia, Benin, Brazil, Chile, European Union (EU), Gabon, Iceland, Indonesia, New Zealand, Norway, Peru, Russian Federation, Samoa, Sierra Leone, United States of America and Uruguay; D. J. Doulman & J. Swan, *A guide to the background and implementation of the 2009 FAO Agreement on Port State Measures to Prevent, Deter and Eliminate Illegal, Unreported and Unregulated Fishing*, FAO Fisheries and Aquaculture Circular No. 1074 (Rome: FAO, 2012), p.24.

456   Canada, France, Ghana, Kenya, Mozambique and Turkey; http://www.fao.org/fishery/topic/166283/en#Efforts.

457   Myanmar and Sri Lanka; http://www.fao.org/fishery/topic/166283/en#Efforts.

three members have ratified it so far.[458] The Agreement will enter into force "thirty days after the date of deposit with the Depositary of the twenty-fifth instrument of ratification, acceptance, approval or accession".[459] In 2012 it was expected that this is "likely to take three to five years."[460]

The new Agreement which is largely based on the FAO Model Scheme on Port State Measures to Combat Illegal, Unreported and Unregulated Fishing[461] is supposed to become the first international binding agreement to stop illegally caught fish from entering international markets through ports.[462] Previously adopted instruments like the Code of Conduct for Responsible Fisheries and in the IPOA-IUU had already encouraged the enhanced implementation of port state measures but there was no binding international instrument on explicit port states measures.[463] The reason for this is that the development and the implementation of port state measures are within the sovereign discretion of each coastal state.[464] According to international law, the coastal State has full sovereignty over its ports.[465] This is also reflected by the Fish Stocks Agreement where the provisions on port state measures are concluded with the requirement that nothing in the respective article "affects the exercise by States of their sovereignty over ports in their territory in accordance with international law."[466] The Port State Measures Agreement is respecting the sovereignty of parties within their areas of national jurisdiction. Nothing shall be construed to affect "the sovereignty of Parties over their internal, archipelagic and territorial waters or their sovereign rights over their continental shelf and in their exclu-

---

458   Chile, European Union, and Norway;
       http://www.fao.org/fishery/topic/166283/en#Efforts.

459   Agreement on Port State Measures to Prevent, Deter and Eliminate Illegal, Unreported and Unregulated Fishing Art.29.1.

460   Doulman & Swan, 2012, p.25.

461   FAO Model Scheme on Port State Measures to Combat Illegal, Unreported and Unregulated Fishing, 2007.

462   FAO, *Report of the Twenty-seventh Session of the FAO Committee on Fisheries*, (2007), Para.68.

463   Code of Conduct for Responsible Fisheries, Art.8.3. International Plan of Action to Prevent, Deter, and Eliminate Illegal, Unreported and Unregulated Fishing (IUU), Paras.52-64.

464   Palma, et al., 2010, p.63.

465   United Nations Convention on the Law of the Sea Arts.25 and 218.

466   Agreement for the Implementation of the Provisions of the United Nations Convention on the Law of the Sea of 10 December 1982 relating to the Conservation and Management of Straddling Fish Stocks and Highly Migratory Fish Stocks, Art.23.4.

sive economic zones"[467] and "the exercise by Parties of their sovereignty over ports in their territory in accordance with international law."[468] In addition, a party to the Agreement is only bound by the provisions of the Agreement itself and not by the measures or decisions of any RFMOs of which it is not a member.[469]

The Port State Measures Agreement basically aims at resolving issues in the regulatory framework for fisheries-related operations in ports, such as the different standards among members of RFMOs in terms of inspection procedures, their information requirements for vessels intending to enter into the port or the penalties imposed.[470] The overall objective of the Agreement is to implement port state measures that effectively ensure the long-term conservation and sustainable use of living marine resources and marine ecosystems.[471] It applies to vessels which seek entry to ports or are already in a port but not flying the flag of the respective port state. Exempted from this requirement are the vessels of neighbouring states that are engaged in artisanal fishing, and container vessels that are not carrying fish or only fish that has been previously landed.[472] Port states can decide not to apply the Agreement to vessels fishing in their EEZs and operating under their authority. In this case, the vessels are subject to measures by the port state itself.[473]

The Agreement calls for a broader cooperation and exchange of information at sub-regional, regional and global levels between relevant states, the FAO and RFMOs.[474] Parties must designate and publicize the ports to which vessels may request entry and provide a list of these ports to the FAO, and they must ensure that every port listed has sufficient capacity to conduct the required inspections.[475] In order to decide about the access to their ports, the port state has to require certain information, as laid down in Annex A of the Agreement.[476] The final decision about access has to be communicated to the vessel or to its repre-

---

467   Agreement on Port State Measures to Prevent, Deter and Eliminate Illegal, Unreported and Unregulated Fishing Art.4.1(a).

468   Ibid. Art.4.1(b).

469   Ibid. Art.4.2.

470   Palma, et al., 2010, p.64.

471   Agreement on Port State Measures to Prevent, Deter and Eliminate Illegal, Unreported and Unregulated Fishing, Art.2.

472   Ibid. Art.3.1.

473   Ibid. Art.3.2.

474   Ibid. Art.6.

475   Ibid. Art.7.

476   Ibid. Art.8.1.

sentative.[477] In cases where entry or use of the port has been denied, the port state has to inform the flag state and, as appropriate, the relevant coastal states, RFMOs and other international organizations.[478] Within its provisions for the use of ports, the Agreement has established a series of reasons for banning a vessel from landing, transshipping, packaging and processing fish that has not been previously landed.[479] This includes a mandatory ban if a flag state does not confirm, within a reasonable period of time from the request of the port state that the fish on board was taken in accordance with the applicable requirements of a relevant RFMO.[480] Based on these provisions, port states can get significant power to enforce international law if they find evidence that vessels are involved in any IUU fishing.[481]

The Agreement further defines when and how inspections should be carried out, and requires the party states to agree on minimum levels for inspection of vessels through RFMOs, the FAO or otherwise.[482] Requirements for the reporting of inspection results are laid down in Annex C.[483] The collected information can be forwarded to other port States or to RFMOs and it can be used to track vessels engaged in IUU fishing.[484] To ensure the efficient electronic exchange and sharing of information, parties are required to cooperate to establish proper mechanisms.[485] The relevant RFMOs have to provide information concerning the measures or decisions they have adopted and implemented.[486]

Flag states are obliged to require their vessels to cooperate with the port state in inspections.[487] Where a party has evidence that a vessel flying its flag is involved in IUU fishing or other activities in support of such fishing, it must request that the port state inspects the vessel or take other measures consistent

---

477 Ibid. Art.9.1.

478 Ibid. Arts.9.3 and 11.3.

479 Ibid. Art.11.1.

480 Ibid. Art.11.1(d).

481 A. Sharp, *The Effectiveness or Not of the New Port State Measures in the Battle to Control Illegal, Unregulated and Unreported Fishing* (Auckland: University of Auckland - Department of Commercial Law, 2010), p.9.

482 Agreement on Port State Measures to Prevent, Deter and Eliminate Illegal, Unreported and Unregulated Fishing Arts.12 and 13 and Annex B.

483 Ibid. Art.14.

484 Sharp, 2010, p.8.

485 Agreement on Port State Measures to Prevent, Deter and Eliminate Illegal, Unreported and Unregulated Fishing Arts.16.1 and 16.2.

486 Ibid. Art.16.5.

487 Ibid. Art.20.1.

with the Agreement.[488] RFMOs, together with the FAO, represent the institutional framework for the development of fair, transparent and non-discriminatory procedures for the identification of such states.[489]

A remaining issue for the Port States Measure Agreement is the poor ratification. During the Resumed Review Conference on the Fish Stocks Agreement, many delegations re-emphasized the importance of the Port State Measures Agreement and called for further ratification, requesting that RFMOs should encourage their members to ratify it, as both IOTC and ICCAT had already done.[490] Accordingly, the Conference recommended that states should be encouraged to become party to the Port States Measures Agreement and to adopt the relevant measures through RFMOs.[491] Although ratification is still in an early phase it has to be recognized that ultimately, the effect of the Port State Measures Agreement is relying on its implementation. Without broad ratification, implementation will remain inconsistent, leaving potential loopholes for IUU fishing.[492]

D. Summary

The analysis has revealed that several international instruments addressing the management of fish, including tuna and tuna-like species, have been developed since the global expansion of industrial fishing in the second half of the twentieth century. In general, these developments were characterized by increasing clarity and specificity over time, while new and more progressive instruments were introduced with the aim of tackling the errors of the past.

It has been shown that the basic framework for the regulation of all fishing activities on tuna and tuna-like species was initially provided by UNCLOS. This treaty defined crucial rights and obligations regarding the different legal zones and different species of fish. Other areas where UNCLOS introduced important provisions include: compliance and enforcement; international obligations for the protection of the environment; scientific research; and the settlement of disputes. Unfortunately, the provisions for the high seas in particular could not

---

488   Ibid. Art.20.2.

489   Ibid. Art.20.3.

490   United Nations General Assembly, *Report of the Resumed Review Conference on the Agreement for the Implementation of the Provisions of the United Nations Convention on the Law of the Sea of 10 December 1982 relating to the Conservation and Management of Straddling Fish Stocks and Highly Migratory Fish Stocks*, (2010) Paras.97-98.

491   Ibid. Annex, Para.III(b)

492   S. Flothmann, et al., Closing Loopholes: Getting Illegal Fishing Under Control, *Scienceexpress - Vol. 328* (2010), p.1236.

prevent the decline of several tuna and tuna-like species. The main causes of this decline were the overutilization of stocks coupled with the inadequate adoption, monitoring and enforcement of effective conservation and management measures, but there were also issues of overcapitalization, excessive fleet size, non-selective gear, vessel reflagging, unreliable databases and a lack of cooperation between states. As a result, during the 1992 Rio Conference, it was recognized that additional instruments were necessary in order to fill the gaps still left after UNCLOS and to clarify any ambiguous provisions.

One of these instruments - the Fish Stocks Agreement - became the most important treaty for the management of straddling and highly migratory fish stocks. The new instrument introduced progressive provisions with regard to conservation and management, the precautionary approach, the compatibility of EEZ and high seas measures, international cooperation, compliance and enforcement and dispute settlement. The Fish Stocks Agreement became an important example of global collaboration and was crucial for the creation of new RFMOs, such as the Western and Central Pacific Fisheries Commission (WCPFC), and the modernizing of older ones, such as the Inter-American Tropical Tuna Commission (IATTC).

Two Review Conferences, in 2006 and then in 2010, assessed the implementation of the Fish Stocks Agreement by both fishing countries and RFMOs. This assessment has identified positive developments regarding the adoption of some conservation and management measures, efforts in data collection and sharing, the development of performance reviews, the cooperation of tuna RFMOs, certain monitoring, control and surveillance operations and the provision of monetary assistance for developing countries. Major issues that were criticized included: continuing overcapacity; the insufficient application of the precautionary and ecosystem approach; a lack of cooperation between coastal states and other states fishing on the high seas; the need for modernization and better cooperation among all RFMOs; poor compliance by flag states, including on IUU fishing; the poor participation of developing states as well as their pending ratification of the Fish Stocks Agreement; and participation in RFMOs by all states involved in the fishery.

Most of the conclusions of the Review Conferences can also be applied to the implementation of other post-Rio instruments, in particular the Compliance Agreement and the Code of Conduct for Responsible Fisheries, but also in part the four International Plans of Action (IPOAs) on seabird bycatch, shark bycatch, overcapacity and IUU fishing. The Compliance Agreement established new requirements for flag states, the marking of vessels and the exchange of information, while the Code of Conduct for Responsible Fisheries addressed conservation, fisheries management, technology, research, port state measures and trade. An evaluation of the implementation of the Code of Conduct for Responsible Fisheries conducted by the FAO detected poor levels of compliance by fishing countries. The level of compliance with the International Plans of Action, at least by some of the RFMOs, including the WCPFC, was much better. Those RFMOs had made real efforts to implement the IPOAs and to establish relevant conservation and management measures. There are now high expectations for the more recent Port States Measures Agreement.

Taken together, these international legal instruments form a comprehensive framework that empowers the relevant policy-makers to manage tuna and tuna-like species sustainably. In particular, the newer instruments are providing a good basis for the development of more specific instruments or conservation and management measures at a regional level. Today, the main challenge for the sustainable management of the fisheries appears to be the effective implementation of these existing instruments and not the suitability of their provisions. The following chapters will analyze how, in practice, the various instruments have been implemented by RFMOs and how they have influenced their performance.

Chapter II

International disputes regarding the sustainable exploitation of tuna and tuna-like species

A. Introduction

After the adoption of the UNCLOS in 1982 there had been two remarkable international disputes on the sustainable exploitation of tuna and tuna like species. In the respective cases on southern bluefin tuna and south-eastern pacific swordfish binding dispute settlement procedures under Part XV of UNCLOS became necessary because the conflicting parties were not able to resolve the respective matters bilaterally. This chapter will present the crucial aspects of these dispute settlement procedures and analyzes their influence on future cases. The disputes over the fishing, importation, and sale of tuna caught with methods that are lethal to dolphins will not be discussed here as the focus of this work is on the conservation and management of tuna and tuna-like species.[493]

The first dispute between Australia and New Zealand on the one side, and Japan on the other was on sustainable exploitation of southern bluefin tuna. Australia and New Zealand requested, after unsuccessful negotiations with Japan, a binding dispute settlement procedure. In a first step, pending the constitution of an Arbitral Tribunal, the International Tribunal for the Law of the Sea delivered an order wherein the judges prescribed provisional measures. In a second step, the Arbitral Tribunal found that the ITLOS had no jurisdiction to rule the merits of the case and requested the parties to resolve the issue by peaceful means under the Convention of the Commission for the Conservation of Southern Bluefin Tuna (CCSBT). This decision caused critique but it became effective and the parties had to solve the issue within the framework of the CCSBT.

The second dispute on swordfish in the Eastern Pacific Ocean between the European Community (EC) and Chile involved proceedings at the ITLOS but also at the World Trade Organisation (WTO). Both parties requested the constitution

---

493 For information on the dolphin disputes see B. Kingsbury, The Tuna-Dolphin Controversy, The World Trade Organization, and the Liberal Project to Reconceptualize International Law, (1994); Dale D. Murphy, The tuna-dolphin wars, *Journal of World Trade - Vol. 40(4)* (2006); E. Trujillo, The WTO Appellate Body Knocks Down U.S. "Dolphin-Safe" Tuna Labels But Leaves a Crack for PPMs, *The American Society of International Law ASIL - Vol. 16(25)* (2012).

of a dispute settlement panel at the WTO and the establishment of a special chamber pursuant to Article 15, Paragraph 2 of the Statute of the ITLOS. However, the actual proceedings were rather short because less than one year after the formal requests both parties agreed on a Provisional Arrangement. They requested to suspend the process for the constitution of the dispute settlement panel at the WTO and before the special chamber. The special chamber agreed to this approach and almost ten years later the dispute could be solved bilaterally.

B.   Dispute on Southern Bluefin Tuna

1.   *History of the dispute*

Southern bluefin tuna is one of the most valuable tuna species and occurs usually throughout the southern ocean south of 30° south.[494] Industrial fishing for this species began in the 1950s and reached the highest catch levels with 81,000 tonnes in 1961.[495] Since then catches declined until in the mid-1980s and it became obvious that the stock was seriously depleted. Therefore, the main fishing countries Australia, Japan and New Zealand initiated informal negotiations leading in 1985 to a first total allowable catch (TAC) of 38,650 tonnes. Despite further significant reductions of the TAC to 11,750 tonnes in 1989 (Japan: 6,065 tonnes, Australia: 5,265 tonnes, New Zealand: 420 tonnes) the stock size continued to decline. In 1997 the stock was estimated to be only between 7-15% of the 1960 level.

In 1994, still before ratifying UNCLOS, Australia, Japan and New Zealand had formalized their cooperation by establishing the Commission for the Conservation of Southern Bluefin Tuna (CCSBT).[496] The objective of the Convention of the CCSBT was "to ensure, through appropriate management, the conservation and optimum utilisation of southern bluefin tuna."[497] The same year the CCSBT adopted the TAC of 11,750 tonnes while maintaining the national shares determined in 1989.

Problems began when Japan, in contrast to the other parties intended to increase its share by trying to add to its TAC first 6000 tonnes and then 3000

---

494   Collette & Nauen, 1983, p.87.

495   Southern Bluefin Tuna Case - Australia and New Zealand v. Japan, Arbitral Award of August 4, 2000, (2000), Para.22.

496   For more detailed information of the CCSBT , see the following chapter.

497   Convention for the Conservation of Southern Bluefin Tuna, 1993, Art.3.

tonnes.[498] Japan further proposed a joint Experimental Fishing Program (EFP) in order to reduce scientific uncertainty regarding the recovery of the stock. Although there was significant uncertainty about the stock status and the migratory patterns of the southern bluefin it might be at least suggested that the actual purpose of the EFP was a greater nominal catch.[499] Due to the opposition by Australia and New Zealand the initial TAC was maintained until 1998. However, a consequence of the dispute was that in the following years the parties were unable to agree on a TAC.

2. Dispute settlement under the Convention of the CCSBT

In 1996 the Commission of the CCSBT had adopted a set of 'objectives and principles for the design and implementation of an experimental fishing program' but without specific catch sizes or specific modalities of execution.[500] Therefore it caused protest by Australia and New Zealand when during a meeting of the CCSBT in 1998, Japan announced an unilateral three-year EFP with an approximate catch of 1,464 mt per year.[501]

As a response to this unilateral action, Australia and New Zealand requested dispute settlement according to Article 16(1) of the Convention of the CCSBT in 1999. Article 16(1) lays down that in case of disputes concerning the interpretation or implementation of the Convention, the relevant Parties are required to "consult among themselves with a view to having the dispute resolved by negotiation, inquiry, mediation, conciliation, arbitration, judicial settlement or other peaceful means of their own choice."[502] Unfortunately the following negotiations among the three parties could not reach mutual agreement and Japan announced to continue its EFP. Thereby, Japan ignored the warning by Australia and New Zealand which announced that they would recognize such behavior as a termination of the negotiations under Article 16(1) of the Convention of

---

498   Southern Bluefin Tuna Case - Australia and New Zealand v. Japan, Arbitral Award of August 4, 2000, (2000), Para.24.

499   M. Haward & A. Bergin, The political economy of Japanese distant water tuna fisheries, *Marine Policy - Vol. 25(2)* (2001), pp.98-99; T. Polacheck, Experimental catches and the precautionary approach: the Southern Bluefin Tuna dispute, *Marine Policy - Vol. 26(4)* (2002), pp.289-290.

500   CCSBT, *Objectives and principles for the design and implementation of an experimental fishing program*, (1996).

501   Southern Bluefin Tuna Case - Australia and New Zealand v. Japan, Arbitral Award of August 4, 2000, (2000), Para.25.

502   Convention for the Conservation of Southern Bluefin Tuna, Art.16.1.

the CCSBT.[503] Therefore, as a result of the continuance of the EFP Australia communicated that in its opinion a full exchange of views on the dispute according to Article 283(1) of UNCLOS had already taken place and it announced further steps under the UNCLOS.[504]

3. *Dispute settlement through the ITLOS*

Article 279 of UNCLOS lays down, that disputes between States Parties have to be settled by peaceful means. If under Part XV Section 1 of UNCLOS no settlement could be reached through negotiation, conciliation or other means, any party can request the submission of the dispute to a court or to a tribunal under Section 2. One possible tribunal is the International Tribunal for the Law of the Sea (ITLOS) which had been established in accordance with Annex VI of UNCLOS.[505] The other means are the International Court of Justice, an arbitral tribunal constituted in accordance with Annex VII and a special arbitral tribunal constituted in accordance with Annex VIII for certain types of disputes including fisheries, protection and preservation of the marine environment or marine scientific research.[506]

On 15th July, 1999 New Zealand and Australia submitted identical notifications to Japan and the ITLOS wherein they stated that Japan´s unilateral experimental fishing was in contravention of UNCLOS.[507] In its request for the prescription of provisional measures two weeks later both parties basically argued that "Japan's current and proposed unilateral actions in relation to southern bluefin tuna taken in the context of a stock at historically low levels, [would] increase the threat to that stock and undermine the disciplines of the accepted scheme for southern bluefin tuna management."[508] In a statement of claim New Zealand and Australia added that Japan had "placed itself in breach of its obligations under international law, specifically articles 64 and 116-119 of UNCLOS, and in relation thereto Article 300 of UNCLOS and the precautionary principle which,

---

503   Southern Bluefin Tuna Case - Australia and New Zealand v. Japan, Arbitral Award of August 4, 2000, (2000), Para.26.

504   Ibid. Para.27.

505   United Nations Convention on the Law of the Sea, Art.287(a).

506   Ibid. Art.287(b-d).

507   K. Leggett, The Southern Bluefin Tuna Cases: ITLOS Order on Provisional Measures, *Review of European Community & International Environmental Law - Vol. 9(1)* (2000), p.76.

508   Southern Bluefin Tuna Cases (New Zealand v. Japan; Australia v. Japan) - Request for the Prescription of Provisional Measures submitted by New Zealand, ITLOS, (1999), Para.15; Southern Bluefin Tuna Cases (New Zealand v. Japan; Australia v. Japan) - Request for the Prescription of Provisional Measures submitted by Australia ITLOS, (1999), Para.15.

under international law, must direct any party in the application and implementation of those articles."[509] The two parties identified violation of Article 119 of UNCLOS which requires the adoption of conservation measures designed to maintain or restore the southern bluefin tuna stocks at levels which can produce the maximum sustainable yield.[510] The additional fishing of the seriously depleted stock was seen as a threat for the recovery of the stock to levels which would permit sustainable harvests approaching maximum sustainable yield.[511]

The notifications of 15th July requested Japan to cease experimental fishing within two weeks and to adopt provisional measures. If Japan would not follow the request the dispute would be decided by an Arbitral Tribunal which would have jurisdiction over this dispute pursuant to Article 288(1) of UNCLOS.[512] Pending the constitution of this Arbitral Tribunal, Australia communicated its intention to seek prescription of provisional measures through the ITLOS.[513] The argumentation was based on Article 290(5) of UNCLOS, stating that "[p]ending the constitution of an arbitral tribunal to which a dispute is being submitted under this section, any court or tribunal agreed upon by the parties or, failing such agreement within two weeks from the date of the request for provisional measures, the International Tribunal for the Law of the Sea [...] may prescribe, modify or revoke provisional measures in accordance with this article if it considers that prima facie the tribunal which is to be constituted would have jurisdiction and that the urgency of the situation so requires. Once constituted, the tribunal to which the dispute has been submitted may modify, revoke or affirm those provisional measures".

---

509   Southern Bluefin Tuna Cases (New Zealand v. Japan; Australia v. Japan) - Annex 2 New Zealand's Statement of Claim dated 15 July 1999, ITLOS, (1999), Para.45; Southern Bluefin Tuna Cases (New Zealand v. Japan; Australia v. Japan) - Annex 2 Australia's Statement of Claim dated 15 July 1999, ITLOS, (1999), Para.45.

510   Southern Bluefin Tuna Cases (New Zealand v. Japan; Australia v. Japan) - Annex 2 New Zealand's Statement of Claim dated 15 July 1999, (1999), Para.58; Southern Bluefin Tuna Cases (New Zealand v. Japan; Australia v. Japan) - Annex 2 Australia's Statement of Claim dated 15 July 1999, (1999), Para.58.

511   Southern Bluefin Tuna Cases (New Zealand v. Japan; Australia v. Japan) - Annex 2 New Zealand's Statement of Claim dated 15 July 1999, (1999), Para.59; Southern Bluefin Tuna Cases (New Zealand v. Japan; Australia v. Japan) - Annex 2 Australia's Statement of Claim dated 15 July 1999, (1999), Para.59.

512   Southern Bluefin Tuna Cases (New Zealand v. Japan; Australia v. Japan) - Request for the Prescription of Provisional Measures submitted by New Zealand, (1999), Para.22; Southern Bluefin Tuna Cases (New Zealand v. Japan; Australia v. Japan) - Request for the Prescription of Provisional Measures submitted by Australia (1999), Para.22.

513   Southern Bluefin Tuna Case - Australia and New Zealand v. Japan, Arbitral Award of August 4, 2000, (2000), Para.28.

As Japan did not follow the notifications, on 30th July, 1999, New Zealand and Australia submitted in accordance with Article 283(1) of UNCLOS two identical requests to the ITLOS. Therein they requested the prescription of provisional measures, including that Japan immediately ceases unilateral experimental fishing; that Japan restricts its catch to its national allocation as last agreed in the CCSBT, subject to the reduction of the catches made in the course of unilateral experimental fishing in 1998 and 1999 and that all parties act consistently with the precautionary principle in fishing for SBT pending a final settlement of the dispute.[514]

Japan, however, raised critique on the procedure chosen by New Zealand and Australia. It argued that the two parties would have come to the wrong forum for resolving what in its view were 'baseless claims'.[515] It accused Australia and New Zealand for having frustrated the functioning of the CCSBT. From Japans point of view, Article 64 did not stipulate specific principles of conservation or concrete conservation measures.[516] Further it was argued that the precautionary principle was not incorporated in UNCLOS and that the status of the principle as a rule of customary international law was not clear.[517] In sum according to the argumentation by Japan, an arbitral tribunal according to Annex VII of UNCLOS would not have the authority prima facie and therefore the ITLOS would not have the authority to prescribe provisional measures.[518] Japan concluded that the claims were still a matter of concern for the CCSBT because the procedures under the Convention of the CCSBT had not been exhausted. It therefore proposed resumed negotiations under the Convention of the CCSBT in order to agree on TAC, annual quota and a joint EFP with the assistance of independent scientific advice.

On 16th August 1999 the ITLOS communicated in an order that it would join the proceedings upon the requests for provisional measures.[519] During the hearings

---

514 Southern Bluefin Tuna Cases (New Zealand v. Japan; Australia v. Japan) - Request for the Prescription of Provisional Measures submitted by New Zealand, (1999), Para.9; Southern Bluefin Tuna Cases (New Zealand v. Japan; Australia v. Japan) - Request for the Prescription of Provisional Measures submitted by Australia (1999), Paras, 8(1)-8(3).

515 Southern Bluefin Tuna Cases (New Zealand v. Japan; Australia v. Japan) - Response of the Government of Japan to Request For Provisional Measures & Counter-Request For Provisional Measures, ITLOS, (1999), Para.1.

516 Ibid. Para.54.

517 Ibid. Para.55.

518 Southern Bluefin Tuna Case - Australia and New Zealand v. Japan, Arbitral Award of August 4, 2000, (2000), Para.34.

519 Southern Bluefin Tuna Cases (New Zealand v. Japan; Australia v. Japan) - Order 1999/4, ITLOS, (1999).

which were held at the ITLOS from 18th to 20th August 1999 all parties basically restated the terms of their requests, respectively its response submitted prior to the ITLOS. In the final order of 27th August 1999 a large majority of the judges prescribed six provisional measures and made two decisions on how to implement these measures.[520]

By 20 votes to 2, the judges ordered that all parties had to ensure that no action is taken which might aggravate or extend the disputes submitted to the arbitral tribunal and that no action is taken which might prejudice the carrying out of any decision on the merits which the arbitral tribunal may render;[521] by 18 votes to 4, that all parties had to ensure that their annual catches would not exceed the TAC of 11,750 tonnes (5,265 tonnes, 6,065 tonnes and 420 tonnes) while taking account of the catches made during the EFP;[522] by 20 votes to 2, to refrain from conducting an EFP involving the taking of a catch of southern bluefin tuna, except with the agreement of the other parties or unless the experimental catch is counted against its annual national allocation;[523] by 21 votes to 1, that the parties resume negotiations in order to reach agreement on conservation and management measures for southern bluefin tuna;[524] and by 20 votes to 2, that the parties should make further efforts to reach agreement with other States and fishing entities engaged in fishing for southern bluefin tuna.[525]

The judges decided by 21 votes to 1 that each party had to submit a report referred to in article 95, paragraph 1, of the Rules of the ITLOS[526] not later than 6 October 1999 and authorizes the President of the ITLOS to request such further reports and information.[527] By 21 votes to 1, it was further decided, in accordance with article 290, paragraph 4, of UNCLOS and Article 94 of the Rules of the ITLOS that the provisional measures prescribed, without delay had to be communicated to all States Parties to the UNCLOS participating in the fishery for southern bluefin tuna.[528]

---

520 Southern Bluefin Tuna Cases (New Zealand v. Japan; Australia v. Japan) - Order 1999/5, ITLOS, (1999).

521 Ibid. Orders 1(a) and1(b).

522 Ibid. Order 1(c).

523 Ibid. Order 1(d).

524 Ibid. Order 1(e).

525 Ibid. Order 1(f).

526 Rules of the Tribunal, ITLOS, (2009).

527 Southern Bluefin Tuna Cases (New Zealand v. Japan; Australia v. Japan) - Order 1999/5, (1999), Order 2.

528 Ibid. Order 3.

Although the judges did not explicitly state that they had applied the precautionary approach the wording of their order may be interpreted like that.[529] This presumably progressive approach was based on weak legal ground because at the time when the order was made the precautionary approach was not accepted as a rule of customary law. After the decision it had been criticized that the ITLOS "in the absence of evidence of an established principle of international law applied common sense and morality rather than positive law."[530]

4. Dispute settlement through the Arbitral Tribunal

In their order from 27th August 1999 the judges also found that an Arbitral Tribunal would prima facie have jurisdiction over the disputes.[531] Therefore, parallel to the final order, an Arbitral Tribunal with five arbitrators was established according to Articles 3(b)-(d) of UNCLOS, Annex VII. From 7th May 2000, hearings on the jurisdiction of the Arbitral Tribunal were held at the World Bank headquarters in Washington. In this hearings Japan contested the jurisdiction of the Arbitral Tribunal arguing that the dispute would only address the interpretation and implementation of the Convention of the CCSBT and not of the UNCLOS.[532] According to Japan the Convention of the CCSBT was not only lex posterior but also lex specialis compared to UNCLOS.[533]

In its final submission Japan declared that the Arbitral Tribunal should adjudge and declare that the case has become moot and should be discontinued. The main argument for that claim was a proposal by Australia in 1999 which had provided an EFP limit of 1500 tonnes.[534] This proposal was interpreted by Japan as an indicator for a general acceptance of the EFP. Further, Japan argued that the Tribunal does not have jurisdiction over the claims made by the applicants in this case and that the claims are not admissible.[535] New Zealand and Austral-

---

529 A. L. Erickson, Out of Stock: Strengthening International Fishery Regulations to Achieve a Healthier Ocean, *North Carolina Journal of International Law and Commercial Regulation - Vol. 34(1)* (2008), p.311.

530 S. Marr, The Southern Bluefin Tuna cases: the precautionary approach and conservation and management of fish resources, *European Journal of International Law - Vol. 11(4)* (2000), p.816.

531 Southern Bluefin Tuna Cases (New Zealand v. Japan; Australia v. Japan) - Order 1999/5, (1999), Para.62.

532 Southern Bluefin Tuna Case - Australia and New Zealand v. Japan, Arbitral Award of August 4, 2000, (2000), Para.38(a).

533 Ibid. Para.38(c).

534 Ibid. Paras.45-46.

535 Ibid. Para.42.

ia made the final submission that the Parties differ on the question whether Japan's EFP and associated conduct is governed by UNCLOS, that a dispute thus exists about the interpretation and application of UNCLOS within the meaning of Part XV, that all the jurisdictional requirements of that Part have been satisfied and that Japan's objections to the admissibility of the dispute are unfounded.[536]

In its answer the Tribunal addressed the submissions. It decided that the case was not moot because among others New Zealand and Australia never had accepted Japan´s offer to reduce the EPF to 1500 tonnes.[537] Further the Tribunal argued that the dispute was rather on the quality of the program than on its quantity. Regarding the question whether the conflict arose solely under the Convention of the CCSBT or also under UNCLOS the Tribunal could not identify any reason why a given act of a State may not violate its obligations under more than one treaty.[538] The conclusion of an implementing convention was not seen as necessarily vacating the obligations imposed by the framework convention upon the parties to the implementing convention.[539] Therefore the Tribunal concluded that it was a case of a single dispute that arose under both conventions.[540]

Therefore, the crucial question was whether the requirements for a compulsory dispute settlement procedure under Part XV Section 2 of UNCLOS were met.[541] The Tribunal communicated that according to Article 281(1) the provisions of Part XV apply only if no settlement by peaceful means has been reached and the parties did not exclude any further procedure.[542] The Tribunal came to the conclusion that Article 16 of the Convention of the CCSBT "exclude[s] any further procedure" within the contemplation of Article 281(1) of UNCLOS.[543] Therefore the final decision was that jurisdiction had to be declined.[544] In its

---

536   Ibid. Para.43.

537   Ibid. Para.46.

538   Ibid. Paras.47-52.

539   Ibid. Para.52.

540   Ibid. Para.54.

541   V. Röben, The Southern Bluefin Tuna Cases: Re-Regionalization of the Settlement of Law of the Sea Disputes?, *Zeitschrift für ausländisches öffentliches Recht und Völkerrecht - Vol. 62* (2002), p.62.

542   Southern Bluefin Tuna Case - Australia and New Zealand v. Japan, Arbitral Award of August 4, 2000, (2000), Paras.55 and 56.

543   Ibid. Para.59.

544   C. Romano, The Southern Bluefin Tuna dispute: Hints of a world to come ... like it or not, *Ocean Development and International Law - Vol. 32(4)* (2001), p.330.

award on 4th August 2000, the Arbitral Tribunal decided four judges to one that it had no jurisdiction to rule on the merits of the dispute.[545] As a consequence it decided that in accordance with Article 290(5) of UNCLOS the provisional measures in force by order of the ITLOS had to be revoked from the day of signature of the award.[546]

This Tribunal had been the first Arbitral Tribunal established under Part XV and Annex VII of UNCLOS.[547] Its final decision was subject to a controversial debate.[548] The main critique was raised on the Tribunal's interpretation of the Convention of the CCSBT.[549] It is doubtful if the provisions of Article 16 can be used as the basis for an exclusion of further procedures as required under Article 281(1) of UNCLOS. Article 16.2 states that any dispute not resolved by peaceful means "shall, with the consent in each case of all parties to the dispute, be referred for settlement to the International Court of Justice or to arbitration". The approach of the Tribunal implied that in the future any claim under UNCLOS that raises issues under a regional agreement might not be adjudicated by UNCLOS.[550] This outcome has the potential to weaken the compulsory procedures under Part XV of UNCLOS. Further it can be questioned if a Tribunal which finds that it has no jurisdiction in a case can have the jurisdiction to revoke provisional measures.[551] Due to these facts it may be concluded that an Arbitral Tribunal might not be the best means to address UNCLOS disputes.[552]

5.   Progress after the decision of the Arbitral Tribunal

Despite the formal critique it has to be recognized that the Arbitral Tribunal has supported the progress already made by the parties and it reminded them that

---

545   Southern Bluefin Tuna Case - Australia and New Zealand v. Japan, Arbitral Award of August 4, 2000, (2000), Para.72.1.

546   Ibid. Para.72.2.

547   Ibid. Para.44.

548   Romano, 2001. pp.331-334; Dean Bialek, Australia and New Zealand v Japan: Southern Bluefin Tuna Case, *Melbourne Journal of International Law - Vol. 1* (2000), pp.157-160; A. Boyle & M. D. Evans, The Southern Bluefin Tuna Arbitration, *The International and Comparative Law Quarterly* (2001), pp.448-451; Röben, 2002, pp.63-67.

549   Romano, 2001, p.331.

550   Röben, 2002, p.66; M. Lodge, *Recommended best practices for regional fisheries management organizations*, Report of an independent panel to develop a model for improved governance by Regional Fisheries Management Organizations (London: Chatham House, 2007), p.81.

551   Röben, 2002, p.67.

552   Boyle & Evans, 2001, p.452.

they have to solve the problem within the framework of the CCSBT.[553] After the decision of the Tribunal, during a special meeting in from 16-18 November 2000, the parties agreed within the framework of the CCSBT to initiate a process to develop a Management Procedure (MP) for the southern bluefin tuna fishery.[554] The MP which was supposed to be adopted in 2004 is a pre-agreed set of rules that can specify changes to the total allowable catch (TAC) based on updated monitoring data.[555] However, in 2004, the adoption date was postponed because of data problems and concerns regarding the validity of the fisheries data. It resulted that since at least the early 1990s there had been substantial under-reporting of Japanese longline catches.[556] This finding hampered the development of the MP and showed the limits of the dispute settlement procedures under Part XV of UNCLOS. In 2011 the MP called 'Bali Procedure' was finally adopted by the extended Commission of the CCSBT.[557] This MP is basically used to set the global TAC and to rebuild the stock.

C.   Dispute on Swordfish in the South-Eastern Pacific Ocean

1.   History of the dispute

Out of the group of billfishes swordfish is probably the most popular food fish. It occurs in tropical, temperate and sometimes cold waters of all oceans.[558] Usually in summer swordfish migrates towards temperate or cold waters for feeding and in autumn it moves back to warm waters for spawning and overwintering. Like most of the other tuna and tuna-like species it migrates through different EEZs and between the EEZs and the high seas.

In the 1990s Chile and the European Community (EC) had controversies on the conservation and sustainable exploitation of swordfish.[559] The controversy was based on Chile's fear that the EC catches on the high seas adjacent to its EEZ

---

553   B. Mansfield, 'Compulsory dispute settlement after the Southern Bluefin Tuna award', *Oceans Management in the 21st Century: Institutional Frameworks and Responses* (Leiden: Martinus Nijhoff, 2004), pp.265-266.

554   CCSBT, *Report of the Special Meeting*, (2000).

555   See also http://www.ccsbt.org/site/management_procedure.php.

556   Kolody, et al., 2008, p.341.

557   CCSBT, *Report of the Special Meeting*, (2011).

558   Nakamura, 1985, p.49.

559   M. A. Orellana, The Swordfish Dispute between the EU and Chile at the ITLOS and the WTO, *Nordic Journal of International Law - Vol. 71* (2002), p.55.

would lead to overexploitation of the stocks within the EEZ.[560] As a consequence Chile implemented Article 165 of the Chilean National Fisheries Law which prohibits unloading and transit of all swordfish catches in Chile's ports by all vessels if catches are made on the high seas bordering Chile's EEZ in contravention of Chilean conservation rules.[561] Chile justified its approach with lack of cooperation by the EC in order to ensure conservation of the highly migratory species according to the requirements of particularly Articles 64 (cooperation in conservation of highly migratory species) and 116-119 conservation of living resources of the high seas) of UNCLOS.

The EC did not agree because by the Chilean approach it saw its economic interests threatened. It complained about limited access to third country markets, particularly regarding the fresh swordfish and it identified a violation of Articles V and XI of the General Agreement on Tariffs and Trade of 1994 (GATT).[562] Article V of the GATT requires freedom of transit for goods through the territory of each contracting party[563] and Article XI lays down that quantitative restrictions on import and export are prohibited subject to import restrictions on agricultural or fisheries products.[564]

Throughout the 1990s both parties communicated bilaterally and initiated certain types of cooperation in order to solve the controversies, but without success. As a consequence, two parallel proceedings to solve the dispute had been initiated by the two parties in 2000. The EC initiated a WTO dispute settlement proceeding to prove consistency with the GATT and Chile invited the EC to establish a formal dispute settlement procedure under Part XV of UNCLOS in order to assess if the dispute was a commercial matter or a matter of conservation.

2. Dispute settlement under the GATT

On 19th April 2000 the EC submitted a formal request to the WTO stating that Chile's measures to prohibit the unloading of swordfish in Chilean ports, established pursuant to Article 165 of Chilean Fishery Law, would be in breach of

---

560 P-T. Stoll & S. Vöneky, The Swordfish Case: Law of the Sea v. Trade, *Zeitschrift für ausländisches öffentliches Recht und Völkerrecht - Vol. 62* (2002), p.21.

561 Chilean National Fishery Law (Ley General de Pesca y Acuicultura) - Consolidated by the Supreme Decree 430 of 28 September 1991, and extended by Decreee 598 of 15 October 1999, 1991.

562 Decision of the Commission of the European Communities - Provisions of the Council Regulation No. 3268/94, Document 300D0296, Para.17, 2000.

563 The General Agreement on Tariffs and Trade, 1994, Art.V.2.

564 Ibid. Art.XI.1 and Art XI.2.c.

Articles V and XI of the GATT.[565] The following consultations between the parties, held on 14th June 2000 enhanced the mutual understanding but did not result in a satisfactory solution of the matter. Therefore, on 6th November 2000, the EC sent to the Chairman of the WTO Dispute Settlement Body a request for the Establishment of a Panel pursuant to Article 6 of the Understanding on Rules and Procedures Governing the Settlement of Disputes (DSU) and Article XXIII of the GATT.[566]

In the following months the parties negotiated bilaterally and reached a solution on the case which was called 'Chile - Measures affecting the Transit and Importation of Swordfish'. On 23 March 2001 the EC circulated a communication from the Permanent Delegation of the European Commission.[567] Therein it was laid down that the EC and Chile had come to a Provisional Arrangement regarding the dispute. In this Arrangement both parties had agreed to suspend the process for the constitution of the panel. The EC maintained the right to revive the proceedings at any time. On 28th March 2001 Chile circulated a communication similar to the one by the EC, confirming the Provisional Arrangement and confirming the intention to suspend the process for the constitution of the panel.[568]

In 2003, 2005 and 2007 the EC and Chile informed the Chairman of the Dispute Settlement Body on the successful implementation of the Arrangement.[569] Both parties reiterated their commitment to further develop the Arrangement, to set-

---

565   Chile - Measures Affecting the Transit and Importation of Swordfish - Request for Consultations by the European Communities of 26 April 2000, WTO, (2000).

566   Chile - Measures Affecting the Transit and Importation of Swordfish - Request for the Establishment of a Panel by the European Communities of 6 November 2000, WTO, (2000).

567   Chile - Measures Affecting the Transit and Importation of Swordfish - Arrangement between the European Communities and Chile - Communication by the European Communities of 23 March 2001, WTO, (2001).

568   Chile - Measures Affecting the Transit and Importation of Swordfish - Arrangement between the European Communities and Chile - Communication from Chile - Addendum of 28 March 2001, WTO, (2001).

569   Chile - Measures Affecting the Transit and Importation of Swordfish - Arrangement between the European Communities and Chile - Communication from Chile and the European Communities - Addendum of 12 November 2003, WTO, (2003); Chile - Measures Affecting the Transit and Importation of Swordfish - Arrangement between the European Communities and Chile - Communication from the European Communities - Addendum of 21 December 2005, WTO, (2005); Chile - Measures Affecting the Transit and Importation of Swordfish - Arrangement between the European Communities and Chile - Communication from the European Communities - Addendum of 13 December 2007, WTO, (2007).

tle the dispute by peaceful means and their decision to maintain the suspension of the process for the constitution of the Panel.

3. *Dispute settlement under the UNCLOS*

On 19th December 2000 Chile and the EC instituted proceedings in the "Case concerning the Conservation and Sustainable Exploitation of Swordfish Stocks in the South-Eastern Pacific Ocean (Chile/European Union)".[570] The government of Chile proposed on behalf of both parties to establish, instead of an arbitrary tribunal, a special chamber pursuant to Article 15, Paragraph 2 of the Statute of the ITLOS.[571] Both parties requested the chamber to decide on a series of issues subject to compulsory procedures entailing binding decisions under Part XV of the UNCLOS.

In particular, Chile requested the chamber to decide whether the EC has complied with its obligations under Articles 116-119 of the UNCLOS on the conservation of swordfish in the high seas adjacent to Chile's EEZ[572] and under Article 64 regarding the direct cooperation with Chile as a coastal State including reporting of catches and other information to the competent international organization and to the coastal State.[573] Chile also requested a decision on whether the EC had challenged the sovereign right and duty of Chile, as a coastal State, to prescribe measures within its national jurisdiction for the conservation of swordfish and to ensure their implementation in its ports, in a non-discriminatory manner, as well as the measures themselves, and whether such challenge would be compatible with the UNCLOS.[574] A further request was whether the obligations arising under Articles 300 and 297.1(b) of the UNCLOS, as well as the general thrust of the UNCLOS in that regard, had been fulfilled in this case by the EC.[575] Article 300 requires the parties to "fulfil in good faith the obligations assumed under th[e] Convention and [to] exercise the rights, jurisdiction and freedoms recognized in th[e] Convention in a manner which would not constitute an abuse of right." According to Article 297.1(b) compulsory procedures entailing binding decisions involve "that a State in exercising [...] freedoms, rights or uses has acted in contravention of this Convention or of laws or

---

570 Case concerning the Conservation and Sustainable Exploitation of Swordfish Stocks in the South-Eastern Pacific Ocean (Chile / European Union) - Order 2000/3, (2000).

571 Ibid., Order 2, Order 2.9 and Order 3.

572 Ibid. Order 2.3(a).

573 Ibid. Order 2.3(b).

574 Ibid. Order 2.3(c).

575 Ibid. Order 2.3(d).

regulations adopted by the coastal State in conformity with this Convention and other rules of international law not incompatible with this Convention".

The EC requested whether the Chilean Decree 598 which was the basis for Chile's unilateral conservation measures is in breach of Articles 87, 89 and 116-119 of the UNCLOS.[576] Another request was whether the "Galapagos Agreement" was negotiated compliant to the provisions of the UNCLOS and whether its substantive provisions are in consonance with Articles 64 and 116 to 119 of the UNCLOS.[577] The EC further requested decision on whether Chile's actions concerning the conservation of swordfish are in conformity with article 300 of the UNCLOS and whether Chile and the EC remain under a duty to negotiate an agreement on cooperation under article 64 of the UNCLOS as well as whether the jurisdiction of the special chamber extends to this issue.[578]

The ITLOS decided to accept the request from the two parties and to form a special chamber with five judges.[579] However, on 9th March 2001 Chile and the EC informed the president of the chamber that they had reached a provisional arrangement concerning the dispute. They therefore requested to suspend the proceedings before the special chamber while reserving the rights to reviving the proceedings at any time. Their request was accepted by the president on 15th March 2001. He adopted the decision by an order[580] which was renewed at the request of the parties in 2003, 2005 and 2007.[581]

On 20th, and 23th October 2008 the two parties submitted letters to the ITLOS stating that they had agreed on a draft text of a new 'Understanding concerning the conservation of swordfish stocks in the South Eastern Pacific Ocean' with the final goal to discontinue the proceedings before the Special Chamber.[582] The

---

576   Ibid. Order 2.3(e).

577   Ibid. Order 2.3(f).

578   Ibid. Order 2.3(g) and 2.3(h).

579   Ibid. Decision of the ITLOS Para.1.

580   Case concerning the Conservation and Sustainable Exploitation of Swordfish Stocks in the South-Eastern Pacific Ocean (Chile / European Union) - Order 2001/1, ITLOS, (2001), Orders 4-6.

581   Case concerning the Conservation and Sustainable Exploitation of Swordfish Stocks in the South-Eastern Pacific Ocean (Chile / European Union) - Order 2003/2, ITLOS, (2003), Order 5; Case concerning the Conservation and Sustainable Exploitation of Swordfish Stocks in the South-Eastern Pacific Ocean (Chile / European Union) - Order 2005/1, ITLOS, (2005); Case concerning the Conservation and Sustainable Exploitation of Swordfish Stocks in the South-Eastern Pacific Ocean (Chile / European Union) - Order 2007/3, ITLOS, (2007).

582   Case concerning the Conservation and Sustainable Exploitation of Swordfish Stocks in the South-Eastern Pacific Ocean (Chile / European Union) - Order 2008/1, ITLOS, (2008).

main element of this document was a more structured framework of fisheries cooperation, catch levels commensurate with the objective of ensuring the sustainability of swordfish as well as safeguarding the marine ecosystem, freezing their fishing effort at the 2008 level or at the maximum historical peak, establish a Bilateral Scientific and Technical Committee, enhance current multilateral consultation and access for EU vessels to designated Chilean ports for landings, transshipments, replenishing or repairs.[583] For that purpose the parties requested to suspend the time-limits for the proceedings for at least one year starting from 1 January 2009 and the chamber agreed.[584]

The last order of this case is dated on 16th December 2009. Therein it is laid down that Chile and the EC had agreed not to seek a further extension of the suspension but to request an order for discontinuance of the case.[585] In the meantime the EC had been replaced by the European Union (EU). Therefore, Chile and now the EU provided a joint communication informing the special chamber that they would be committed to the signature, ratification or approval, and implementation of and compliance with the new Understanding agreed between negotiators for both Parties on 16 October 2008.[586]

The special chamber followed the communication of the parties and placed on record, pursuant to article 105, paragraph 2, of the Rules of the ITLOS, the discontinuance, by agreement of the Parties, of the proceedings initiated on 20 December 2000 by Chile and the European Community and ordered that the case was removed from the List of cases.[587] On 28th May 2010 the two parties also notified the WTO Dispute Settlement Body and the Council for Trade in Goods about the removal from the list.[588] In addition, they communicated that they

---

583 Understanding concerning the conservation of swordfish stocks in the South Eastern Pacific Ocean, 2010. Order 12.

584 Case concerning the Conservation and Sustainable Exploitation of Swordfish Stocks in the South-Eastern Pacific Ocean (Chile / European Union) - Order 2008/1, (2008) Order 8.

585 Case concerning the Conservation and Sustainable Exploitation of Swordfish Stocks in the South-Eastern Pacific Ocean (Chile / European Union) - Order 2009/1, ITLOS, (2009). Order 8.

586 Understanding concerning the conservation of swordfish stocks in the South Eastern Pacific Ocean.

587 Case concerning the Conservation and Sustainable Exploitation of Swordfish Stocks in the South-Eastern Pacific Ocean (Chile / European Union) - Order 2009/1, (2009). Order 14.

588 Chile - Measures Affecting the Transit and Importation of Swordfish - Joint Communication from the European Union and Chile - Addendum of 28 May 2010, WTO, (2010).

had unconditionally agreed that neither party shall further exercise any procedural right accruing to it under the DSU in the case 'Chile-Measures Affecting the Transit and Importation of Swordfish'.

D. Summary

The two cases have shown that international jurisdiction dealing with fisheries targeting tuna and tuna-like species is highly complex. In this complex environment the dispute settlement procedures under the UNCLOS but also under the GATT have proven to be a useful instrument to guide negotiations between conflicting parties. Although in both cases, in the end, the dispute had been solved through negotiations which are located formally outside of the framework of UNCLOS it is remarkable that the procedures were applied consistently and turned out important to clarify the competences of the institutions involved. Generally it can be recognized that all parties have fully cooperated with the institutions even though they questioned the legitimacy of their establishment. This is a positive finding with regard to possible future cases involving tuna and tuna-like species.

Despite this positive outcome, the decision of the Arbitral Tribunal in the Southern Bluefin Case revealed that the dispute settlement procedures under Part XV of UNCLOS leave room for interpretation. At first instance the judges of the ITLOS prescribed six provisional measures and made two decisions on how to implement these measures. The Arbitral Tribunal as the court of second instance revoked the measures because Article 16 of the Convention of the CCSBT excluded any further procedure within the contemplation of Article 281(1) of UNCLOS. The effect of this decision is unclear. On the positive side the decision of the Tribunal has strengthened the authority of RFMOs. However, it implied also that in the future any claim under UNCLOS that raises issues under a regional agreement might not be adjudicated by UNCLOS. This outcome has the potential to weaken the compulsory procedures under Part XV of UNCLOS. Despite this formal critique on the decision of the Arbitral Tribunal it has to be recognized that the decision has supported the progress already made by the parties and it reminded them that they have to solve the problem within the framework of the CCSBT.

Also in the Swordfish Case it can be recognized that the initiation of formal dispute settlement procedures had a positive effect on the willingness of the parties to reach a solution. Even the fact that two jurisdictional forums with potentially contradictory or incompatible views had been chosen did not result to be a problem. The opposite was the case. The procedures facilitated bilateral negotiations and forced the parties to reach a mutual solution. In the end agreement of the parties led to the discontinuance of the proceedings.

Chapter III

The five tuna RFMOs – Differences and Similarities

A.   Introduction

As outlined in the first chapter, there are several provisions within international treaties that require regional cooperation in the management of those fisheries catching tuna and tuna-like species. The most important provisions for such cooperation are laid down in UNCLOS and the Fish Stocks Agreement. For the management of transboundary and straddling stocks, UNCLOS states that all relevant fishing states "shall seek, either directly or through appropriate subregional or regional organizations, to agree upon the measures necessary to coordinate and ensure the conservation and development of such stocks".[589] With respect to highly migratory species, the states concerned "shall cooperate directly or through appropriate international organizations with a view to ensuring conservation and promoting the objective of optimum utilization of such species throughout the region, both within and beyond the exclusive economic zone."[590] Very similar provisions can be found in the Fish Stocks Agreement, which declares that member states "shall, in accordance with the Convention, pursue cooperation in relation to straddling fish stocks and highly migratory fish stocks either directly or through appropriate subregional or regional fisheries management organizations or arrangements".[591]

In practice, this 'cooperation' is conducted through Regional Fisheries Management Organizations (RFMOs). These are regional fisheries bodies with a management mandate. Overall, there are currently 20 of these bodies.[592] Five of them specialize in the management of tuna and tuna-like species (tuna RFMOs), covering their whole migratory range with jurisdiction both on the high seas and in the EEZs. Essentially, the tuna RFMOs adopt conservation and management measures, collect data, carry out scientific research and monitor fishing operations. Named according to either the area or the species they manage, they are:

- the Inter-American Tropical Tuna Commission (IATTC);

---

589   United Nations Convention on the Law of the Sea, Arts.63.1 and 63.2.

590   Ibid. Art.64.1.

591   Agreement for the Implementation of the Provisions of the United Nations Convention on the Law of the Sea of 10 December 1982 relating to the Conservation and Management of Straddling Fish Stocks and Highly Migratory Fish Stocks, Art.8.1.

592   For a list of all fisheries bodies, see http://www.fao.org/fishery/rfb/search/en.

- the International Commission for the Conservation of Atlantic Tunas (ICCAT);

- the Commission for the Conservation of Southern Bluefin Tuna (CCSBT);

- the Indian Ocean Tuna Commission (IOTC); and

- the Western and Central Pacific Fisheries Commission (WCPFC).

The five tuna RFMO had been established over a long time span between 1950 and 2004. Hence, the content of the constituent instruments, which in the case of IATTC, ICCAT IOTC and WCPFC are Conventions and in the case of the IOTC an Agreement, reflect the stage of development in international fisheries law at the time of their adoption. The more recently established constituent instruments contain provisions which reflect the more recent developments while the older ones often lack relevant provisions. With respect to the time of establishment the Inter American Tropical Tuna Commission (IATTC) is the oldest tuna RFMO.[593] Its original Convention was adopted on the 31st May, 1949.[594] However, the 1949 Convention was recently replaced by the so-called 'Antigua Convention' which came into force on the 27th August, 2010.[595] The International Commission for the Conservation of Atlantic Tunas (ICCAT) is more than 15 years younger than the IATTC, but it is still, a well-established tuna RFMO when compared to CCSBT, IOTC and WCPFC. On the 21st March, 1969, the Convention of the International Commission for the Conservation of Atlantic Tunas entered into force after it had been signed at the Conference of Plenipotentiaries on the Conservation of Atlantic Tunas in Rio de Janeiro, Brazil, in 1966.[596] The ICCAT Convention has been amended twice since, in 1984 and in 1992,[597] but in the areas of conservation and management, these amendments

---

593   For further information on the history of the IATTC, see: R. L. Allen, *The Inter American Tropical Tuna Commission*, Symposium on World Tuna Fisheries Commemorating the 50th Anniversary of the Establishment of the Inter American Tropical Tuna Commission (La Jolla: IATTC, 2001), pp.21-29.

594   Convention for the Establishment of an Inter-American Tropical Tuna Commission, 1949.

595   Convention for the Strengthening of the Inter-American Tropical Tuna Commission established by the 1949 Convention between the United States of America and the Republic of Costa Rica "Antigua Convention".

596   International Convention for the Conservation of Atlantic Tunas.

597   In 1984, Arts. XIV, XV and XVI (signature, ratification, entry into force) of the Convention were modified: ICCAT, *Conference of Plenipotentiaries of the States Parties to the international convention for the conservation of Atlantic tunas*, (1984), Para.I. In 1992, when paragraph 2 of Article X (budget) was modified: ICCAT, *Conference of Plenipotentiaries of the Contracting Parties to the International Convention for the Conservation of Atlantic Tunas*, (1992), Art.1.

are not comparable with the fundamental changes made by the IATTC to its Convention. The Commission for the Conservation of Southern Bluefin Tuna (CCSBT) is one of the two tuna RFMOs that were established just before the adoption of the Fish Stocks Agreement. The Convention was signed in May 1993, and one year later it entered into force. The Indian Ocean Tuna Commission (IOTC), like the CCSBT, can trace its origins to before the Fish Stocks Agreement. The Agreement for the Establishment of the IOTC was approved by the FAO Council at its 105th Session, on the 25th November, 1993, and it entered into force on the 27th March, 1996. The Western and Central Pacific Fisheries Commission (WCPFC) is the most recently established tuna RFMO. Its Convention was adopted on the 4th September, 2000, and entered into force on the 19th June, 2004.[598]

This chapter will give an overview of the five tuna RFMOs. It is intended to assess their current performance in the context of the developments in international fisheries law. A further aim is the comparison of the tuna RFMOs with regard to the implementation of the requirements laid down in their respective constituent instruments.

B.   Evolution of catches and stock status

1.   *Evolution of catches*

In 2010, the total catch of the principal market tuna species skipjack, yellowfin, bigeye albacore and bluefin which was made in all areas of competence of the tuna RFMOs remained with about 4.3 million metric tonnes (mt) approximately at the same level as in previous years since 2002.[599] However, the global trend is not representative for the development in all areas. Three types of tuna RFMOs can be identified with regard to the development of catches. One type is represented by the two oldest RFMOs, IATTC and ICCAT, established in 1950 and 1969 respectively. In the Convention areas of these organizations, comparatively high catch levels were reached already in the 1970s. A second type is represented by the much younger CCSBT which was established in 1994. Despite its recent establishment it has to be considered that the highest catch levels for southern bluefin tuna were reached already in the 1960s. The third type includes the two youngest RFMOs, IOTC and WCPFC, which were established in 1996 and 2004 respectively, and which have both shown continuously increasing catch rates from their foundation until their highest catch levels were reached in recent years.

---

598   Convention on the Conservation and Management of Highly Migratory Fish Stocks in the Western and Central Pacific Ocean.

599   FAO, 2012, p.53.

Figure 3.1   Annual catches WCPFC

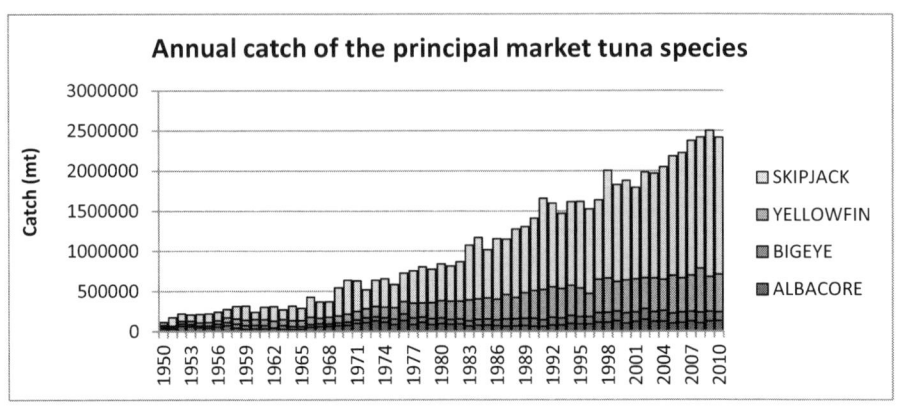

Source: WCPFC[600], IATTC[601] and FAO[602]

The Convention area of the WCPFC is the most productive area with regard to the total catches of tuna and tuna like species (Figure 3.1).[603] Since 1961, the catches have shown a continuous increase. In 2010, the total catch of these species was 2,414,994 mt.[604] This catch represented 84 percent of the total Pacific Ocean tuna catch (2,875,909 mt), and more than half of the estimated tuna catch worldwide. In 2010, the total catch was less than in 2009 but it was still the second highest annual catch ever, recorded after a continuous increase over the last fifty years. Skipjack (1,706,166 mt) accounted for most of the catches followed by yellowfin (470,161 mt), albacore (129,670 mt) and bigeye (108,997 mt) tuna. The principal fishing gears in the WCPFC are purse seine (75 percent), longline (10 percent) and pole and line (8 percent).[605]

---

600   Secretariat of the Pacific Community, *WCPFC Tuna Fishery Yearbook* (Noumea: Western and Central Pacific Fisheries Commission, 2010); WCPFC, *Summary Report - Seventh Regular Session of the Scientific Committee*, (2011).

601   IATTC, *The fishery for tunas and billfishes in the Eastern Pacific Ocean in 2010* (La Jolla: IATTC, 2011), pp.27-28; IATTC, *Fishery Status Report No. 1 (Tunas and Billfishes in the Eastern Pacific Ocean in 2002)* (La Jolla: IATTC, 2003), pp.33-34.

602   http://www.fao.org/fishery/statistics/tuna-catches/query/en.

603   Yellowfin tuna, skipjack tuna, bigeye tuna, albacore tuna.

604   WCPFC, *Summary Report - Seventh Regular Session of the Scientific Committee*, (2011).

605   International Seafood Sustainability Foundation, *Status of the world fisheries for tuna: Management of tuna stocks and fisheries*, Technical Report 2012-07 (Washington: ISSF, 2012), p.6.

Figure 3.2   Annual catches IOTC

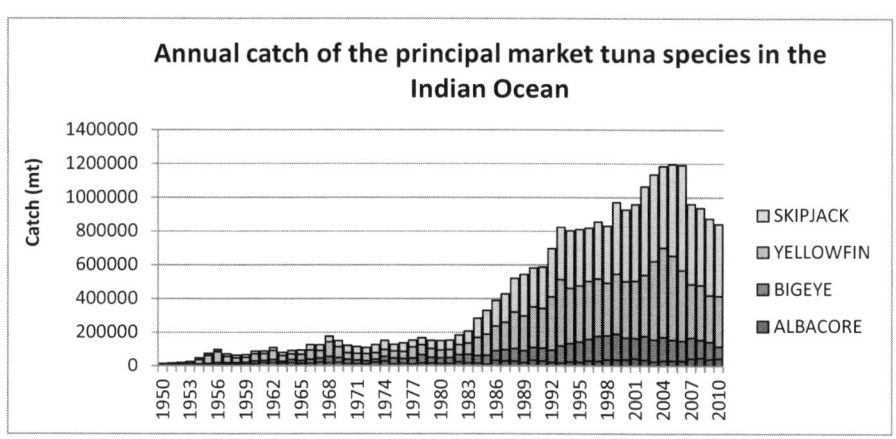

Source: IOTC[606] and FAO[607]

Currently the area of competence of the Indian Ocean is the second most productive area (Figure 3.2). In 2010, the Scientific Committee of the IOTC reported for the principal market tuna species,[608] a decrease of the catches.[609] The total catch, at 842,993 mt, was significantly lower than the all-time high of 1,195,696 mt just five years earlier, in 2005. The catch composition was skipjack (428,719 mt), yellowfin (299,074 mt), bigeye (71,489 mt) and albacore tuna (43,711 mt). The main gear types in the IOTC are gillnet (43 percent), purse seine (33 percent), longline (16 percent) and pole and line (10 percent).[610]

606   IOTC, *Report of the Fourteenth Session of the IOTC Scientific Committee*, (2011), pp.78-181.

607   http://www.fao.org/fishery/statistics/tuna-catches/query/en.

608   Yellowfin tuna, bigeye tuna, skipjack tuna and albacore tuna.

609   IOTC, *Report of the Fourteenth Session of the IOTC Scientific Committee*, (2011), pp.78-181.

610   Foundation, 2012, p.9.

Figure 3.3   Annual catches IATTC

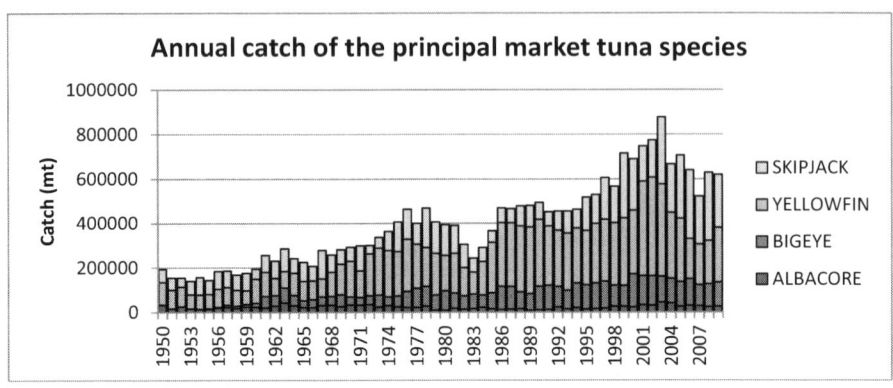

Source: Data from IATTC[611] and the FAO[612]

The third most productive area is the Convention area of the IATTC (Figure 3.3). According to the information provided by the Scientific Advisory Committee of the IATTC, in 2009 the total catch weight of the principal market tuna species[613] was 619,251 metric tonnes (mt).[614] This represented a significant decrease from the all-time high in 2003 of 877,054 mt. The catch was composed by yellowfin (245,963 mt), skipjack (238,863 mt), bigeye (108,006 mt) and albacore tuna (26.415 mt). The principal fishing gears are purse seine (91 percent) and longline (8 percent).[615]

---

611   IATTC, 2011, pp.27-28; IATTC, 2003, pp.33-34.

612   http://www.fao.org/fishery/statistics/tuna-catches/query/en.

613   Yellowfin tuna, skipjack tuna, bigeye tuna and albacore tuna.

614   IATTC, 2011, pp.26-29.

615   Foundation, 2012, p.4.

Figure 3.4   Annual catches ICCAT

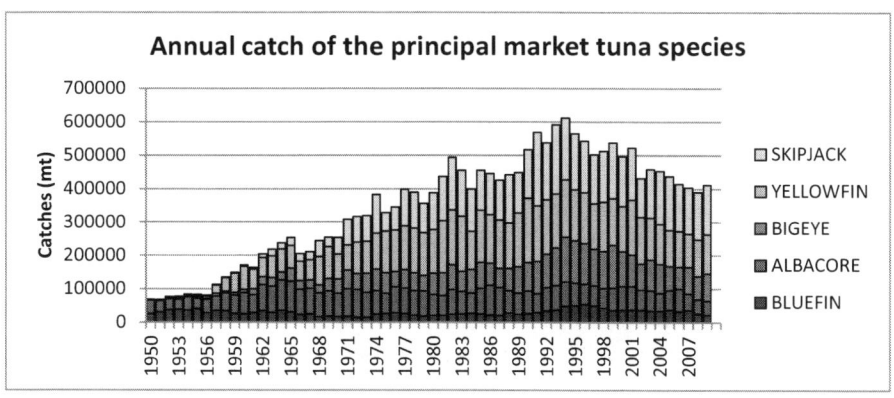

Source: Data from ICCAT[616] and FAO[617]

The convention area of the ICCAT is the fourth most productive area (Figure 3.4). Catches reached their highest levels during the early 1990s. Catch data from the Standing Committee on Research and Statistics shows that, in 2009, the total catch weight of the principal market tuna species[618] (at 411,721 mt) was significantly less than the all-time high of 611,893 mt in 1994.[619] These catches are divided between skipjack (148,653 mt), yellowfin (117,340 mt), bigeye (81,813 mt), albacore (42,235 mt) and bluefin tuna (21,680 mt). The main fishing gears in the ICCAT are purse seine (54 percent), pole and line (19 percent) and longline (20 percent).[620]

616   ICCAT, *Report for Biennial Period, 2010-11 PART II (2011) - Vol. 2 (Standing Committee on Research & Statistics)*, (2012) pp.1-102.

617   http://www.fao.org/fishery/statistics/tuna-catches/query/en.

618   Yellowfin tuna, bigeye, tuna, skipjack tuna, albacore tuna and bluefin tuna.

619   ICCAT, *Report for Biennial Period, 2010-11 PART II (2011) - Vol. 2 (Standing Committee on Research & Statistics)*, (2012), pp.11-102.

620   Foundation, 2012, pp.7-8.

Figure 3.5   Annual catches CCSBT

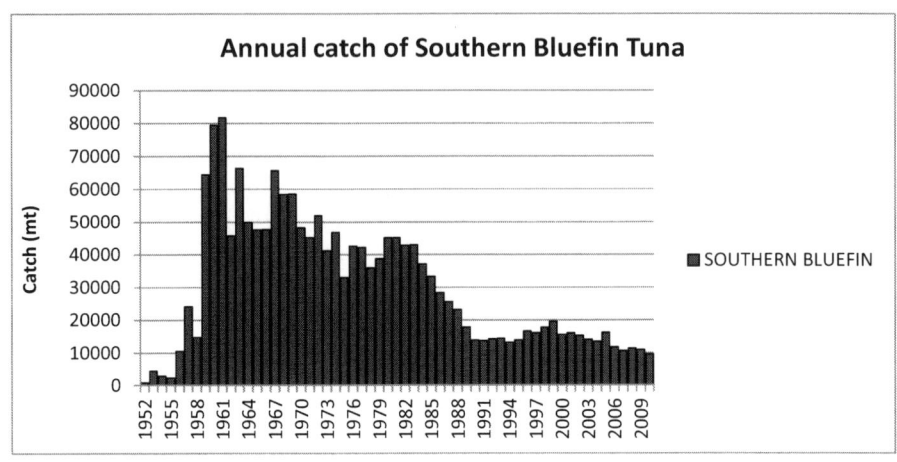

Source: Data from CCSBT[621]

Overall catches in the convention area of the CCSBT are very small in comparison to the other four tuna RFMOs (Figure 3.5). Catch data provided by the Scientific Committee of the CCSBT shows that in 2010 the reported southern bluefin tuna catch, at 9,547 mt, was the lowest catch since 1955.[622] This catch represents only a fraction of the record catches from the 1960s to the early 80s, with the all-time high of 81,750 mt reached in 1961. In the CCSBT the two main gear types are longline (58 percent) and purse seine (42 percent).[623]

2.   Stock status

A pertinent point of comparison between the tuna RFMOs is the stock status of tuna and tuna-like species within their geographic areas of competence. This subchapter will show the results of stock assessments that have been conducted by or for the scientific bodies of the tuna RFMOs.[624] Although it cannot always be excluded that those assessments are to some extend politically influenced, there is no doubt that they provide the best data available.[625]

---

621   CCSBT, *Report of the Sixteenth Meeting of the Scientific Committee* (2011) Para.19 and Attachment 4.

622   Ibid. Para.19 and Attachment 4.

623   Foundation, 2012, p.10.

624   Further information on the scientific bodies will be provided in the next subchapter.

625   T. Polacheck, Politics and independent scientific advice in RFMO processes: A case study of crossing boundaries, *Marine Policy, Vol. 36(1)* (2012).

In the chapters three and four, the terms 'Maximum Sustainable Yield' 'over-fishing', 'overfished' and 'sustainable' will be frequently used. The FAO defines the 'Maximum Sustainable Yield (MSY) as "the highest theoretical equilibrium yield that can be continuously taken (on average) from a [fish] stock under existing (average) environmental conditions without affecting significantly the reproduction process."[626] The two main indicators which are used in stock assessments are B_{MSY} (biomass corresponding to MSY) and F_{MSY} (the fishing mortality rate which, if applied constantly, would result in MSY).[627] Tables 3.1 -3.5 will show the stock status in the areas of competence of the five tuna RFMOs. In accordance with the practice of the Scientific Committee of the WCPFC the terms 'overfishing' and 'overfished' will be used as follows. For a stock with current fishing mortality (F) below F_{MSY} and current biomass (B) above B_{MSY} there is no overfishing occurring, nor is the stock being overfished. If either fishing mortality is above F_{MSY} or the current biomass is below B_{MSY} either overfishing is occurring or the stock has been overfished. When the fishing mortality is above F_{MSY} and current biomass is below B_{MSY} then overfishing is occurring and the stock has also been overfished. The management is described as 'sustainable' when the current fishing mortality is below the fishing mortality that can produce MSY and when the current biomass is above the biomass that can produce MSY.

The assessments which focus almost exclusively on the economically most important tuna and tuna-like species, show clearly that in the areas of competence of WCPFC and IOTC comparatively few stocks are overfished and less overfishing is occurring. In contrast, IATTC, ICCAT and CCSBT are struggling with the management. In particular, the stocks managed by ICCAT and CCSBT are often overfished and overfishing is still occurring. Generally, it can be noted that, at present, the latter organizations have to rebuild several of their stocks while IOTC and WCPFC are facing the different challenge of maintaining their comparatively good position.

---

626   http://www.fao.org/fi/glossary/default.asp.

627   Caddy & Mahon, 1995.

Table 3.1   Stock status WCPFC

	No overfishing F < Fmsy Not overfished	Overfishing F > Fmsy Not overfished	No overfishing F < Fmsy Overfished	Overfishing F > Fmsy Overfished
Skipjack tuna	X			
Yellowfin tuna	X			
Albacore tuna (N)	X			
Albacore tuna (S)	X			
Bigeye tuna		X		
Pacific bluefin	X			
Swordfish (N)	X			
Swordfish (S)	X			
Striped marlin (SW)	(X)			
Striped marlin (N)				X

Source: Unterweger, based on WCPFC data

According to the latest stock assessments discussed during the Scientific Committee Meetings of the WCPFC in 2011 it can be recognized that eight out of ten assessed stocks are neither overfished nor is overfishing occurring (Table 3.1).[628] Only two stocks are in critical conditions. The striped marlin in the north pacific is overfished and overfishing is occurring[629] while for bigeye tuna overfishing is occurring but the stock is not overfished yet[630]. More action seems to be needed in order to rebuild these stocks. The Scientific Committee found that current catch levels of bigeye tuna are unlikely to be sustainable in the long term, and that the stock is approaching an overfished state. In particular a reduction of catches of juvenile bigeye was seen as necessary in order to increase MSY levels and to obtain greater overall yields. With regard to striped marlin in the north Pacific it was found that immediate reduction of fishing mortality would be necessary and that catch limits adopted in 2010 would have to be reviewed.[631]

628   WCPFC, *Summary Report - Seventh Regular Session of the Scientific Committee*, (2011).

629   WCPFC, *Summary Report - Sixth Regular Session of the Scientific Committee*, (2010), p.65.

630   WCPFC, *Summary Report - Seventh Regular Session of the Scientific Committee*, (2011) p.26.

631   Ibid. p.69.

Table 3.2   *Stock status IOTC*

	No overfishing F < Fmsy Not overfished	Overfishing F > Fmsy Not overfished	No overfishing F < Fmsy Overfished	Overfishing F > Fmsy Overfished
Skipjack tuna	X			
Yellowfin tuna	X			
Bigeye tuna	X			
Albacore tuna		X		
Swordfish	X			
Swordfish (SW)			X	

Source: Unterweger, based on IOTC data

In 2011 the Scientific Committee of the IOTC revealed that in the Indian Ocean only two out of six assessed stocks are in critical conditions (Table 3.2).[632] For albacore tuna overfishing is occurring, and swordfish in the south West Indian Ocean is overfished.[633] The Committee stated that maintaining or increasing effort in the albacore fishery would probably result in further declines in biomass, productivity and catch per unit effort. With regard to swordfish in the southwest Indian Ocean it was recognized that there had been already a decrease in catch and effort in the previous years. With catches at current levels the risk of exceeding MSY-based reference points was therefore low. For yellowfin and bigeye tuna it was found that recent declines in longline and purse seiner effort have lowered the pressure on the stocks and that maintaining current fishing mortality would not reduce the population to an overfished state.[634]

---

632   IOTC, *Report of the Fourteenth Session of the IOTC Scientific Committee*, (2011)

633   Ibid. pp.78 and 149

634   Ibid. pp.87 and 107.

Table 3.3   Stock status IATTC

	No overfishing F < Fmsy Not overfished	Overfishing F > Fmsy Not overfished	No overfishing F < Fmsy Overfished	Overfishing F > Fmsy Overfished
Yellowfin			X	
Skipjack	X	(X)		
Bigeye		X		
Albacore (N)	X			
Albacore (S)	X			
Pacific bluefin		X		
Swordfish (SE)	X			
Striped marlin	X			

Source: Unterweger, based on IATTC data

In 2011 the Scientific Advisory Committee of the IATTC has shown that three, or possibly four of the eight assessed stocks in the Eastern Pacific Ocean are currently overfished (yellowfin tuna)[635] or overfishing is occurring (bigeye[636], pacific bluefin[637] and possibly skipjack tuna)[638] (Table 3.3). It was communicated that the current levels of fishing mortality (2008-2010) of yellowfin tuna would allow the spawning biomass to rebuild, and remain above the level corresponding to MSY. For bigeye tuna a recovery trend was recognized due to resolutions initiated in 2004 but uncertainty about recruitment and biomass level did not allow a precise prediction. With regard to pacific bluefin tuna it has been criticized that the IATTC was failing to adopt a resolution that would allow the recovery of the stock. The stock status of skipjack tuna is not certain but it is possible that the exploitation rate is approaching or above the level associated with MSY.

---

635   Aires-da-Silva & Maunder, 2011, p.18.

636   Aires-da-Silva & Maunder, 2011, p.3.

637   ISC, *Report of the Pacific Bluefin Tuna Working Group Workshop*, (2010).

638   Maunder, 2011, p.2.

114

Table 3.4   Stock status ICCAT

	No overfishing F < Fmsy Not overfished	Overfishing F > Fmsy Not overfished	No overfishing F < Fmsy Overfished	Overfishing F > Fmsy Overfished
Yellowfin			X	
Bigeye	X			
Skipjack	X			
Albacore (N)			X	(X)
Albacore (S)			X	
Bluefin (W)				X
Bluefin (E+Med)				X
Blue marlin				X
White marlin				X
Sailfish				X
Swordfish (N)	X			
Swordfish (S)	X			
Swordfish (Med)				X

Source: Unterweger, based on ICCAT data

Despite the ICCAT's long experience in managing tuna and-tuna like species it was concluded by the Standing Committee on Research and Statistics that only four of the 13 assessed stocks are neither overfished nor is overfishing occurring (Table 3.4).[639] Three stocks (yellowfin[640] and albacore tuna in the north and south Atlantic[641]) are overfished, and for six or seven stocks both overfishing is occurring and the stock is overfished (bluefin tuna in all parts of the Atlantic,[642] blue marlin,[643] white marlin,[644] sailfish,[645] swordfish in the Mediterranean[646] and possibly albacore tuna in the north Atlantic). For bluefin tuna in the eastern pacific and the Mediterranean it was found despite significant uncertainty in scientific information that that previously established recommendations appeared to have resulted in reductions in catch and fishing mortality rates.[647] For western Atlantic bluefin tuna there is also uncertainty but it was suggested that

639   ICCAT, *Report for Biennial Period, 2010-11 PART II (2011) - Vol. 2 (Standing Committee on Research & Statistics)*, (2012).

640   Ibid. pp.12-13.

641   Ibid. pp.59-60.

642   Ibid. pp.78-84

643   Ibid. p.104.

644   Ibid. p.104.

645   Ibid. pp.118-119.

646   Ibid. p.128.

647   Ibid. p.81.

spawning biomass is slowly rebuilding.[648] For both albacore stocks declining trends in stock size have been identified while the assessment of yellowfin tuna has shown that current catch levels have the potential to lead to a biomass that supports the MSY by 2016. For blue and white marlin as well as for sailfish there is remaining concern although it seems that the declining trend has partially stabilized. Swordfish in the Atlantic has recovered in the last years but the stock in the Mediterranean needs further reductions in fishing mortality.

Table 3.5   Stock status CCSBT

| | No overfishing | Overfishing | No overfishing | Overfishing |
| | $F < Fmsy$ | $F > Fmsy$ | $F < Fmsy$ | $F > Fmsy$ |
	Not overfished	Not overfished	Overfished	Overfished
Southern Bluefin			X	

Source: Unterweger, based on CCSBT data

The latest stock assessments of the Scientific Committee of the CCSBT indicate that fishing mortality had declined significantly from 2005 to 2010 (Table 3.5).[649] According to this, it was concluded that there is no longer overfishing occurring. The decline in fishing mortality is related to the reductions in catches, two reductions in the global TAC (in 2006 and 2009) as well as higher recruitment to the stock in the early 2000s. Despite this progress, the stock is still overfished. The maintenance of the current fishing mortality is needed in order to rebuild the stock to the biomass that can support harvest of the MSY.

C.   Legal framework

1.   Areas of competence

All RFMOs have determined specific geographic areas of competence.[650] The respective areas usually comprise the occurrence of particular species or species groups. In tuna RFMOs this is basically the migratory range of tuna and tuna-like species. The Fish Stocks Agreement contains explicit provisions that require States to agree on the area of application while taking into account the cooperation requirements in the management of straddling and highly migratory stocks as well as the characteristics of subregions or regions.[651] Today, with the excep-

---

648   Ibid. p.85.

649   CCSBT, *Report of the Sixteenth Meeting of the Scientific Committee* (2011), Para.101.

650   For a further information on all RFMOs see http://www.fao.org/fishery/rfb/search/en

651   Agreement for the Implementation of the Provisions of the United Nations Convention on the Law of the Sea of 10 December 1982 relating to the Conservation and Management of Straddling Fish Stocks and Highly Migratory Fish Stocks, Arts.9.1(b) and 7.1.

tion of territorial and archipelagic waters, all maritime areas where tuna and tuna-like species occur are covered by the 'convention areas'[652] or the 'area of competence'[653] of tuna RFMOs.[654] Each tuna RFMO has determined the extension of its area. Some of them have agreed on specific coordinates while others just refer to particular oceans and parts of it or to the migration route of a particular tuna species. The global coverage can be seen in Figures 3.6-3.10. It has to be considered that the illustrations also include the territorial sea and archipelagic waters which are not included in the convention areas or area of competence.

Figure 3.6   *IATTC Convention area*

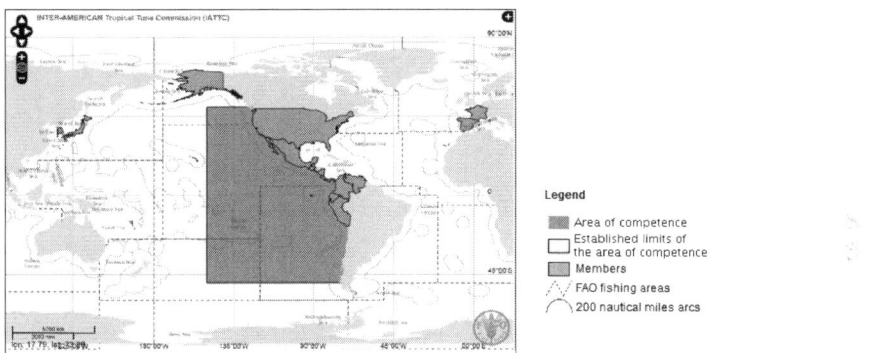

Source: FAO

The area of application of the Antigua Convention (Figure 3.6) comprises roughly the eastern half of the Pacific Ocean. According to Article III of the Convention, the boundaries on the landward side are the "coastline of North, Central, and South America".[655] On the seaward side, the area is limited by "the 50°N parallel from the coast of North America to its intersection with the 150°W meridian; the 150°W meridian to its intersection with the 50°S parallel; and the 50°S parallel to its intersection with the coast of South America."

---

652   IATTC, ICCAT, IOTC and WCPFC call it convention area.

653   IOTC calls it area of competence.

654   According to Articles 2 and 49 of UNCLOS coastal and archipelagic States have full souvereignity over the resources within the territorial and archipelagic waters.

655   Convention for the Strengthening of the Inter-American Tropical Tuna Commission established by the 1949 Convention between the United States of America and the Republic of Costa Rica "Antigua Convention" Art.III.

Figure 3.7   WCPFC Convention area

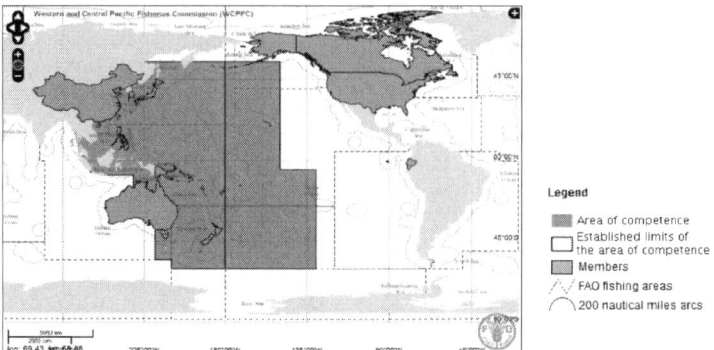

Source: FAO

The Convention area of the WCPFC (Figure 3.7) roughly covers all the waters of the Pacific Ocean that are relevant for tuna fisheries and not covered by the Convention area of the IATTC. Outer limits are provided only for the southern and eastern border where the area is extending "[f]rom the south coast of Australia due south along the 141° meridian of east longitude to its intersection with the 55° parallel of south latitude; thence due east along the 55° parallel of south latitude to its intersection with the 150° meridian of east longitude; thence due south along the 150° meridian of east longitude to its intersection with the 60° parallel of south latitude; thence due east along the 60° parallel of south latitude to its intersection with the 130° meridian of west longitude; thence due north along the 130° meridian of west longitude to its intersection with the 4° parallel of south latitude; thence due west along the 4° parallel of south latitude to its intersection with the 150° meridian of west longitude; thence due north along the 150° meridian of west longitude."[656] For the northern and the western part of the Convention area there are no coordinates. It is only stated that the Convention "shall be applied throughout the range of the stocks, or to specific areas within the Convention Area, as determined by the Commission."[657]

---

656   Convention on the Conservation and Management of Highly Migratory Fish Stocks in the Western and Central Pacific Ocean, Art.3.1.

657   Ibid. Art.3.3.

Figure 3.8   ICCAT Convention area

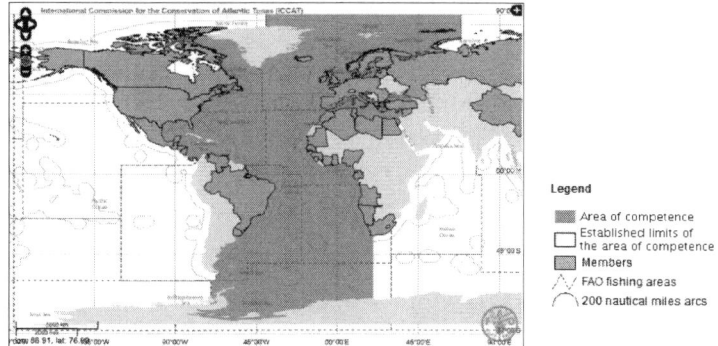

Within the Convention of the ICCAT (Figure 3.8) there are no specific coordinates. Article I only lays down that the Convention area covers "all waters of the Atlantic Ocean, including the adjacent Seas." As indicated by the map, this is a huge area, including all Atlantic waters between the two poles, as well as semi-enclosed areas such as the Mediterranean Sea, Baltic Sea and Black Sea.

Figure 3.9   IOTC area of competence

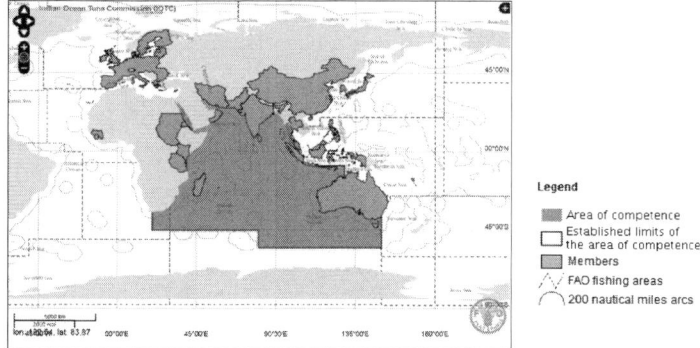

The IOTC is using the FAO Statistical Areas for the determination of its area of competence (Figure 3.9). According to Article II of the Agreement for the Establishment of the Indian Ocean Tuna Commission, this area is comprising the Indian Ocean, which is represented by FAO statistical areas 51 and 57[658] and the

---

658   http://www.fao.org/fishery/area/search/en.

"adjacent seas, north of the Antarctic Convergence, insofar as it is necessary to cover such seas for the purpose of conserving and managing stocks that migrate into or out of the Indian Ocean."[659]

Figure 3.10   CCSBT Convention area

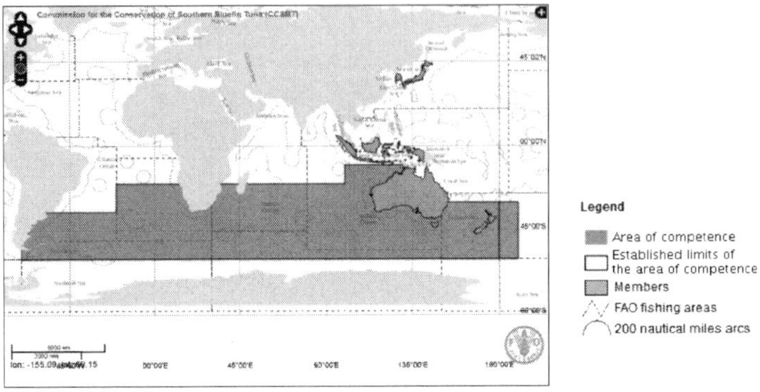

Source: FAO

The Convention of the CCSBT does not provide any boundaries or geographical limitations. Instead, reference is made to the southern bluefin tuna which is managed by the CCSBT.[660] In some areas where vessels are fishing for this tuna species the Convention areas of ICCAT, IOTC and WCPFC there is an overlap of responsibility (Figure 3.10). A Memorandum of Understanding (MOU) between CCSBT and WCPFC clarifies that the CCSBT has the conservation and management mandate for southern bluefin tuna in the relevant area of overlap.[661] With ICCAT and IOTC, there are MOUs for the regulation of transshipment, but there is no formal agreement with regard to the management mandate.[662]

2.   *Species coverage*

For all tuna RFMOs it is crucial to know to which species its conservation and management mandate applies. Articles 63 and 64 of UNCLOS require coopera-

---

659   Agreement for the Establishment of the Indian Ocean Tuna Commission, Art.II.

660   Convention for the Conservation of Southern Bluefin Tuna, Art.1.

661   Memorandum of Understanding between WCPFC and CCSBT, 2006.

662   Memorandum of Understanding between the CCSBT and ICCAT Secretariats for Transshipment at sea by Large-Scale Fishing Vessels, 2009; Memorandum of Understanding between the CCSBT and IOTC Secretariats for Monitoring Transshipment at Sea by Large-Scale Tuna Longline Fishing Vessels, 2009.

tion in conservation and management of transboundary and straddling stocks as well as of highly migratory species.[663] A specific list was only provided for highly migratory species in Annex I.[664] This list includes the economically most important tuna and tuna-like species. The Fish Stocks Agreement did not provide any further guidance with regard to the inclusion or exclusion of particular species.[665] Most of the tuna RFMOs developed their own provisions that are more specific than those of UNCLOS.

The IOTC is the only tuna RFMO which has developed an own list in Annex B of its Agreement.[666] The list names all the relevant principal market tuna species as well as other tuna and billfish species set out in Annex I of UNCLOS but it also includes other region-specific tuna-like species like longtail tuna (Thunnus tongol) or indo-pacific king mackerel (Scomberomorus guttatus). All species set out in this Annex are covered by the Agreement.[667]

No particular species are named in the Antigua Convention of the IATTC. The covered species are "stocks of tunas and tuna-like species and other species of fish taken by vessels fishing for tunas and tuna-like species in the Convention Area".[668] Although this includes non target species as well, it is stated in Article VII that when fulfilling its conservation and management functions, the Commission must give "priority to tunas and tuna-like species".[669]

The Convention of the ICCAT predates the UNCLOS. The Preamble of its Convention states that conservation requirements apply to "the populations of tuna and tuna-like fishes", further defined as the Scombriformes.[670] These species, as well as "other species of fishes exploited in tuna fishing in the Convention area as are not under investigation by another international fishery organization", must be studied in order to carry out the objectives of the Commission.[671]

---

663 United Nations Convention on the Law of the Sea, Arts.63 and 64.

664 Ibid. Annex I.

665 Agreement for the Implementation of the Provisions of the United Nations Convention on the Law of the Sea of 10 December 1982 relating to the Conservation and Management of Straddling Fish Stocks and Highly Migratory Fish Stocks Arts.3 and.8.1.

666 Agreement for the Establishment of the Indian Ocean Tuna Commission, Annex B.

667 Ibid. Art.III.

668 Convention for the Strengthening of the Inter-American Tropical Tuna Commission established by the 1949 Convention between the United States of America and the Republic of Costa Rica "Antigua Convention" Art.I.1.

669 Ibid. Art.VII.1.

670 International Convention for the Conservation of Atlantic Tunas, Preamble and Art.I.

671 ICCAT, International Convention for the Conservation of Atlantic Tunas, Art.IV(1).

Explicit reference to UNCLOS is only made by the WCPFC Convention. It incorporates Annex I when defining 'highly migratory fish stocks'.[672] In addition the Commission is entitled to determine other species as 'highly migratory'.[673] The Convention applies to "all stocks of highly migratory fish within the Convention Area except sauries[674] [...] throughout the range of the stocks, or to specific areas within the Convention Area, as determined by the Commission."[675]

The Convention of the CCSBT is the only constituent instrument of a tuna RFMO that applies only to one target species. Article I of the Convention of the CCSBT states that the Convention "shall apply to southern bluefin tuna (Thunnus maccoyii)." For scientific purposes the Convention further requires the collection of scientific information of 'ecologically related species'.[676]

3. *Organizational structure*

The general organizational structure is quite similar in all tuna RFMOs. They are composed by a Commission which is the executive body, subsidiary bodies to support the work of the Commission and an independent Secretariat with permanent staff. With delegates from each member country the Commissions meet at least once a year and have a series of different functions. The main function is the adoption of conservation and management measures. These measures address catch and effort limits, compliance, membership, finance and administration, work programmes, subsidiary bodies, research priorities, relations to other management and advisory bodies, implementation of international instruments, observers at meetings and dispute settlement.[677] Across all tuna RFMOs the measures are named differently. IATTC, CCSBT and IOTC

---

672 Convention on the Conservation and Management of Highly Migratory Fish Stocks in the Western and Central Pacific Ocean, Art.1(f).

673 Ibid.

674 Sauries (Scomberesocidae) have been exempted because Japan had the opinion that these are no highly migratory species; Serdy, 2004, p.237.

675 Convention on the Conservation and Management of Highly Migratory Fish Stocks in the Western and Central Pacific Ocean, Art.3.3.

676 Convention for the Conservation of Southern Bluefin Tuna, Arts.5.2, 5,3, 8.1(a) and 9.2(c).

677 J. Swan, *Decision-making in Regional Fishery Bodies or Arrangements- the evolving role of RFBs and international agreement on decision-making processes.*, FAO Fisheries Circular - No. 995 (Rome: FAO, 2004), p.10; T. L. McDorman, Implementing Existing Tools: Turning Words Into Actions Decision-Making Processes of Regional Fisheries Management Organisations (RFMOs), *The International Journal of Marine and Coastal Law - Vol. 20* (2005), p.425.

distinguish between binding resolutions and non-binding recommendations.[678] In contrast, in the ICCAT recommendations are binding but resolutions are not binding.[679] Also different is the naming in the WCPFC with binding conservation and management measures and non-binding resolutions.[680]

Sound conservation and management measures require comprehensive scientific advice from competent bodies. Therefore, subject to their particular historic, geographic and organizational characteristics, every tuna RFMO has developed one or more subsidiary bodies that provide scientific advice to the Commission. The principal scientific body in CCSBT, IOTC and WCPFC are the so called Scientific Committees. In the IATTC it is the Scientific Advisory Committee and in the ICCAT the Standing Committee on Research and Statistics. The main difference between these committees is how the scientific information is produced. In ICCAT, IOTC and CCSBT scientific advices are the outcome of the effort of various working groups which are composed by national scientists from contracting and cooperating parties.[681] The ICCAT has further established specific panels for different species groups.[682] These panels can directly propose recommendations to the Commission. In contrast, in the IOTC the working groups on species groups are not expected to advise the Commissions. The IOTC has the possibility to establish for particular stocks sub-commissions with wider competences than the working parties.[683] So far such sub-commissions have not been established. The special feature of the CCSBT is that national scientists are supported by an independent expert panel which facilitates consensus or provides own views.[684] The IATTC follows a slightly different approach. Information is provided by working groups but its Secretariat has wider competences than those of every other tuna RFMO. With proper scientific staff, the Secretariat coordinates the scientific information and carries out proper re-

---

678   Convention for the Strengthening of the Inter-American Tropical Tuna Commission established by the 1949 Convention between the United States of America and the Republic of Costa Rica "Antigua Convention", Art.VII(c)-(m); Convention for the Conservation of Southern Bluefin Tuna, Art.8.3 and 8.5; Agreement for the Establishment of the Indian Ocean Tuna Commission, Art.V.2(c) and V.3.

679   International Convention for the Conservation of Atlantic Tunas, Art.VII.

680   Convention on the Conservation and Management of Highly Migratory Fish Stocks in the Western and Central Pacific Ocean, Art.10.1(a) and 10.1(c).

681   M. Aranda, et al., *A report review of the tuna RFMOs: CCSBT, IATTC, IOTC, ICCAT and WCPFC*, EU FP7 project n°212188 TXOTX (Pasaia: AZTI Tecnalia, 2010), p.48.

682   International Convention for the Conservation of Atlantic Tunas, Art.VI.

683   Agreement for the Establishment of the Indian Ocean Tuna Commission, Art.XII.2.

684   Aranda, et al., 2010, p.14.

search.[685] Another different model is applied by the WCPFC where the Scientific Committee gets its scientific data from the Secretariat of the Pacific Community and the International Scientific Committee which are both external science providers.[686] Further, the Northern Committee of the WCPFC, which is composed by member countries of the northern part of the convention area, is entitled to make recommendations to the Commission.[687]

Subsidiary bodies have also been established by the Commissions in order to review and monitor compliance with the conservation and management measures. In all tuna RFMOs so called Compliance Committees (IOTC, CCSBT, WCPFC), Conservation and Management Measures Compliance Committee (ICCAT) or Committee for the Review of Implementation of Measures Adopted by the Commission (IATTC) have been established. These committees provide advice on compliance to the Commission and meet, like the scientific bodies, usually prior to the annual Commission meeting.[688]

All tuna RFMOs have further established independent Secretariats that employ permanent staff including an Executive Director and different experts. The Secretariats fulfill all tasks under the authority of the Commissions. The main tasks usually include: receiving and transmitting communications; organizing meetings; facilitating the compilation and dissemination of catch and effort data; preparing reports; supporting agreements on monitoring, control and surveillance; coordinating scientific advice; publishing decisions as well as administration and finance.[689] Some Secretariats play a more proactive role than others. As mentioned before, the Secretariat of the IATTC has comparatively wide scientific competences and a great number of employees. The Secretariat of the WCPFC has recently shown its competences with regard to compliance.[690] It is largely involved in the developed of a compliance monitoring scheme which is intended to improve compliance of the members.

---

685 Lodge, 2007, p.32

686 Convention on the Conservation and Management of Highly Migratory Fish Stocks in the Western and Central Pacific Ocean, Art.13.

687 Ibid. Art.11.7.

688 Agreement for the Establishment of the Indian Ocean Tuna Commission, Art.XII.5; Convention for the Conservation of Southern Bluefin Tuna, Art.8.10; Convention on the Conservation and Management of Highly Migratory Fish Stocks in the Western and Central Pacific Ocean, Art.11.1; International Convention for the Conservation of Atlantic Tunas, Art.IX.3; Convention for the Strengthening of the Inter-American Tropical Tuna Commission established by the 1949 Convention between the United States of America and the Republic of Costa Rica "Antigua Convention", Art.X.

689 Lodge, 2007, p.109.

690 WCPFC, *Conservation and Management Measure for Compliance Monitoring Scheme*, (2012).

4. State participation in RFMOs

In the recent past, participation was together with the allocation of fishing opportunities and illegal unreported and unregulated (IUU) fishing one of the main issues in RFMOs.[691] A major problem was that prior to the Fish Stocks Agreement RFMOs did not have the possibility to substantially and actively intervene in IUU fishing activities of non-member states.[692] According to UNCLOS on the high seas the flag states had exclusive jurisdiction over the vessels flying their flag.[693] One of the progressive achievements of the Fish Stocks Agreement was therefore the linkage between participation, allocation and IUU fishing.[694] States fishing in the area of competence of any RFMO have to cooperate by becoming members of the respective RFMO, or by agreeing to apply the conservation and management measures established by the RFMO.[695] This is not an explicit obligation to become a member but it requires all states with an interest in the fish resources to comply with the conservation and management measures that have been established by the RFMOs.[696] Although this requirement applies only to parties to the Fish Stocks Agreement it has been a remarkable step forward as it substantiated the duty to cooperate.

Today all tuna RFMOs are composed by members or contracting parties and so called cooperating non-members or cooperating non-contracting parties (Table 3.6). The terminology varies among the tuna RFMOs. With 48 Contracting Parties and five Cooperating Non-Contracting Parties, ICCAT is the biggest tuna RFMO in terms of members.[697] The CCSBT is, with only six members and three cooperating non-members, currently the smallest tuna RFMO.[698] The IOTC has 31 members and two Cooperating Non-Contracting Parties.[699] The IATTC has 21 members and two cooperating non-parties.[700] The WCPFC has 25 members,

---

691   E. J. Molenaar, Participation, Allocation and Unregulated Fishing: The Practice of Regional Fisheries Management Organisations, *International Journal of Marine and Coastal Law - Vol.18* (2003).

692   Meltzer & Fuller, 2009, p.63.

693   United Nations Convention on the Law of the Sea, Art.94.

694   Molenaar, 2003, p.458.

695   Agreement for the Implementation of the Provisions of the United Nations Convention on the Law of the Sea of 10 December 1982 relating to the Conservation and Management of Straddling Fish Stocks and Highly Migratory Fish Stocks, Art.8.3.

696   Ibid. Art.8.4.

697   http://www.iccat.int/en/contracting.htm.

698   http://www.ccsbt.org/site/origins_of_the_convention.php.

699   http://www.iotc.org/English/info/comstruct.php.

700   https://www.iattc.org/HomeENG.htm.

seven participating territories and 11 Cooperating Non-members.[701] Participating territories exist only in the WCPFC. These overseas territories of former colonies are entitled to participate fully in the work of the Commission, to attend all meetings and they are also allowed to speak at the meetings. [702]

Table 3.6   Participation in tuna RFMOs

	ICCAT	IATTC	CCSBT	IOTC	WCPFC
Albania	CP				
Algeria	CP				
Angola	CP				
American Samoa					PT
Australia			M	M	M
Barbados	CP				
Belize	CP	M		M	CNM
Bolivia		CNM			
Brazil	CP				
Canada	CP	M			M
Cape Verde	CP				
China	CP	M		M	M
Chinese Taipei[703]	CNCP	M	M		M
Colombia	CNCP	M		M	
Comoros				M	
Cook Islands		CNM			M
Costa Rica		M			
Côte d'Ivoire	CP				
Croatia	CP				

701   http://www.wcpfc.int/.

702   Convention on the Conservation and Management of Highly Migratory Fish Stocks in the Western and Central Pacific Ocean Arts.43.1 and 43.3.

703   In the CCSBT Chinese Taipei is named 'Fishing Entity of Taiwan'.

	ICCAT	IATTC	CCSBT	IOTC	WCPFC
Curaçao	CNCP				
Democratic People's Republic of Korea					CNM
Ecuador		M			CNM
Egypt	CP				
El Salvador	CNCP	M			CNM
Equatorial Guinea	CP				
Eritrea				M	
European Union	CP	M	CNM	M	M
Federated States of Micronesia					M
Fiji				M	M
France	CP	M			M
French Polynesia					PT
Gabon	CP				
Ghana	CP				
Guatemala	CP	M			
Guinea	CP			M	
Guam					PT
Guatemala					
Honduras	CP				
Iceland	CP				
India				M	
Indonesia			M	M	CNM
Islamic Republic of Iran				M	
Japan	CP	M	M	M	M
Kenya				M	
Kiribati		M			M
Libya	CP				

	ICCAT	IATTC	CCSBT	IOTC	WCPFC
Madagascar				M	
Malaysia				M	
Maldives				M	
Mauritania	CP				
Mexico	CP	M			CNM
Morocco	CP				
Mozambique				M	
Namibia	CP				
Nauru					M
New Zealand			M		M
New Caledonia					PT
Nicaragua	CP	M			
Nigeria	CP				
Niue					M
Northern Mariana Islands					PT
Norway	CP				
Palau					M
Panama	CP	M			CNM
Pakistan				M	
Papua New Guinea					M
Peru		M			
Philippines	CP		CNM	M	M
Republic of Korea	CP	M	M	M	M
Republic of Marshall Islands					M
Russia	CP				
Saint Vincent/Grenadines	CP				
Samoa					M

	ICCAT	IATTC	CCSBT	IOTC	WCPFC
Sao Tome and Principe	CP				
Senegal	CP			CNCP	CNM
Seychelles				M	
Sierra Leone	CP			M	
South Africa	CP		CNM	CNCP	
Solomon Islands					M
Sri Lanka				M	
St Kitts and Nevis					CNM
Sudan				M	
Sultanate of Oman				M	
Suriname	CNCP				
Syrian Arab Republic	CP				
Tanzania				M	
Thailand				M	CNM
Trinidad and Tobago	CP				
Tokelau					PT
Tonga					M
Tunisia	CP				
Turkey	CP				
Tuvalu					M
United Kingdom	CP			M	
United States of America	CP	M			M
Uruguay	CP				
Vanuatu	CP	M		M	M
Venezuela	CP	M			
Vietnam					CNM
Wallis and Futuna					PT
Yemen				M	

Source: Unterweger, based on information provided by the websites of the tuna RFMOs.

CP= Contracting Party; CNCP = Cooperating Non-Contracting Party; M = Member; CNM = Cooperating Non-Member; PT = Participating Territory

5. *Accession*

The Fish Stocks Agreement states that States "having a real interest in the fisheries concerned" are entitled to become members of an RFMO.[704] Although it is not clear what exactly is meant by the term 'real interest' it is assumed that these are the "States fishing for the stocks on the high seas and relevant coastal States" mentioned in Article 8.3.[705] Overall it can be concluded that this provision does not exclude anybody from participating. Another important question is how nature and extent of participatory rights for new members or participants are determined. Article 11 is providing a comprehensive list of criteria for the determination of participatory rights.[706] The criteria include status of the stocks, existing level of fishing efforts, historic fishing activity; contribution to conservation, needs of coastal states and the interests of developing states.

In practice all tuna RFMOs are generally open to new members but the organizations differ with regard to possible accession. In ICCAT and CCSBT the new member's signature and a declaration on compliance is sufficient for accession. The great number of members of the ICCAT shows that this RFMO is very open for new members. According to its Convention any State or Regional Economic Integration Organizations (REIOs) which is a member of the United Nations or of any Specialized Agency of the United Nations can become a member by signing the Convention.[707] In view of the provisions of the Convention, the CCSBT is open as well. Accession is possible for every State, whose vessels are fishing for southern bluefin tuna, or any other coastal State through whose EEZ southern bluefin tuna migrates.[708] This Convention becomes effective for a new member when it deposits its instrument of accession. REIOs are not mentioned in the Convention text. This might lead to the conclusion that they are not per-

---

704   Agreement for the Implementation of the Provisions of the United Nations Convention on the Law of the Sea of 10 December 1982 relating to the Conservation and Management of Straddling Fish Stocks and Highly Migratory Fish Stocks, Art.8.3.

705   E. J. Molenaar, The Concept of "Real Interest" and Other Aspects of Co-operation through Regional Fisheries Management Mechanisms, *The International Journal of Marine and Coastal Law - Vol.15(4)* (2000), pp.475-532.

706   Agreement for the Implementation of the Provisions of the United Nations Convention on the Law of the Sea of 10 December 1982 relating to the Conservation and Management of Straddling Fish Stocks and Highly Migratory Fish Stocks, Art.11.

707   International Convention for the Conservation of Atlantic Tunas, Art.XIV.1.

708   Convention for the Conservation of Southern Bluefin Tuna, Art.18.

mitted to join.[709] However, the European Union as a REIO became a cooperating non-member in 2004.

IOTC, WCPFC and IATTC require formal consultation and a qualified decision of the current members. The membership requirements in the IOTC are more restrictive than in the ICCAT. Generally, membership is open to Members and Associate Members of FAO which are Indian Ocean Coastal States or REIOs, or states or REIOs with vessels fishing in the area of competence.[710] A non-member of the FAO which is a member of the United Nations, or of any Specialized Agency of the United Nations can be admitted by a two-thirds majority vote of the Members if it is an Indian Ocean Coastal State or REIO, or a state or REIO with vessels fishing in the area of competence.[711] After its adoption, the WCPFC Convention was open for signature and accession by the participants of the negotiation process.[712] All the current members were participants of that process. However, States and REIOs with a real interest in the fisheries can become members if they are invited with the agreement of all contracting parties.[713] After its adoption in 2003 the Antigua Convention of the IATTC was open for signature.[714] Entitled to become members were all parties of the 1949 Convention, coastal states within the Convention area, states and REIOs which had participated in the negotiations of the Antigua Convention and states fishing in the Convention area in the four years prior to the adoption of the new Convention.[715] So far 14 members have signed[716] and 16 members have ratified[717] the Antigua Convention.[718] Two original members have neither signed

---

709 E. Meltzer, Global Overview of Straddling and Highly Migratory Fish Stocks; Maps and Charts Detailing RFMO Coverage and Implementation, *The International Journal of Marine and Coastal Law - Vol. 20* (2005), p.596.

710 Agreement for the Establishment of the Indian Ocean Tuna Commission, Art.IV.1.

711 Ibid. Art.IV.2.

712 Convention on the Conservation and Management of Highly Migratory Fish Stocks in the Western and Central Pacific Ocean, Art.34.1.

713 Ibid. Art.35.2.

714 Convention for the Strengthening of the Inter-American Tropical Tuna Commission established by the 1949 Convention between the United States of America and the Republic of Costa Rica "Antigua Convention", Art.XXVII.1.

715 Ibid. Art.XXVII.1(a-d).

716 Canada, China, Chinese Taipei, Costa Rica Ecuador, El Salvador, the European Union, France, Guatemala, Mexico, Nicaragua, Peru, the United States and Venezuela.

717 Belize, Canada, China, Chinese Taipei, Costa Rica, El Salvador, the European Union, France, Guatemala, Japan, Kiribati, Korea, Mexico, Nicaragua and Panama.

718 See also http://www.iattc.org/IATTCdocumentationENG.htm.

nor ratified or acceded to the Convention yet.[719] The Convention is open to accession by any State or REIO that meets the requirements for signature, states fishing for fish stocks covered by the Convention, following consultations with the Parties and states that are invited to accede on the basis of a decision by the Parties.[720]

A general problem of all RFMOs is the fact that existing RFMO members are usually not interested in providing new members with shares of the total catch.[721] If this is taken into account then RFMOs are rather restrictive or closed.[722] In practice RFMOs are often increasing the total catch in order to accommodate new members.[723] One example is the CCSBT who allocated in 2003 additional catches to the Fishing entity of Taiwan instead of reducing the respective share of the original members.[724]

6.   Decision making

Decision making procedures in RFMOs are often seen as a reason for the lack of sustainability in their conservation and management.[725] Especially the pressure of states that seek to maintain or improve fishing opportunities for their own fleets leads to poor decision making by tuna RFMOs.[726] The principal problems are the decision making procedure itself and possible objections by certain members. Neither UNCLOS nor the Fish Stocks Agreement address decision making processes explicitly.[727] The few provisions within the Fish Stocks

---

719   Colombia and Vanuatu.

720   Convention for the Strengthening of the Inter-American Tropical Tuna Commission established by the 1949 Convention between the United States of America and the Republic of Costa Rica "Antigua Convention", Art.XXX.

721   A. Willock & M. Lack, *Follow the leader - Learning from experience and best practice in regional fisheries management organizations* (Sydney: WWF International and TRAFFIC International, 2006), p.6.

722   E. J. Molenaar, 'Regional Fisheries Management Organizations: Issues of Participation, Allocation and Unregulated Fishing', *Oceans Management in the 21st Century: Institutional Frameworks and Responses* (Leiden: Martinus Nijhoff, 2004), p.85.

723   Lodge, 2007, p.36.

724   CCSBT, *Report of the Performance Review Working Group Part 1 - Self Assessment*, (2008), p.20.

725   Willock & Lack, 2006, p.34.

726   Allen, 2010, p.6.

727   McDorman, 2005, p.423.

Agreement are very general and open to interpretation.[728] Parties are only required to agree through RFMOs "on decision-making procedures which facilitate the adoption of conservation and management measures in a timely and effective manner."[729] As a consequence, tuna RFMOs have developed different approaches of decision making.

Generally, all constituent instruments of the tuna RFMOs endeavor consensus based decisions but there are also some which provide majority based voting procedures.[730] It is not totally clear which approach is resulting in better decisions. Although the consensus based decision making procedures guarantee that all members are willing to comply with the decision, the problem is that the outcome of such procedures is often the lowest denominator.[731] Consensus based decisions are further blamed for having a tendency to defer management decisions until a crisis is approaching.[732] Majority based decision making procedures have the potential to provide more progressive decisions but they can be weakened if there is the possibility of objections.[733]

a)   Decision making procedure

In CCSBT and in IATTC consensus based decision making is mandatory. During the Commission meetings of the CCSBT every party has one vote and decisions have to be taken by a unanimous vote of the parties that are present at the meeting.[734] IATTC specifies this general requirement[735] and puts an extra emphasis on the need for consensus in decisions on allocation of catch, capacity or effort.[736]

---

728   Agreement for the Implementation of the Provisions of the United Nations Convention on the Law of the Sea of 10 December 1982 relating to the Conservation and Management of Straddling Fish Stocks and Highly Migratory Fish Stocks, Arts. 6.3(a), 10(j), 12.1 and 28.

729   Ibid. Art.10(j).

730   Allen, 2010, p.8.

731   Meltzer & Fuller, 2009, p.192.

732   M.W. Lodge & S. N. Nandan, Some Suggestions towards Better Implementation of the United Nations Agreement on Straddling Fish Stocks and Highly Migration Fish Stocks of 1995, *International Journal of Marine & Coastal Law - Vol. 20* (2005), p.376.

733   Meltzer & Fuller, 2009, p.193.

734   Convention for the Conservation of Southern Bluefin Tuna, Art.7.

735   Convention for the Strengthening of the Inter-American Tropical Tuna Commission established by the 1949 Convention between the United States of America and the Republic of Costa Rica "Antigua Convention", Art.IX.1.

736   Ibid. Art.IX.3(b) and Art.VII.1(l).

In contrast, in the other three tuna RFMOs there exists the possibility to decide on certain decisions by a qualified majority. In general, decisions in the ICCAT have to be taken by a majority of the Contracting Parties of the Commission.[737] An exception is provided for the adoption of recommendations to maintain tuna and tuna-like species at levels which permit the maximum sustainable catch.[738] These recommendations have to be adopted by at least two-thirds of all Contracting Parties. The basic requirements on decision making in the IOTC are very similar to those of the ICCAT. According to its Agreement the Commission of the IOTC can adopt binding conservation and management measures by a two-thirds majority of the members present and voting.[739] Recommendations on conservation and management of the stocks for furthering the objectives of the Agreement require only a simple majority of the members present and voting.[740] The WCPFC applies a mixture of mandatory consensus based decision making and majority voting.[741] Consensus is required for the most important decisions like the allocation of catch and effort.[742] Questions of procedure can be taken by simple majority. A special feature of the WCPFC is chambered voting on questions of substance. These questions include conservation and management measures and must be taken by a three fourths majority provided that this majority includes a three-fourths majority of the members of the South Pacific Forum Fisheries Agency[743] and a three-fourths majority of nonmembers of the South Pacific Forum Fisheries Agency.[744] In no circumstances proposal shall be defeated by two or fewer votes in either chamber.

b)   *Objections*

Only the constituent instruments of the three tuna RFMOs which provide voting procedures (ICCAT, IOTC, WCPFC) allow objections or seeking of review.

---

737   International Convention for the Conservation of Atlantic Tunas, Art.III.3; ICCAT, *Rules of procedure*, (2012), Para.9.2.

738   International Convention for the Conservation of Atlantic Tunas, Art.VIII, Para1(b)(i)

739   Agreement for the Establishment of the Indian Ocean Tuna Commission, Art.IX.1.

740   Ibid. Art.IX.8.

741   Convention on the Conservation and Management of Highly Migratory Fish Stocks in the Western and Central Pacific Ocean, Art.20.2.

742   Ibid. Art.10.4.

743   More detailed information on the South Pacific Forum Fisheries Agency in the next chapter.

744   Convention on the Conservation and Management of Highly Migratory Fish Stocks in the Western and Central Pacific Ocean, Art.20.2.

The fact that in CCSBT and IATTC all decisions and resolutions are taken by consensus of all members prevents the occurrence of disputes and objections.[745]

The ICCAT established the most comprehensive provisions on objection procedures. If a Contracting Party objects to a recommendation within the six months, the recommendation does not become effective for an additional 60 days.[746] Any other Contracting Party can present an objection prior to the expiration of the additional 60 days period, or within 45 days of the date of the notification of an objection made by another Contracting Party.[747] Except for those Contracting Parties that have presented an objection the recommendation becomes effective at the end of the extended period or periods for objection.[748] A recommendation which has met with an objection presented by up to one-fourth of the Contracting Parties has no effect to the objecting parties.[749] A recommendation which has met with objection from more than one-fourth but less than the majority of the Contracting Parties the recommendation becomes effective for the Contracting Parties that have not objected.[750] In the case that objections have been presented by a majority of the Contracting Parties the recommendation does not become effective.[751] Any Contracting Party objecting to a recommendation may at any time withdraw that objection.[752]

In the IOTC any Member of the Commission has the possibility to object to a conservation and management measure within 120 days from the date of notification by the Commission.[753] If there is an objection from a member, any other Member may also object to the measure within 60 days after the initial 120-day objection period. A Member of the Commission that has objected to a measure is not bound by that measure. An objection can be withdrawn at any time by the objecting member. If objections are made by more than one third of Members, none of the Members are bound by that measure.[754]

Slightly different is the situation in the WCPFC which provides the most progressive approach to avoid objections. A member that does not agree with a

---

745   Swan, 2004, p.67.

746   International Convention for the Conservation of Atlantic Tunas, Art.VIII.3(a).

747   Ibid. Art.VIII.3(b).

748   Ibid. Art.VIII.3(c).

749   Ibid. Art.VIII.3(d-e).

750   Ibid. Art.VIII.3(f).

751   Ibid. Art.VIII.3(g).

752   Ibid. Art.VIII.4.

753   Agreement for the Establishment of the Indian Ocean Tuna Commission, Art.IX.5.

754   Ibid. Art.IX.6.

certain decision can seek a review of the decision.[755] In the next step a review panel evaluates the need to modify, amend or revoke the decision.[756] If it determines that there is no need the decision becomes binding. That means that in such a case a respective member could be bound by the decision even if it does not agree. It is not clear if that would succeed in practice. So far there had been no such case.

7. Settlement of disputes

In RFMOs basically three possible types of disputes can arise: Disputes on the interpretation or application of the constituent instrument of an RFMO; disputes on the meaning of UNCLOS or the Fish Stocks Agreement; or disputes on the compatibility between conservation and management measures adopted for the EEZ by a coastal state or for the high seas by a tuna RFMO.[757] The settlement of disputes and its possible application in RFMOs is a good example for progressive development in international fisheries law. Already UNCLOS introduced an innovative and far reaching compulsory dispute settlement system.[758] According to Part XV disputes have to be settled by peaceful means.[759] The preferred manner is always through negotiation, conciliation or other means.[760] If that is not possible any party can request a compulsory dispute settlement procedure through an international tribunal or court.[761] The dispute settlement system of UNCLOS has been already applied in two important disputes on the conservation and management of tuna and tuna-like species.[762] The southern bluefin case revealed that the UNCLOS is not necessarily lex superior to the constituent instruments of RFMOs. In its award the Arbitral Tribunal had concluded that the provisions on dispute settlement laid down in the Convention of the CCSBT implicitly excluded further provisions under UNCLOS and redirected the dispute to the CCSBT.

After the ratification of the Fish Stocks Agreement the Tribunal might have come to a different decision because the new agreement had laid down in Article 30.2 that the provisions of Part XV of the UNCLOS "apply mutatis mutandis

---

755 Convention on the Conservation and Management of Highly Migratory Fish Stocks in the Western and Central Pacific Ocean, Art.20.6.

756 Ibid. Art.20.8.

757 Lodge, 2007, p.78.

758 T. Koh, *A Constitution for the Oceans* (Montego Bay: United Nations, 1983).

759 United Nations Convention on the Law of the Sea, Art.279.

760 Ibid. Arts.283-284.

761 Ibid. Art.286.

762 More detailed information is provided in the previous chapter.

to any dispute between States Parties to this Agreement concerning the interpretation or application of a subregional, regional or global fisheries agreement relating to straddling fish stocks or highly migratory fish stocks to which they are parties".[763] According to that, Part XV applies now to all parties of the Fish Stocks Agreement regardless the constituent instruments of the RFMOs.[764] Despite this progress it has to be considered that there are still some RFMO members which are not parties to the Fish Stocks Agreement yet. In a possible dispute, those countries might refer to the decision of the Arbitral Tribunal in the southern bluefin case. For that reason it is useful to have a closer look at the constituent instruments of the tuna RFMOs. Explicit provisions on dispute settlement are included in the instruments of WCPFC, IATTC, IOTC and CCSBT. No dispute settlement procedures can be found in the Convention of the ICCAT which is the only constituent instrument that predates the UNCLOS.

Among all tuna RFMOs the dispute settlement procedure of the WCPFC comes closest to the provisions of the Fish Stocks Agreement. Article 31 of the WCPFC Convention refers directly to Part VIII of the Fish Stocks Agreement which applies, "mutatis mutandis, to any dispute between members of the Commission, whether or not they are also Parties to the Agreement."[765]

The Antigua Convention of the IATTC does not explicitly refer to the dispute settlement provisions of the Fish Stocks Agreement nor does it mention UNCLOS or the international courts. It is only stated that members shall cooperate "to prevent disputes" and "if a dispute is not settled [...] the members in question shall consult among themselves as soon as possible in order to settle the dispute through any peaceful means they may agree upon, in accordance with international law."[766] Unique among the tuna RFMOs is the provision that disputes of a technical nature which cannot be resolved by the conflicting members may be referred, "by mutual consent, to a non-binding ad hoc expert panel constituted within the framework of the Commission".[767] This panel "shall confer with the members concerned and shall endeavor to resolve the dispute expeditiously without recourse to binding procedures for the settle-

---

763 Agreement for the Implementation of the Provisions of the United Nations Convention on the Law of the Sea of 10 December 1982 relating to the Conservation and Management of Straddling Fish Stocks and Highly Migratory Fish Stocks, Art.30.2.

764 Lodge, 2007, p.80.

765 Convention on the Conservation and Management of Highly Migratory Fish Stocks in the Western and Central Pacific Ocean, Art.31.

766 Convention for the Strengthening of the Inter-American Tropical Tuna Commission established by the 1949 Convention between the United States of America and the Republic of Costa Rica "Antigua Convention", Arts.XXV.1 and 2.

767 Ibid. Arts.XXV.3, sent.1.

ment of disputes."[768] According to these provisions the IATTC currently lacks a compulsory dispute settlement mechanism. This is a weak point of the comparatively new Antigua Convention but in the future it might become less relevant if there is increasing ratification of the Fish Stocks Agreement with its superimposing provisions.[769] Also conflicting States which are Parties to UNCLOS have to at least comply with UNCLOS dispute settlement requirements, which to some extent are compulsory.

In the IOTC all disputes regarding the interpretation or application of the Agreement which are not settled by the Commission have to be referred for settlement to a conciliation procedure adopted by the Commission.[770] The results of this procedure, while not binding in character, shall become the basis for renewed consideration by the parties concerned of the matter out of which the disagreement arose.[771] The Agreement refers to the international courts. A dispute which is not settled after this procedure "may be referred to the International Court of Justice in accordance with the Statute of the International Court of Justice, unless the parties to the dispute agree to another method of settlement."[772] However, the Agreement of the IOTC does not make reference to a compulsory dispute settlement mechanism. This is considered as a major gap.[773] However, as for the IATTC relevance of this problem will decrease if more countries become parties to the Fish Stocks Agreement

The dispute settlement procedure under the Convention of the CCSBT had been discussed already extensively in the previous chapter. Therefore, the only question which is raised at this point is whether Article 16 on dispute settlement in the Convention of the CCSBT can be used as the basis for an exclusion of compulsory dispute settlement under UNCLOS. This might be doubted especially because Article 16.2 is stating that any dispute not resolved by peaceful means "shall, with the consent in each case of all parties to the dispute, be referred for settlement to the International Court of Justice or to arbitration".

---

768   Ibid. Arts.XXV.3, sent.2.

769   A. Serdy, 'International Fisheries Law and the Transferability of Quota: Principles and Precedents', *Conservation and Management of Transnational Tuna Fisheries* (Iowa: Wiley-Blackwell, 2010), pp.117-118.

770   Agreement for the Establishment of the Indian Ocean Tuna Commission, Art.XXIII, sent.1.

771   Ibid. Art.XXIII, sent.2.

772   Ibid. Art.XXIII. sent.3.

773   IOTC, *Report of the IOTC Performance Review Panel* (Victoria: IOTC, 2009), p.18.

8. Regulation of total catch and fishing mortality

A basic task for all tuna RFMOs is the regulation of total catch and fishing mortality. In international fisheries management two principal approaches are used for that purpose: Input controls and output controls. Input controls are regulating the amount of effort which can be put in a fishery and output controls are regulating the catch which can be taken from a fishery.[774] An additional approach is the use of technical measures which usually influence the efficiency of the fishing gear.

The UNCLOS and the Fish Stocks provide limited guidance on which approach should be used in tuna fisheries.[775] The provisions of UNCLOS regarding the conservation of straddling stocks and highly migratory species do not make any preference for a certain type of control.[776] For the conservation of the living resources of the high seas UNCLOS established minimum requirements for the determination of the total allowable catch (TAC) and other conservation measures but without indicating preference for catch or effort based regulations.[777] The Fish Stocks Agreement contains more detailed provisions on catch and effort regulations.[778] However, it makes no preference either. It only lays down that, in fulfilling their obligation to cooperate, states have to "agree, as appropriate, on participatory rights such as allocations of allowable catch or levels of fishing effort."[779]

This limited guidance through UNCLOS and the Fish Stocks Agreement but also differences among the tuna RFMOs regarding the time of their establishment, available scientific information, species composition or states involved in the fishery have led to different strategies of regulating total catch and fishing mortality. Generally it has to be recognized that the first tuna RFMOs IATTC

---

774 FAO Fishery Resources Division and Fishery Policy and Planning Division, *Fisheries management*, FAO Technical Guidelines for Responsible Fisheries - No. 4 (Rome: FAO, 1997), pp.45-51.

775 M. C. Engler Palma, *Allocation of Fishing Opportunities in Regional Fisheries Management Organizations: A Legal Analysis in the Light of Equity* (Halifax: Dalhousie University, 2010), p.77.

776 United Nations Convention on the Law of the Sea, Art.63.2 and Art.64.1.

777 They have to be designed "to maintain or restore populations of harvested species at levels which can produce the maximum sustainable yield, as qualified by relevant environmental and economic factors."

778 Agreement for the Implementation of the Provisions of the United Nations Convention on the Law of the Sea of 10 December 1982 relating to the Conservation and Management of Straddling Fish Stocks and Highly Migratory Fish Stocks, Arts.5(h) and (j), Art.6.6.

779 Ibid. Art.10(b).

and ICCAT initially mainly used output control. When, due to increasing fishing capacity, overfishing became a problem both shifted to input control and technical measures as well. In contrast, the younger IOTC and WCPFC were established when, on an international level, overfishing was already a problem. These tuna RFMOs show a clear preference for input control and technical measures. The CCSBT plays a special role. Prior to its establishment there was already a long history of using output control to manage the southern bluefin tuna fishery. Also after the establishment this remained the dominating approach to regulate catch and fishing mortality.

The IATTC was the first tuna RFMO which adopted a conservation and management measure for tuna and tuna-like species. This early resolution focused on output control and determined catch limits for yellowfin tuna.[780] The 1949 Convention of the IATTC did not provide any guidance on specific types of control. The new Antigua Convention requires that conservation and management goals have to be achieved, among others, "through the setting of the total allowable catch of such fish stocks as the Commission may decide and/or the total allowable level of fishing capacity and/or level of fishing effort for the Convention Area as a whole."[781] The current resolution for the most important species yellowfin, bigeye and skipjack tuna requires a mixture of input control, output control and technical measures.[782] The purse seine fishery is managed with capacity allocations and time closures while the longline fishery is managed by national quotas.[783] A special characteristic of the IATTC is a closed regional vessel register for purse seine vessels.[784] The objective of this register which is unique among the tuna RFMOs is to freeze capacity by limiting the number of vessels that are allowed to fish in the convention area.[785]

The Convention of the ICCAT does not state any preference for input or output control. The Commission is only required to „make recommendations designed to maintain the populations of tuna and tuna-like fishes [...] at levels which will permit the maximum sustainable catch.[786] Due to the time of its establishment

---

780   Joseph, et al., 2010, p.14.

781   Convention for the Strengthening of the Inter-American Tropical Tuna Commission established by the 1949 Convention between the United States of America and the Republic of Costa Rica "Antigua Convention", Art.7.1(c).

782   IATTC, *Resolution on a Multinational Program for the Conservation of Tuna in the Eastern Pacific Ocean in 2011-2013*, (2011).

783   D. J. Agnew, et al., *Allocation issues for WCPFC tuna resources* (London: Marine Resources Assessment Group Ltd, 2006), pp.32-33.

784   IATTC, *Resolution on fleet capacity*, (2001).

785   Joseph, et al., 2010, p.24.

786   International Convention for the Conservation of Atlantic Tunas, Art.8.1(a).

the ICCAT made similar experiences like the IATTC.[787] Initially the management system was mainly based on output control before input control and technical measures were adopted to reduce overfishing.[788] The current management approach of the ICCAT for bigeye and bluefin tuna is still based on TAC and national quota allocations[789] but also on effort limits[790] and technical measures like time/area closures.[791] Yellowfin tuna is managed exclusively with effort limits since 1993.[792]

The CCSBT is one of the tuna RFMOs that has been created when global overfishing of tuna and tuna-like species and especially of southern bluefin tuna already had become a problem. However, already long before establishing the CCSBT the principal fishing countries had used TACs and national quotas to manage the southern bluefin tuna fishery. The Convention of the CCSBT reflects this preference for output control by stating that the Commission "shall decide upon a total allowable catch and its allocation among the Parties unless the Commission decides upon other appropriate measures".[793] Today, the CCSBT is the only tuna RFMOs which manages its stocks exclusively with TACs and country allocations.[794]

In the Convention of the WCPFC, there is no preference given for input or output control. The Commission has to "determine the total allowable catch or total level of fishing effort within the Convention Area for such highly migratory

---

787 Joseph, et al., 2010, p.15.

788 ICCAT, *Recommendation by ICCAT on a Yellowfin Size Limit*, (1972); ICCAT, *Recommendation by ICCAT Concerning a Limit on Bluefin Tuna Size and Fishing Mortality*, (1974).

789 ICCAT, *Recommendation by ICCAT on a multi-year conservation and management program for bigeye tuna*, (2004), Paras.3-7; ICCAT, *Recommendation Amending the Recommendation by ICCAT to Establish a Multi-annual Recovery Plan for Bluefin Tuna in the Eastern Atlantic and Mediterranean*, (2010), Paras. 4-20.

790 ICCAT, *Recommendation by ICCAT on a multi-year conservation and management program for bigeye tuna*, (2004), Paras.1-2; ICCAT, *Recommendation Amending the Recommendation by ICCAT to Establish a Multi-annual Recovery Plan for Bluefin Tuna in the Eastern Atlantic and Mediterranean*, (2010), Paras.41-49.

791 ICCAT, *Recommendation by ICCAT on a multi-year conservation and management program for bigeye tuna*, (2004), Paras.8-12. ICCAT, *Recommendation Amending the Recommendation by ICCAT to Establish a Multi-annual Recovery Plan for Bluefin Tuna in the Eastern Atlantic and Mediterranean*, (2010),Paras.21-25.

792 ICCAT, *Recommendation by ICCAT on supplemental regulatory measures for the management of Atlantic yellowfin tuna*, (1993), Para.1.

793 Convention for the Conservation of Southern Bluefin Tuna, Art.8.3(a).

794 CCSBT, *Resolution on the Allocation of the Global Total Allowable Catch*, (2011).

fish stocks as the Commission may decide."[795] For that purpose the Commission is entitled to adopt measures on the quantity of any species or stocks which may be caught; the level of fishing effort; limitations of fishing capacity; areas and periods in which fishing may occur; minimum size of fish and other species; fishing gear and technology; as well as particular sub-regions or regions.[796] In practice, the Commission basically requires its members to remain below the catch and effort levels of previous reference years. For the most important target species bigeye and yellowfin tuna there are effort limits for purse seine fisheries[797] and catch limits for longline fisheries.[798] The respective provisions are accompanied by technical measures like Fish Aggregating Device (FAD) or area closures.[799] To date, no TAC has been determined for any species.

The Agreement of the IOTC does not provide guidance on a preferred method to regulate total catch and fishing mortality. Very generally it is only laid down that the Commission has to "adopt [...] conservation and management measures, to ensure the conservation of the stocks covered by th[e] Agreement and to promote the objective of their optimum utilization throughout the Area."[800] In practice the IOTC is mainly focusing on capacity limits[801] and technical measures like area closures.[802] Only one resolution is setting a catch limit by stating that all CPCs have to limit their catch of bigeye tuna to recent levels of catch reported by the Scientific Committee.[803] A performance report of a re-

---

795   Convention on the Conservation and Management of Highly Migratory Fish Stocks in the Western and Central Pacific Ocean, Art.10.1(a).

796   Ibid. Art.10.2.

797   WCPFC, *Conservation and Magement Measures for Bigeye and Yellowfin Tuna in the Western and Central Pacific Ocean*, (2008), Para.10.

798   Ibid. Paras.31-38.

799   Ibid. Paras.17.b, and 19; WCPFC, *Conservation and Management Measure for the Eastern High-Seas Pocket Special Management Area*, (2010).

800   Agreement for the Establishment of the Indian Ocean Tuna Commission.

801   IOTC, *Resolution 03/01 On the limitation of fishing capacity of Contracting Parties and Cooperating non-Contracting Parties*, (2003); IOTC, *Resolution 06/05 On the limitation of fishing capacity, in terms of number of vessels, of IOTC contracting parties and co-operating non contracting parties (superseded by Resolution 09/02)*, (2006); IOTC, *Resolution 07/05 Limitation of fishing capacity of IOTC Contracting Parties and Cooperating non-Contracting Parties in terms of number of longline vessels targeting swordfish and albacore (superseded by Resolution 09/02)*, (2007); IOTC, *Resolution 09/02 On the implementation of a limitation of fishing capacity of Contracting Parties and Cooperating non-Contracting Parties*, (2009).

802   IOTC, *Resolution 10/01 For the Conservation and Management of Tropical Tunas Stocks in the IOTC Area of Competence*, (2010).

803   IOTC, *Resolution 05/01 On conservation and management measures for bigeye tuna*, (2005) Para.1.

view panel criticized the current conservation and management and elaborated that other approaches like catch limits, total allowable catch (TAC) or total allowable effort (TAE) should be explored in the future.[804] As a consequence, currently the IOTC is developing allocation criteria which are intended to establish, among others, catch or fishing capacity limits.[805]

9. *Allocation criteria*

The allocation of catch and effort is described as "the process of providing temporary or permanent access, use or presumptive rights to fish."[806] It is probably the most controversial issue in RFMOs because every country is interested to get the greatest possible share.[807] A sensitive task for the RFMOS is to balance conservation interests with the economic and social interests when allocating catch and effort. Unfortunately the international legal instruments provide limited guidance on that.[808] The UNCLOS does not contain any allocation criteria at all and the Fish Stocks Agreement requires agreement "on participatory rights such as allocations of allowable catch or levels of fishing effort" but without stating how these rights should be allocated or on which criteria the allocation should be the based on.[809] Article 11 of the Fish Stocks Agreement provides only a list of allocation criteria for new members and participants. These quite detailed criteria include: stock status; historical catches; contribution to conservation and management; provision of scientific data; needs of coastal fishing communities and coastal states as well as the interests of developing states.[810] Some guidance is further given by Article 25 which states that developing should get facilitated access to high seas fisheries.[811] The tuna RFMOs have developed their own criteria to allocate catch and effort. Most of them are using allocation criteria similar to those provided by Article 11 of the Fish Stocks

---

804   IOTC, 2009, pp.24-25.

805   IOTC, *Report of the Technical Committee Meeting on Allocation Criteria*, (2011).

806   R. Q. Grafton, et al., *The Economics of Allocation in Tuna Regional Fisheries Management Organizations (RFMOs)*, Economics and Environment Network Working Paper, EEN0612 (Canberra: Australian National University, 2006), p.4.

807   Miyake, et al., 2010, Executive summary; Q. Grafton, et al., 'The Economics of Allocation in Tuna Regional Fisheries Management Organizations', *Conservation and Management of Transnational Tuna Fisheries* (Iowa: Blackwell Publishing, 2010), p.156.

808   Lodge & Nandan, 2005, p.374.

809   Agreement for the Implementation of the Provisions of the United Nations Convention on the Law of the Sea of 10 December 1982 relating to the Conservation and Management of Straddling Fish Stocks and Highly Migratory Fish Stocks, Art.10(b).

810   Ibid. Arts.11.

811   Ibid. Art.25.1(b).

Agreement.[812] The most common criteria to calculate country allocations are historical catches which are comparatively fair and easy to quantify.[813]

The Convention of the CCSBT provides criteria which have to be considered when catches are allocated.[814] These criteria include: scientific evidence; need for orderly and sustainable development of the fisheries; interests of Parties through whose EEZ southern bluefin tuna migrates; interests of Parties engaged in fishing; contribution to conservation and scientific research as well as any other factors. In the recent past, allocation was still mainly based on political negotiations and on the unpublished scheme that has been negotiated between the original members in 1986.[815] However, in 2011 the CCSBT has adopted a management procedure for the setting of the TAC[816] and a resolution on the allocation of the global total allowable catch.[817] Therein it is stated that, if there are changes to the TAC, Member's allocation will increase or decrease consistent with its nominal percentage level.[818] Cooperating Non-Members receive a fixed amount of the TAC.[819]

The Convention of the ICCAT does not provide any allocation criteria or guidance with regard to the development of such criteria. However, in 2001 the ICCAT adopted a non-binding document defining criteria for the allocation of fishing possibilities.[820] These criteria which are grouped by past/present fishing activity, stock status, status of the qualifying participants and compliance/data submission/scientific research are very extensive and inclusive. Soon after the adoption it became obvious that the effect of the new criteria was overestimated when the members were not able to adopt management measures for Eastern and Mediterranean bluefin tuna and Southern swordfish.[821] Particularly the inclusiveness makes it difficult for the ICCAT to weight the criteria.[822] Therefore,

---

812 Engler Palma, 2010, p.71.

813 A. Cox, *Quota Allocation in International Fisheries*, OECD Food, Agriculture and Fisheries Papers No. 22 (Paris: OECD Publishing, 2009), p.15.

814 Convention for the Conservation of Southern Bluefin Tuna, Art.8.4.

815 Australia, New Zealand and Japan; See also: Agnew, et al., 2006, p.32.

816 CCSBT, *Resolution on the Adoption of a Management Procedure*, (2011).

817 CCSBT, *Resolution on the Allocation of the Global Total Allowable Catch*, (2011).

818 Ibid. Paras.6-7.

819 Ibid. Paras.8.

820 ICCAT, *Criteria for the allocation of fishing possibilities*, (2001).

821 Engler Palma, 2010, pp.70-71.

822 Cox, 2009, p.15.

in practice decisions on allocation are made through political negotiations while the criteria are rarely applied in their entire scope.[823]

The Convention of the WCPFC is providing a comprehensive list with guidelines for the development of allocation criteria.[824] These guidelines are almost as inclusive as the criteria of the ICCAT and show a strong emphasis on the interests of small island developing states.[825] Based on these guidelines the Commission has to "develop, where necessary, criteria for the allocation of the total allowable catch or the total level of fishing effort for highly migratory fish stocks in the Convention Area."[826] So far the WCPFC has not established any formal criteria nor has there been any formal allocation process.[827] The limitations of fishing effort and catches are an implicit form of allocation mainly based on historical catches and the interests of small island developing states.[828]

The Antigua Convention of the IATTC states - quite similar as the WCPFC Convention - that its Commission shall "where necessary, develop criteria for, and make decisions relating to, the allocation of total allowable catch, or total allowable fishing capacity, including carrying capacity, or the level of fishing effort, taking into account all relevant factors."[829] In contrast to the WCPFC there are no guidelines for the establishment of such criteria. In practice the criteria which are used for the allocation of purse seine effort are based on a resolution on fleet capacity.[830] They include national catches between 1985 and 1998, historical EEZ catches, tuna landings and contribution to the IATTC conserva-

---

823   OECD, *Strengthening the International Commission for the Conservation of Atlantic Tunas (ICCAT)*, Strengthening Regional Fisheries Management Organisations (Paris: OECD Publishing, 2009), p.59.

824   Convention on the Conservation and Management of Highly Migratory Fish Stocks in the Western and Central Pacific Ocean, Art.10.3.

825   Ibid. Arts.10.3(d), (g), (h), (i), (j).

826   Ibid. Art.10.1(g).

827   T. Henriksen, et al., *Law and politics in ocean governance: the UN Fish Stocks Agreement and regional fisheries management regimes* (Leiden ; Boston: Martinus Nijhoff Publishers, 2006), pp.180-181.

828   H. Parris & A. Lee, 'Allocation Models in the Western and Central Pacific Fisheries Commission', *Navigating Pacific Fisheries Legal and Policy Trends in the Implementation of International Fisheries Instruments in the Western and Central Pacific Ocean* (Wollongong: Australian National Centre for Ocean Resources and Security, 2009), p.256.

829   Convention for the Strengthening of the Inter-American Tropical Tuna Commission established by the 1949 Convention between the United States of America and the Republic of Costa Rica "Antigua Convention", Art.7.1(f).

830   IATTC, *Resolution on fleet capacity*, (1998), Art.1.

tion programme. Catch quotas in the longline fishery are based on the stock status and historical catches.[831]

The legal framework of the IOTC does not provide any allocation criteria. Previous limitations of fishing capacity can be interpreted as an implicit form of allocation based on historical catches but there are no criteria which go beyond that.[832] Currently the IOTC is discussing options for the allocation of fishing opportunities including the development of allocation criteria.[833] In 2010 the Commission was required to "adopt an allocation quota system or any other relevant measure for the yellowfin and bigeye tunas at its plenary session in 2012."[834] Due to technical problems the decision has been postponed to 2013.[835]

10. Application of the precautionary approach

Awareness of problems like overfishing and IUU fishing led in the 1990s to increasing importance of the precautionary approach in international fisheries management.[836] The precautionary approach is basically based on the Principle 15 of the Rio Declaration on Environment and Development. The core idea is that "[w]here there are threats of serious or irreversible damage, lack of full scientific certainty shall not be used as a reason for postponing cost-effective measures to prevent environmental degradation.[837] The Code of Conduct for Responsible Fisheries and the Fish Stocks Agreement provide the formal basis for the application of the precautionary approach in international fisheries.[838]

---

831   Agnew, et al., 2006, p.33.

832   IOTC, *Approaches to Allocation Criteria in Other Tuna Regional Fishery Management Organizations* (Nairobi, Kenya: IOTC, 2011).

833   IOTC, *Report of the Technical Committee Meeting on Allocation Criteria*, (2011).

834   IOTC, *Resolution 10/01 for the conservation and management of tropicaltunas stocks in the IOTC area of competence*, (2010), Para.13.

835   IOTC, *Report of the Sixteenth Session of the Indian Ocean Tuna Commission*, (2012).

836   FAO, *Precautionary approach to capture fisheries and species introductions*, FAO Technical Guidelines for Responsible Fisheries - Vol.2 (Lysekil: FAO, 1996); R. Hilborn, et al., The Precautionary Approach and risk management: can they increase the probability of successes in fishery management?, *Canadian Journal of Fisheries & Aquatic Sciences - Vol. 58* (2001); Eli P. Fenichel, et al., Real options for precautionary fisheries management, *Fish and Fisheries - Vol. 9(2)* (2008).

837   Rio Declaration on Environment and Development, 1992, Principle 15.

838   Agreement for the Implementation of the Provisions of the United Nations Convention on the Law of the Sea of 10 December 1982 relating to the Conservation and Management of Straddling Fish Stocks and Highly Migratory Fish Stocks, Art.6; Code of Conduct for Responsible Fisheries, Art.6.5.

The Fish Stocks Agreement was the first internationally binding instrument that explicitly required the application in order to protect tuna and tuna-like species.[839] A main pillar of the approach is that "[t]he absence of adequate scientific information shall not be used as a reason for postponing or failing to take conservation and management measures."[840]

Target and limit reference points are key criteria for successful application of the precautionary approach.[841] Target reference points indicate a "state of a fishing and/or resource which is considered to be desirable and at which management action, whether during development or stock rebuilding, should aim."[842] Limit reference points indicate "a state of a fishery and/or resource which is considered to be undesirable and which management action should avoid."[843] The Code of Conduct and the Fish Stocks Agreement contain detailed provisions on reference points.[844] Specific guidelines for the application of precautionary reference points are provided in Annex II of the Fish Stocks Agreement.[845] Those guidelines require, among others, that:

- two types of precautionary reference points should be used: conservation, or limit, reference points and management, or target, reference points;[846]

- fishery management strategies shall ensure that the risk of exceeding limit reference points is very low and that target reference points are not exceeded on average;[847] and that

---

839   Agreement for the Implementation of the Provisions of the United Nations Convention on the Law of the Sea of 10 December 1982 relating to the Conservation and Management of Straddling Fish Stocks and Highly Migratory Fish Stocks, Art.6.1.

840   Ibid. Art.6.2.

841   M. L. Mooney-Seus & A. A. Rosenberg, *Regional fisheries management organizations: progress in adopting the precautionary approach and ecosystem-based management* (London: Chatham House, 2007), p.2.

842   Caddy & Mahon, 1995, p.8.

843   Ibid. p.8.

844   For further information see the first chapter.

845   Agreement for the Implementation of the Provisions of the United Nations Convention on the Law of the Sea of 10 December 1982 relating to the Conservation and Management of Straddling Fish Stocks and Highly Migratory Fish Stocks, Annex II.

846   Ibid. Annex II.2.

847   Ibid. Annex II.5.

- the fishing mortality rate which generates maximum sustainable yield should (F_{MSY}) be regarded as a minimum standard for limit reference points.[848]

Despite the fact that all tuna RFMOs are interested in long term conservation and sustainable use of tuna and tuna-like species the application of the precautionary approach is rather poor.[849] In the past, overfishing was often happening due to a failure in adopting precautionary measures in the face of uncertainty.[850] Overall it has to be recognized that many of the stocks of tuna and tuna-like species are overexploited or even depleted.[851] Moderately exploited stocks like the skipjack tuna do rather benefit from their resilience than from precautionary management practices.[852] There are few examples where precautionary measures have been taken for tuna and tuna-like species. The conservation and management measure for albacore tuna in the South Pacific is seen as the only example where as a result of the precautionary approach a tuna RFMO has explicitly identified optimum catches at levels less than the MSY.[853] The measure has been adopted in 2005 and its provisions on effort control reflect the biological uncertainties.[854]

Among the constituent instruments of the tuna RFMOs only those recently adopted by IATTC and WCPFC are making explicit reference to the precautionary approach. The IATTC's Antigua Convention provides the overall objective "to ensure the long-term conservation and sustainable use of the fish stocks covered by th[e] Convention, in accordance with the relevant rules of international law."[855] For that purpose it calls for the application of the precautionary approach in accordance with Code of Conduct and/or the Fish Stocks Agreement.[856] In particular, the members are required to "be more cautious when information is uncertain, unreliable or inadequate."[857] Formal limit and target

---

848   Ibid. Annex II.7.

849   Aranda, et al., 2010, pp.10-11.

850   Willock & Lack, 2006, p.12.

851   FAO, 2012, pp.53-54.

852   K. M. Schaefer, 'Reproductive biology of tunas', *Tuna: Physiology, ecology and evolution* (San Diego: Academic Press, 2001).

853   Allen, 2010, p.25.

854   WCPFC, *Conservation and Management Measure for South Pacific Albacore*, (2005).

855   Convention for the Strengthening of the Inter-American Tropical Tuna Commission established by the 1949 Convention between the United States of America and the Republic of Costa Rica "Antigua Convention" Art.II.

856   Ibid. Arts.IV.1 and VII.1(m).

857   Ibid. Arts.IV.2.

reference points are still under development and have not been adopted yet.[858] Informal reference points are based on the requirement of the 1949 Convention to maintain stocks at levels that support maximum sustainable yield.[859] In the past, the IATTC has informally used fishing mortality that produces maximum sustainable yield (F_{MSY}) as target reference point by trying to adjust fishing effort to levels that correspond to F_{MSY}.[860] However, it has been noted that this is not consistent with the Fish Stocks Agreement where F_{MSY} is a limit reference point.[861] Spawning biomass corresponding to maximum sustainable yield (B_{MSY}) was used as an informal reference point as well but it is not clear if it represented a target or limit reference point.[862]

Pursuant to its Article 2 the objective of the Convention of the WCPFC is "to ensure, through effective management, the long-term conservation and sustainable use of highly migratory fish stocks in the western and central Pacific Ocean".[863] Regarding the adoption of the precautionary approach the Convention of the WCPFC requires its members to "apply the guidelines set out in Annex II of the [Fish Stocks] Agreement, which shall form an integral part of th[e] Convention". [864] However, despite these progressive provisions, the application of the precautionary approach had been rather limited in the past.[865] This refers in particular to the adoption of reference points. Since 2009 approaches for identification of appropriate reference points were discussed on the basis of a working paper but there was no formal adoption.[866] The MSY was only considered as an informal limit reference point due to the fact that during scientific meetings stock assessment outputs are usually given in relation to the B_{MSY} and

---

858   Mooney-Seus & Rosenberg, 2007, pp.43-54.

859   Convention for the Establishment of an Inter-American Tropical Tuna Commission, Art.II.5.

860   M. N. Maunder, *Reference points, decision rules, and management strategy evaluation for tunas and associated species in the eastern Pacific Ocean - IATTC Stock Assessment Report 2013* (La Jolla: IATTC, 2013), pp.107-108.

861   Agreement for the Implementation of the Provisions of the United Nations Convention on the Law of the Sea of 10 December 1982 relating to the Conservation and Management of Straddling Fish Stocks and Highly Migratory Fish Stocks, Annex II.5.

862   Maunder, 2013, pp.107-108.

863   Convention on the Conservation and Management of Highly Migratory Fish Stocks in the Western and Central Pacific Ocean, Art.2.

864   Ibid. Art.6.1(a).

865   Willock & Lack, 2006, p.13.

866   C. Davies & M. Basson, *Approaches for identification of appropriate reference points and implementation of MSE within the WCPO* (Manila: WCPFC, 2009).

F_{MSY}.[867] At its last meeting in 2012 the Commission finaly adopted, at least limit reference points for the most important target species.[868] The agreed limit for bigeye, yellowfin and skipjack tuna is the fishing mortality at a level no greater than F_{MSY}.[869] The Commission further envisaged the adption of target reference points for the near future.[870]

Within the Articles of the Agreement of the IOTC there is no explicit objective stated. Only the Preamble describes that the parties desire to cooperate "with a view to ensuring the conservation of tuna and tuna-like species in the Indian Ocean, and promoting their optimum utilization and the sustainable development of the fisheries."[871] However, some of the most recent developments with regard to the adoption of the precautionary approach in tuna fisheries have taken place in the IOTC. In 2012 the IOTC has formally adopted a resolution on the Implementation of the Precautionary Approach.[872] This resolution establishes general principles that guide the application of the precautionary approach in the context of the IOTC.[873] Among other measures, the Commission recommended provisional (non-binding) target and limit reference points for albacore, bigeye, skipjack and yellowfin tuna as well as for swordfish.[874] The target reference points for all of these species are B_{MSY} and F_{MSY}.[875] Depending on the species, the limit reference points are set at a biomass reduction to 40%–50% of the MSY level (between 40%-50% of B_{MSY}) or a fishing pressure that exceeds by 30–50% the level that would produce the MSY (between 30%-50% above

---

867   Mooney-Seus & Rosenberg, 2007, pp.132-141; R. Campbell, *Identifying possible Limit Reference Points for the key target species in the WCPFC* (Tonga: WCPFC, 2010), p.3.

868   WCPFC, *Conservation and Management Measure for Bigeye, Yellowfin and Skipjack Tuna in the Western and Central Pacific Ocean*, (2012), Para.1.

869   Ibid.Paras. 2-4.

870   Ibid. Para.1.

871   Agreement for the Establishment of the Indian Ocean Tuna Commission, Preamble.

872   IOTC, *Resolution 12/01 On The Implementation Of The Precautionary Approach*, (2012).

873   IOTC, *Report of the Sixteenth Session of the Indian Ocean Tuna Commission*, (2012), p.16.

874   IOTC, *Recommendation 12/14 On Interim Target And Limit Reference Points*, (2012).

875   Ibid. Para.1.

F_{MSY}).[876] For future conservation and management measures the Commission has committed itself to develop permanent species-specific reference points and harvest control rules.[877]

The objective of the CCSBT Convention, according to Article 3, is "to ensure, through appropriate management, the conservation and optimum utilisation of southern bluefin tuna."[878] In the past there was limited application of the precautionary approach.[879] Until 2008 the only reference point was the management objective to return spawning stock biomass to 1980 levels by 2020.[880] Then it was recognized that this objective was not achievable.[881] A Management Procedure adopted in 2011 provided measures and criteria which are precautionary and conform to the precautionary approach.[882] The Management Procedure is intended to guide the setting of the TAC in order to ensure that the spawning stock biomass achieves an interim rebuilding target.[883] This target is 20% of the original spawning stock biomass by 2035.[884] Currently the CCSBT is using only these target reference points.

No objectives are named in the Articles of the ICCAT Convention. It is only in the Preamble that parties to the Convention express their desire "to co-operate in maintaining the populations [of tuna and tuna-like species] at levels which will permit the maximum sustainable catch for food and other purposes".[885] The ICCAT has not formally adopted the precautionary approach yet but it is discussing it intensively and many of its principles have been adopted.[886] Al-

---

876   Limit reference points for each species: Albacore tuna: 40% of Bmsy, 40% above Fmsy; Bigeye tuna: 50% of Bmsy, 30% above Fmsy; Skipjack tuna: 40% of Bmsy, 50% above Fmsy; Yellowfin tuna: 40% of Bmsy, 40% above Fmsy; Swordfish: 40% of Bmsy; 40% above Fmsy; ibid. Para.1; IOTC, *Report of the Sixteenth Session of the Indian Ocean Tuna Commission,* (2012).

877   IOTC, *Recommendation 12/14 On Interim Target And Limit Reference Points,* (2012), Para.2.

878   Convention for the Conservation of Southern Bluefin Tuna, Art.3.

879   Aranda, et al., 2010, pp.15-16; CCSBT, *Report of the Independent Expert Part 2,* (2008), p.4.

880   CCSBT, *Report of the Seventh Meeting of the Stock Assessment Group,* (2006).

881   CCSBT, *Report of the Performance Review Working Group Part 1 - Self Assessment,* (2008), p.47.

882   P. de Bruyn, et al., The Precautionary Approach to fisheries management: How this is taken into account by tuna regional fisheries management organisations (RFMOs), *Marine Policy - Vol. 38* (2013), p.399.

883   CCSBT, *Resolution on the Adoption of a Management Procedure,* (2011), Art.2.

884   Ibid. Art.6(i).

885   International Convention for the Conservation of Atlantic Tunas, Preamble.

886   de Bruyn, et al., 2013, p.400; ICCAT, *Report of the 3rd Meeting of the Working Group on the Future of ICCAT,* (2012), p.2.

ready in 1998 an Ad Hoc Working Group on the precautionary approach was established but it has met only sporadically.[887] In 2010 and in 2011 a further application of the precautionary approach has been recommended by the ICCAT Working group on stock Assessment Methods.[888] Officially the ICCAT is using F_{MSY} and B_{MSY} as target reference points.[889] MSY is considered rather as a target reference point than as limit reference point.[890]

11. Monitoring, compliance and enforcement

Already UNCLOS had provided comprehensive provisions on monitoring, compliance and enforcement through RFMOs.[891] The Fish Stocks Agreement, as one of the instruments adopted after the 1992 Rio Conference, specified these provisions.[892] The most important provisions cover flag state duties, international cooperation on compliance and enforcement and port state enforcement.[893] The tuna RFMOs have developed a variety of measures that implement these provisions. Traditional tools which are used by all tuna RFMOs are vessel register, vessel monitoring systems and regional observer programmes. In contrast, the more recently developed port states measures and trade measures have been established only in some of the tuna RFMOs.

a)   Vessel register

Tuna RFMOs share the need to know which vessels are active within their areas of competence. This knowledge is crucial for all subsequent compliance and enforcement measures. Therefore, all flag states have to establish a national record of the authorized fishing vessels.[894] These records must be transmitted to

---

887   ICCAT, *Report of the ICCAT Ad Hoc Working Group Meeting on the Precautionary Approach*, (1998).

888   ICCAT, *Report of the 2010 ICCAT Working Group on Stock Assessment Methods*, (2010); ICCAT, *Report of the 2011 Joint Meeting of the ICCAT Working Group on Stock Assessment Methods and Bluefin Tuna Species Group to Analyze Assessment Methods Developed Under the GBYP and Electronic Tagging*, (2011).

889   Mooney-Seus & Rosenberg, 2007, pp.64-73; Hedley, et al., 2003, p.14.

890   G. D. Hurry, et al., *Report of the independent Review - ICCAT* (Madrid: ICCAT, 2008), p.39.

891   United Nations Convention on the Law of the Sea, Arts.62.4 and 117.

892   Another related instrument is the Compliance Agreement.

893   Agreement for the Implementation of the Provisions of the United Nations Convention on the Law of the Sea of 10 December 1982 relating to the Conservation and Management of Straddling Fish Stocks and Highly Migratory Fish Stocks, Arts.18-23.

894   Ibid. Art.18.3.

the Secretariats of the respective tuna RFMOs.[895] They collect all national records and establish so called 'positive vessel lists' or 'white lists' of all authorized vessels and publish them on their websites.[896] In addition the WCPFC has an Interim Register of Non-Member Fish Carrier and Bunker Vessels.[897] A joint global list of all registered tuna fishing vessels is provided on the website of 'Tuna-org' which is an informal framework for sharing information from tuna RFMOs.[898]

WCPFC, IATTC, ICCAT and IOTC have further developed an 'IUU vessel list' or 'black list' with all vessels that have been involved in illegal, unreported and undocumented fishing.[899] The list has to include a series of previous and current information about the vessel, including names of vessel; flags of vessel; names of owners, including beneficial owners, and owner's place of registration; operators of the vessel; call signs of the vessel; lloyd or International Maritime Organization (IMO) numbers; photographs of the vessel; the date of when a vessel was first included on the IUU List; and a summary of activities which

---

895 WCPFC, *Conservation and Management Measure - Record of Fishing Vessels and Authorization to Fish*, (2004); IATTC, *Resolution (amended) on a Regional Vessel Register*, (2011); ICCAT, *Recommendation by ICCAT Concerning the Establishment of an ICCAT Record of Vessels 20 Meters in Length Overall or Greater Authorized to Operate in the Convention Area*, (2011); IOTC, *Resolution 07/02 Concerning the establishment of an IOTC Record of Vessels Authorised to operate in the IOTC area*, (2007); CCSBT, *Resolution on amendment of the Resolution on "Illegal, Unregulated and Unreported Fishing (IUU) and Establishment of a CCSBT Record of Vessels over 24 meters Authorized to Fish for Southern Bluefin Tuna"*, (2008).

896 https://www.iattc.org/VesselRegister/VesselList.aspx?List=RegVessels&Lang=ENG; http://www.iccat.int/en/vesselsrecord.asp; http://www.ccsbt.org/site/authorised_vessels.php; http://www.iotc.org/English/record/search3.php;   http://www.wcpfc.int/record-fishing-vessel-database.

897 http://www.wcpfc.int/vessels#Register.

898 http://www.tuna-org.org.

899 http://www.wcpfc.int/vessels#IUU; https://www.iattc.org/VesselRegister/IUU.aspx?Lang=ENG; http://www.iccat.int/en/IUU.asp; http://www.iotc.org/English/iuu/search.php.

justify inclusion of the vessel on the list.[900] Despite these comprehensive requirements it has to be recognized that in practice the information provided on the websites is rather limited. Especially the IMO number which is crucial for identification purposes is often missing.

The ICCAT has currently 30 vessels listed. For most of them little information is available especially on previous owner and operator. Nine of the vessels have been listed for fishing in the Mediterranean sea during closed season, two lacked the use of VMS, two caught tuna species illegally, one was fishing without being on the ICCAT record of vessels, one was involved in illegal transshipment, one had been fishing in Brazilian waters without license, one without quota and others are listed due to illegal activities in the areas of competence of other tuna RFMOs. The IATTC lists 14 vessels but with less detailed information as provided by the ICCAT. Information on the reason for being listed is not provided at all. The IOTC has listed four vessels. Information is also incomplete. For two vessels the reasons for listing are mentioned, but unspecific by solely referring to respective resolutions or recommendations. The WCPFC has listed only three vessels but with almost all information required by the respective conservation and management measure. One vessel was fishing on the high seas of the Convention Area without being on the WCPFC Record of Fishing Vessels, one was fishing without nationality and another one was fishing in the EEZ of the Republic of the Marshall Islands without permission and in contravention of Republic of the Marshall Islands's laws and regulations.

b) Vessel Monitoring System

Vessel Monitoring Systems (VMS) help RFMOs and coastal states to monitor position, course and speed of all active fishing vessels that are fishing in their area of competence.[901] Technically, VMS have developed from conventional radio, radar or sonar based types to more modern systems that use satellite

---

900   ICCAT, *Recommendation by ICCAT Further Amending Recommendation 09-10 Establishing a List of Fishing Vessels Presumed to be Engaged in Illegal, Unreported and Unregulated (IUU) Fishing Activities in the ICCAT Convention Area*, (2011), Annex 1, IOTC, *Resolution 11/03 On Establishing A List Of Vessels Presumed To Have Carried Out Illegal, Unreported And Unregulated Fishing In The IOTC Area of Competence*, (2011), Annex I.A; WCPFC, *Conservation and Management Measure to establish a List of Vessels presumed to have carried out Illegal, Unreported and Unregulated fishing activities in the WCPO*, (2010), Para.19; IATTC, *Resolution to establish a list of vessels presumed to have carried out illegal, unreported and unregulated fishing activities in the eastern pacific ocean*, (2005), Para.2.

901   FAO, *Fishing operations: Vessel monitoring systems*, FAO Technical Guidelines for Responsible Fisheries - No. 1 (Rome: FAO, 1998), p.4.

based data.[902] In the meantime all tuna RFMOs have established vessel monitoring systems (VMS) which are satellite based. Tracking units fitted to each vessel regularly transmit the information through a communications satellite to a land earth station.

According to Article 62.4 of UNCLOS coastal states are entitled to require foreign fishing states which are fishing in their EEZ to provide among others vessel position reports.[903] Vessels of RFMO members which are fishing on the high seas have to comply with the VMS requirements of the respective RFMO.[904] The Fish Stocks Agreement requires all flag states to collect and share, in a timely manner, complete and accurate data concerning fishing activities, including vessel position and to develop and implement vessel monitoring systems in accordance with national, subregional, regional or global programmes.[905]

The first VMS had been established by the ICCAT. In 2003 a recommendation required the implementation of a VMS for all commercial fishing vessels exceeding 20 meters between perpendiculars or 24 meters length overall.[906] No specific area of application is determined in the recommendation. The IATTC followed in 2004 with a similar resolution for vessels with 24 meters or more in length.[907] A specific area of application is determined neither. In 2006 the IOTC adopted a resolution for the establishment of a its satellite based VMS "for all vessels greater than 15 meters in length overall [...] which operate in the IOTC Area and which fish on the high seas.[908] The "fisheries jurisdiction of any coastal state" is explicitly exempted.[909] The VMS of the WCPFC applies to all vessels 24 meters in length which are fishing on the high seas of the Convention area.[910] A difference to other tuna RFMOs is the fact that the VMS is only activated in the southern and central part of the Convention Area.[911] Due to the

---

902   Ibid. p.9; Palma, et al., 2010, p.144.

903   United Nations Convention on the Law of the Sea, Art.62.4(e).

904   Palma, et al., 2010, p.221.

905   Agreement for the Implementation of the Provisions of the United Nations Convention on the Law of the Sea of 10 December 1982 relating to the Conservation and Management of Straddling Fish Stocks and Highly Migratory Fish Stocks, Arts.5(j) and 18.3(g)(iii).

906   ICCAT, *Recommendation by ICCAT concerning minimum standards for the establishment of a Vessel Monitoring System in the ICCAT Convention area*, (2003), Para.1.

907   IATTC, *Resolution on the establishment of a Vessel Monitoring System (VMS)*, (2004), Para.1.

908   IOTC, *Resolution 06/03 On establishing a Vessel Monitoring System*, (2006), Para.1.

909   Ibid. Para.1.

910   WCPFC, *Conservation and Management Measure - Commission Vessel Monitoring System*, (2006), Para.5(a) and (b).

911   Ibid. Para.2.

lack of willingness of some member states the VMS is not yet activated north of 20°N and west of 175°E.[912] In the CCSBT the VMS is based on a resolution that has been adopted in 2008.[913] According to this resolution vessels fishing for southern bluefin tuna have to adopt and implement a satellite-linked VMS that complies with the systems of ICCAT, IOTC and WCPFC as well as with the VMS of the Commission for the Conservation of Antarctic Marine Living Resources.[914]

c)   *Regional observer programmes*

Observers which are on board of a fishing or carrier vessel have the possibility to monitor all fishing operations that are carried out at sea.[915] Among others they can document how, where and when fishing operations are carried out. Already Article 62.4 of UNCLOS had laid down that observers or trainees on board of foreign flagged vessels have to be part of the terms and conditions for access to coastal state EEZs.[916] More specific are the provisions in Article 18.3 of the the Fish Stocks Agreement where observer programmes are explicitly named and form an integral part of the flag state duties.[917] The main requirement is "the implementation of national observer programmes and subregional and regional observer programmes in which the flag State is a participant, including requirements for such vessels to permit access by observers from other States to carry out the functions agreed under the programmes".[918] Similar provisions can be found in the Code of Conduct[919] and in the IPOA-IUU[920].

All tuna RFMOs have regional observer programmes that generate data on fisheries science and vessel compliance. The IATTC has developed an observer programme for the purse seine and the longline fishery. Since 1994, this programme requires 100 percent observer coverage of purse seine vessels larger

---

912   Ibid. Para.3.

913   CCSBT, *Resolution on establishing the CCSBT Vessel Monitoring System*, (2008).

914   Ibid. Para.1.

915   FAO, *Guidelines for developing an at-sea fishery observer programme*, FAO Technical Paper - No. 414 (Rome: FAO, 2002).

916   United Nations Convention on the Law of the Sea, Art.62.4(g).

917   Lodge, 2007, pp.47-48.

918   Agreement for the Implementation of the Provisions of the United Nations Convention on the Law of the Sea of 10 December 1982 relating to the Conservation and Management of Straddling Fish Stocks and Highly Migratory Fish Stocks, Art.18.3(g)(ii).

919   Code of Conduct for Responsible Fisheries, Arts.7.7.3; 8.1.4; and 8.4.3.

920   International Plan of Action to Prevent, Deter, and Eliminate Illegal, Unreported and Unregulated Fishing (IUU), Paras 24.4; 47.4; and 80.9.

than 363 mt capacity.[921] Seventy percent of these observers are employed by the IATTC and managed by the Secretariat.[922] Thirty percent are from national observer programmes.[923] In 2011 a further resolution had been adopted that requires, from 2013, at least five percent observer coverage also for longline vessels which are greater than 20 metres length overall.[924] No observers are currently on smaller purse seine and longline vessels, or on vessels using other gear types.[925]

The WCPFC provided already in its Convention comprehensive provisions on the development of a regional observer programme.[926] A conservation and management measure adopted in 2007 required the implementation of a regional observer programme including a five percent observer coverage in each fishery until June 2012.[927] Mandatory 100 percent observer coverage for the purse seine vessels fishing in the area between 20° north and 20° south is already in place since 2009.[928] Observers are sourced from the Regional Observer Programme. The current implementation of the five percent coverage in the longline fishery is expected to become a challenge.[929]

With a regional observer programme in the bluefin tuna fishery and national observers in the bigeye tuna fishery the ICCAT is covering only a very small portion of all ICCAT fisheries.[930] The Regional Observer Programme for Bluefin Tuna was adopted in 2008 and amended in 2010. The comprehensive programme requires with regard to the eastern Atlantic and Mediterranean bluefin tuna fishery an observer coverage of 100 percent on all purse seine vessels irre-

921 Meltzer, 2005, p.599.

922 International Seafood Sustainability Foundation, *KOBE III Bycatch Joint Technical Working Group: Harmonisation of Purse-seine Data Collected by Tuna-RFMO Observer Programmes*, ISSF Technical Report 2012-12 (Washington: ISSF, 2012).

923 Meltzer, 2005, p.599.

924 IATTC, *Resolution on scientific observers for longline vessels*, (2011), Para.1.

925 Aranda, et al., 2010.

926 Convention on the Conservation and Management of Highly Migratory Fish Stocks in the Western and Central Pacific Ocean, Art.28.

927 WCPFC, *Conservation and Management Measure for the Regional Observer Programme*, (2007), Para.2; Attachment K, Annex C, Para.6.

928 WCPFC, *Conservation and Magement Measures for Bigeye and Yellowfin Tuna in the Western and Central Pacific Ocean*, (2008), Para.28.

929 WCPFC, *Summary Report - Seventh Regular Session of the Technical Compliance Committee*, (2011), p.32.

930 D. A. Russell & D. L. Vander Zwaag, *Recasting Transboundary Fisheries Management Arrangements in Light of Sustainability Principles* (Leiden: Martinus Nijhoff, 2010), p.297.

spective of their length during all the annual fishing season from 2013 and during all transfer of bluefin tuna to the cages and all harvest of fish from the cage.[931] The second fishery where observers are mandatory is the bigeye tuna fishery. National observers are required on at least 5% of the longline vessels over 24 meters.[932]

The actual regional observer scheme of the IOTC is based on a resolution that had been adopted in 2011.[933] The observers can be either national or non-national of the flag State.[934] The basic requirement is that "at least 5 percent of the number of operations/sets for each gear type by the fleet of each CPC while fishing in the IOTC area of competence of 24 meters overall length and over, and under 24 meters if they fish outside their EEZs shall be covered by this observer scheme."[935] Without determining a timeline it is further stated that coverage of the artisanal fishing vessels should progressively increase towards 5% of the total levels of vessel activity.[936]

Since 2003 a set of scientific observer standards are in place in the CCSBT as well. These standards provide the framework for the operation of the CCSBT Scientific Observer Program.[937] This program which aims on catch and effort monitoring has established a target observer coverage of 10 percent in the purse seine and longline fishery fisheries.[938]

For at-sea transshipment observer coverage is mandatory in all tuna RFMOs. The basic requirement of the respective measures is the prohibition of transshipment if there is no observer on the receiving (carrier) vessel. The first organization that adopted such a measure was the ICCAT. The ICCAT Regional Observer Programme for At-Sea Transshipments is working since 2007 and applied initially to large-scale longline vessels (greater than 20 m) of participating parties or entities.[939] Only those parties or entities that participate are allowed to

---

931   ICCAT, *Recommendation Amending the Recommendation by ICCAT to Establish a Multi-annual Recovery Plan for Bluefin Tuna in the Eastern Atlantic and Mediterranean*, (2010), Para 91.

932   ICCAT, *Recommendation by ICCAT on a multi-year conservation and management program for bigeye tuna*, (2004), Para.15.

933   IOTC, *Resolution 11/04 On a regional observer scheme*, (2011).

934   Ibid. Para.5.a.

935   Ibid. Para.2.

936   Ibid. Para.4.

937   CCSBT, *CCSBT Scientific Observer Program Standards*, (2003), Para.2.

938   Ibid. Para.5.

939   ICCAT, *Recommendation by ICCAT establishing a programme for transhipment*, (2006), Para.3.

transship at sea. All others are required to transship at port.[940] Participating carrier vessels must have an observer on board when receiving at sea transshipments[941] and the longline vessel must be authorized to transship.[942] The same measure has been adopted by IATTC,[943] WCPFC,[944] IOTC[945] and CCSBT.[946]

d)   Port state measures

Insufficient flag state control is a remaining problem in international fisheries as it leads to IUU fishing.[947] Controlling the vessels when they are using port facilities has been therefore identified as an additional powerful tool to ensure compliance.[948] All vessels involved in fishing operations have to seek port access for landing and transshipment of fish, refueling, repairing or other activities.[949] General requirements for the development of port states measures have been provided by the Fish Stocks Agreement.[950] It further lays down that port states are entitled to inspect documents, fishing gear and catch on board fishing vessels which are voluntarily in its ports or at offshore terminals.[951] More specific provisions are included in the more recently established FAO Model Scheme on Port State Measures to Combat Illegal, Unreported and Unregulated

---

940   Ibid. Para.1.

941   Ibid. Para.15.

942   Ibid. Para.11.

943   IATTC, *Resolution (amended) on establishing a program for transipment by large scale vessels*, (2011), Para.16.

944   WCPFC, *Conservation and Management Measure on Regulation of Transshipment*, (2009), Para.13.

945   IOTC, *Resolution 11/05 On establishing a programme for transipment by large-scale fishing vessels*, (2011), Para.16.

946   CCSBT, *Resolution on Establishing a Program for Transshipment by Large-Scale Fishing Vessels*, (2008).

947   Palma, et al., 2010, p.157.

948   FAO Model Scheme on Port State Measures to Combat Illegal, Unreported and Unregulated Fishing.

949   Lodge, 2007, pp.54.

950   Agreement for the Implementation of the Provisions of the United Nations Convention on the Law of the Sea of 10 December 1982 relating to the Conservation and Management of Straddling Fish Stocks and Highly Migratory Fish Stocks, Art.18.3(g)(ii).

951   Ibid. Arts.23.2

Fishing as well as in the Port States Measures Agreement.[952] The non-binding FAO Model Scheme provides principles and guidelines for the negotiation and adoption of regional memoranda of understanding, for the adoption of resolutions or recommendations within RFMOs or for measures adopted at the national level.[953] The binding Port States Measures Agreement is, so far, on an international level the most progressive approach to adopt harmonized port states measures. It is intended to reach similar standards among all RFMO members with regard to inspection procedures, information to be provided by the vessel or regarding penalties.[954] The Agreement is not in force yet but will enter into force thirty days after the date of deposit with the Depositary of the twenty-fifth instrument of ratification, acceptance, approval or accession.[955]

All tuna RFMOs have some sort of port states measures in place. Depending on the time when the respective measures have been adopted there are variations with regard to the implementation of the international requirements. Among the tuna RFMOs the IOTC has developed the most progressive port states measures. A first programme of inspection in port had been established in 2001 and superseded in 2005.[956] The respective program basically includes the core provisions of the Fish Stocks Agreement.[957] Port states were entitled, among others, to inspect documents, fishing gear and catch on board of the fishing vessels.[958] The latest resolution on port state measures was adopted in 2010.[959] They are based on the Port States Measures Agreement and include most of its sub-

---

952 International Plan of Action to Prevent, Deter, and Eliminate Illegal, Unreported and Unregulated Fishing (IUU), Paras.52-64; FAO, *Model Scheme on Port State Measures to Combat Illegal, Unreported and Unregulated Fishing* (Rome: FAO, 2007); Agreement on Port State Measures to Prevent, Deter and Eliminate Illegal, Unreported and Unregulated Fishing.

953 FAO, *Report of the Technical Consultation to Review Port State Measures to Combat Illegal, Unreported and Unregulated Fishing, FAO Fisheries Report - No. 759*, (2004), Para.16.

954 Palma, et al., 2010, p.63.

955 Agreement on Port State Measures to Prevent, Deter and Eliminate Illegal, Unreported and Unregulated Fishing, Art.29.1.

956 IOTC, *Resolution 02/01 Relating to the establishment of an IOTC programme of inspection in port*, (2001); IOTC, *Resolution 05/03 relating to the establishment of an IOTC programme of inspection in port*, (2005).

957 IOTC, *Resolution 05/03 relating to the establishment of an IOTC programme of inspection in port*, (2005). Paras.2 and 3.

958 Ibid. Para.3.

959 IOTC, *Resolution 10/11 on Port State Measures to prevent, deter and eliminate illegal, unreported and unregulated fishing*, (2010).

stantive duties.[960] Particular the requirements on designation and capacity of ports, information provided prior to entry into port as well as the flag state duties are identical to those in the Port States Measures Agreement.[961] However, sometimes the scope of the some measures of the resolution lacks clarity because of overlaps with other resolutions that contain port related measures as well.[962] Another weak point is its application only to the ports within the area of competence of the IOTC.[963]

In 1998 ICCAT has established a port inspection scheme which requires that inspections are carried out by appropriate authorities in order to monitor compliance with conservation and management measures.[964] According to this scheme inspectors are entitled to "examine the fish, fishing gear, fish samples, and all relevant documents, including fishing logbooks and cargo manifest".[965] In addition there are port state measures in other recommendations.[966] Despite the fact that most of the recommendations have been adopted prior to the Port States Measures Agreement it can be recognized that most of the measures conform to its obligations. However it is criticized that due to the fact that port state measures are not systemized there is no systematic and comprehensive approach to port controls.[967]

The legal framework of the WCPFC lacks such a systematic and comprehensive approach as well. The WCPFC has not developed a port state scheme but its

---

960 Fabra, et al., 2011, pp.18-20.

961 IOTC, *Resolution 10/11 on Port State Measures to prevent, deter and eliminate illegal, unreported and unregulated fishing*, (2010), Paras.5, 6 and 17.

962 IOTC, *Resolution 01/03 Establishing a Scheme to promote compliance by Non-Contracting Party vessels with resolutions established by IOTC*, (2001); IOTC, *Resolution 05/03 relating to the establishment of an IOTC programme of inspection in port*, (2005); IOTC, *Resolution 09/03 On establishing a list of vessels presumed to have carried out illegal, unregulated and unreported fishing in the IOTC area*, (2009).

963 IOTC, *Resolution 10/11 on Port State Measures to prevent, deter and eliminate illegal, unreported and unregulated fishing*, (2010), Para.3.

964 ICCAT, *Recommendation by ICCAT for a Revised ICCAT Port Inspection Scheme*, (1997), Para.1.

965 Ibid. Para.3.

966 ICCAT, *Recommendation by ICCAT establishing a programme for transhipment*, (2006); ICCAT, *Recommendation Amending the Recommendation by ICCAT to Establish a Multi-annual Recovery Plan for Bluefin Tuna in the Eastern Atlantic and Mediterranean*, (2010); ICCAT, *Recommendation by ICCAT concerning the ban on landings and transhipments of vessels from non-Contracting Parties identified as having committed a serious infringement*, (1998)

967 Fabra, et al., 2011, pp.15-17.

Convention reflects, like the IOTC scheme, the provisions of the Fish Stocks Agreement. Generally it is stated that a port state has the right and the duty to take measures "to promote the effectiveness of subregional, regional and global conservation and management measures."[968] Port states are entitled, among others, to inspect documents, fishing gear and catch on board of fishing vessel that voluntarily enters one of its ports or offshore terminals.[969] Port state measures are further included in several conservation and management measures.[970] Generally it can be recognized that the obligations on the use of ports and the denial of entry are conform to the obligations of the Port States Measures Agreement but there are no provisions regarding the information provided prior to entry.[971] Weak points are further the insufficient obligations on inspections or the designation and capacity of ports.[972]

So far IATTC and CCSBT are the tuna RFMOs with the least developed systems of port state control.[973] They have neither a port inspection scheme nor a port inspection programme in place and in comparison to the Port States Measures Agreement they are particularly weak when it comes to denial of entry, information provided prior to entry and designation and capacity of ports.[974] In addition there are no explicit port state related provisions included in their Convention text. Some resolutions of the IATTC contain port states related provision but they are rather limited.[975] The same applies to the CCSBT where only

---

968 Convention on the Conservation and Management of Highly Migratory Fish Stocks in the Western and Central Pacific Ocean, Art.27.1.

969 Ibid. Art.27.2.

970 WCPFC, *Conservation and Management Measure - WCPFC Record of Fishing Vessels and Authorization to Fish*, (2009); WCPFC, *Conservation and Management for Sharks*, (2009); WCPFC, *Conservation and Management Measure on Regulation of Transshipment*, (2009); WCPFC, *Conservation and Magement Measures for Bigeye and Yellowfin Tuna in the Western and Central Pacific Ocean*, (2008); WCPFC, *Conservation and Management Measure to Establish a List of Vessels presumed to Have Carried out Illegal, Unreported and Unregulated Fishing Activities in the WCPO*, (2007).

971 Fabra, et al., 2011, pp.21-23.

972 Ibid. p.23.

973 Ibid. pp.9-14.

974 Ibid. pp.9-14.

975 IATTC, *Resolution (amended) on establishing a program for transipment by large scale vessels*, (2011); IATTC, *Resolution (amended) on establishing a list of longline fishing vessels over 24 meters (LSTLFVs) authorized to operate in the eastern pacific ocean*, (2011); IATTC, *Resolution to establish a list of vessels presumed to have carried out illegal, unreported and unregulated fishing activities in the eastern pacific ocean*, (2005); IATTC, *Resolution on the conservation of sharks caught in association with fisheries in the eastern pacific ocean*, (2005).

some provisions are included in resolutions.[976] Therefore, in 2011 a strategic plan of the Commission recommended the adoption of explicit port state measures.[977]

e)   Trade-related measures

Another approach to deal with the limited success of internationally agreed instruments are trade-related measures.[978] These measures which have been described as, 'the heart of internationally agreed marked related measures', are increasingly applied by the tuna RFMOs.[979] Their main purpose is to reduce opportunities and incentives for IUU fishing and to improve information on fishing mortality.[980] Most of the relevant international legal instruments contain provisions for the use of trade-related measures.[981] The Fish Stocks Agreement requires RFMOS to take measures to deter activities of vessels that are undermining the effectiveness of subregional or regional conservation and management measures.[982] Port states are entitled to adopt regulations that empower relevant national authorities to prohibit landings and transshipments if the catch that has been taken undermining the effectiveness of subregional, regional or global conservation and management measures on the high seas.[983] Explicit provisions that facilitate the establishment of trade measures are further provided by the IPOA-IUU. States are encouraged to identify vessels engaged in

976   CCSBT, *Resolution on action plans to ensure compliance with Conservation and Management Measures*, (2009); CCSBT, *Resolution on Establishing a Program for Transshipment by Large-Scale Fishing Vessels*, (2008). CCSBT, *Resolution on amendment of the Resolution on "Illegal, Unregulated and Unreported Fishing (IUU) and Establishment of a CCSBT Record of Vessels over 24 meters Authorized to Fish for Southern Bluefin Tuna"*, (2008).

977   CCSBT, *Strategic Plan for the Commission for the Conservation of Southern Bluefin Tuna*, (2011), p.5.

978   Schneider, 2000.

979   Palma, et al., 2010, p.193.

980   Lack, 2007. p.1.

981   Ibid. p.5.

982   Agreement for the Implementation of the Provisions of the United Nations Convention on the Law of the Sea of 10 December 1982 relating to the Conservation and Management of Straddling Fish Stocks and Highly Migratory Fish Stocks, Art.17.4.

983   Ibid. Art.23.3.

IUU fishing and to adopt and implement trade related measures.[984] Such measures could include multilateral catch documentation and certification.[985]

Tuna RFMOs are using a variety of possible trade related measures against vessels and states that have been involved in IUU fishing. The most important measures that are currently in place include catch and trade documentation schemes, prohibition of landing and transshipment from vessel registered on IUU vessel lists as well as trade restrictive measures.[986]

aa) Catch and trade documentation schemes

Catch and trade documentation schemes are two different trade related methods to document all catches and their trade. Catch documentation schemes are monitoring landed catch. All catches are covered from the point of first capture by a flag State through international trade routes including imports, exports and re-exports, and potential farming operations to the State of final destination.[987] Trade documentation schemes or statistical document programs are monitoring only the part of the catch that enters international trade.[988] While IATTC, ICCAT CCSBT and IOTC have catch or/and trade documentation schemes in place the WCPFC has not developed any scheme yet.

The first trade based scheme that had been adopted by a tuna RFMO was the Statistical Document Program of ICCAT for bluefin tuna in 1992.[989] This program was followed by similar ones for swordfish and bigeye tuna.[990] In 2006 the Commission further recommended the development of pilot projects to investigate the feasibility of electronic systems to improve the statistical document

---

984   International Plan of Action to Prevent, Deter, and Eliminate Illegal, Unreported and Unregulated Fishing (IUU), Para.66.

985   Ibid. Para.69.

986   R.G. Tarasofsky, *Regional Fisheries Organizations and the World Trade Organization: Compatibility or Conflict?*, 2003), pp.2-6; Lack, 2007. pp.1-2; Lodge, 2007, p.58.

987   S. Clarke, *Best Practice Study of Fish Catch Documentation Schemes* (Brisbane: Marine Resources Assessment Group Ltd - Asia Pacific, 2010), p.1.

988   Lack, 2007. p.8.

989   ICCAT, *Recommendation by ICCAT Concerning the ICCAT Bluefin Tuna Statistical Document Program*, (1992).

990   ICCAT, *Recommendation by ICCAT establishing a Swordfish Statistical Document Program*, (2001); ICCAT, *Recommendation by ICCAT concerning the ICCAT Bigeye Tuna Statistical Document Program*, (2001); ICCAT, *Recommendation by ICCAT concerning the amendment of the forms of the ICCAT bluefin/bigeye/swordfish statistical documents*, (2003).

programs.[991] The two main problems of the Bluefin Tuna Statistical Document Program were that domestic consumption of bluefin tuna could not be detected and quantities of tuna caged for farming purposes could not be adequately determined.[992] Therefore, in 2007 the Commission adopted a recommendation on an ICCAT Bluefin Tuna Catch Documentation Program to overcome these shortcomings.[993] The new scheme was intended to ensure that all catches were reported regardless whether they were used for export, domestic consumption or farming.[994] Since then the scheme was amended twice in order to make some refinements and to incorporate further guidance towards enhanced clarity and better implementation.[995] The latest recommendation was adopted in 2010. It requires the development of an electronic Bluefin Tuna Catch Documentation System to improve the current programme through the treatment of shipments, the ability to detect fraud and deter IUU shipments and the facilitation of automated links between the various actors involved, including exporting and importing authorities.[996]

In 2009 the CCSBT adopted a resolution on the Implementation of a catch documentation scheme in order to track and validate legitimate product flow from catch to the point of first sale on domestic or export markets.[997] This scheme was replacing the Statistical Document Programme which had been operated since 2000. One reasons for the catch documentation scheme was the reduction of overharvest which had been conducted by one of the CCSBT members in

---

991   ICCAT, *Recommendation by ICCAT on an Electronic Statistical Document Pilot Program,* (2006).

992   ICCAT, *Expert Panel on CITES Listing Proposals - Regional Fisheries Bodies,* CoP14 Doc. 68 Annex - Annex 1 - Summary of measures taken historically by ICCAT for bluefin tuna (Madrid: ICCAT, 2010), p.6.

993   ICCAT, *Recommendation by ICCAT on an ICCAT Bluefin Tuna Catch Documentation Program* (2007).

994   ICCAT, 2010, p.7.

995   ICCAT, *Recommendation by ICCAT on an ICCAT Bluefin Tuna Catch Documentation Program* (2008); ICCAT, *Recommendation by ICCAT Amending the Recommendation 08-12 on an ICCAT Bluefin Tuna Catch Documentation Program,* (2009).

996   ICCAT, *Recommendation by ICCAT on an Electronic Bluefin Tuna Catch Document Programme (eBCE),* (2010).

997   CCSBT, *Resolution on the Implementation of a CCSBT Catch Documentation Scheme,* (2006).

previous years.[998] In 2012, the 2009 resolution was modified in order to develop further documentation forms and reports.[999]

IOTC and IATTC have no catch documentation schemes but similar statistical document programmes for bigeye tuna which are based on resolutions from 2001 and 2003.[1000] The main purpose of these programmes was to use trade data in order to reduce uncertainty on the catch of bigeye tuna and to eliminate IUU fishing operations.[1001] All bigeye tuna which are caught in their areas of competence and imported into the territory of a Contracting Party have to be accompanied by a Bigeye Tuna Statistical Document or a Bigeye Tuna Re-export Certificate in accordance with specific requirements laid down in two annexes.[1002]

So far the WCPFC has developed neither a trade documentation scheme nor a catch documentation scheme. In the past there was a continuous communication on such schemes but problems like mixed species catches, landings of fish outside the Convention Area in non-member States, charter vessels controlled by coastal States, and limited ability and capacity of Pacific Island Countries made progress difficult.[1003] A proposal for a catch documentation scheme provided by Papua New Guinea and the European Union was again discussed during the annual session in 2012.[1004] All Members, Cooperating Non-members and participating territories supported the establishment of an Intersessional Working Group in order to develop a catch documentation scheme.[1005]

---

998   WWF, *WWF Position – 9th Regular Session of the Western and Central Pacific Fisheries Commission (WCPFC)* (Manila: WWF, 2012), p.9.

999   CCSBT, *Resolution on the Implementation of a CCSBT Catch Documentation Scheme*, (2012), Appendixes 1,2 and 3.

1000 IOTC, *Resolution 01/06 Concerning the IOTC bigeye tuna statistical document programme*, (2001); IATTC, *Resolution on IATTC bigeye tuna statistical document program*, (2003),

1001 IOTC, *Resolution 01/06 Concerning the IOTC bigeye tuna statistical document programme*, (2001), Preamble; IATTC, *Resolution on IATTC bigeye tuna statistical document program*, (2003), Preamble.

1002 IOTC, *Resolution 01/06 Concerning the IOTC bigeye tuna statistical document programme*, (2001), Para.1, Annex 1, Annex 2; IATTC, *Resolution on IATTC bigeye tuna statistical document program*, (2003), Para.1, Annex 1, Annex 2.

1003 Clarke, 2010, p.67.

1004 WCPFC, *Joint FFA/ EU proposal: WCPFC Catch Documentation Scheme Proposed Intercessional Working Group Operations and Terms of Reference*, (2012)

1005 WCPFC, *Summary Report - Ninth Regular Session of the Commission*, (2012), p.47.

bb) Trade restrictive measures

Trade restrictive measures are sanctions against flag states that failed to take action against vessels that undermined the conservation and management measures established by the RFMOs.[1006] The intention of these sanctions is to exclude non-compliant countries from market access and to provide incentives for future compliance. Although sanctions can be a useful tool to improve compliance it has to be considered that there are challenges like the need for a comprehensive black list and the possible shift to other markets.[1007] Another problem is the consistency between trade restrictive measures and other international obligations like those of the WTO.[1008] Import prohibitions are prima facie incompatible with the WTO but there are exceptions if there is a multilateral basis for the measure, if the decision making process is transparent and taken after alternative measures have failed and if the interests of the main trading partners have been taken into account.[1009]

ICCAT, IATTC IOTC and WCPFC require their members and cooperating non-members to prohibit importations of tuna and tuna-like species that have been caught by vessels included in a IUU vessel list.[1010] However, the respective provisions do not imply specific trade restrictive measures. Action plans for the adoption of trade restrictive measures have been developed so far only by ICCAT and CCSBT.[1011] These RFMOs required their members to imposed trade bans on certain non-compliant countries. In 1994 ICCAT was the first tuna

---

1006 Lack, 2007, p.32.

1007 Le Gallic, 2008; OECD, *Transition to Responsible Fisheries: Economic and Policy Implications*, (Paris: OECD Publishing, 2000).

1008 Tarasofsky, 2003.

1009 R. Tarasofsky, *Enhancing the Effectiveness of Regional Fisheries Management Organizations through Trade and Market Measures* (London: Chatham House, 2007), p.5.

1010 ICCAT, *Recommendation by ICCAT Further Amending Recommendation 09-10 Establishing a List of Fishing Vessels Presumed to be Engaged in Illegal, Unreported and Unregulated (IUU) Fishing Activities in the ICCAT Convention Area*, (2011), Para.9; IATTC, *Resolution to establish a list of vessels presumed to have carried out illegal, unreported and unregulated fishing activities in the eastern pacific ocean*, (2005), Para.9(e); IOTC, *Resolution 11/03 On Establishing A List Of Vessels Presumed To Have Carried Out Illegal, Unreported And Unregulated Fishing In The IOTC Area of Competence*, (2011), Para.13(e); WCPFC, *Conservation and Management Measure to establish a List of Vessels presumed to have carried out Illegal, Unreported and Unregulated fishing activities in the WCPO*, (2010), Para.22(e).

1011 ICCAT, *Recommendation by ICCAT Regarding Belize and Honduras Pursuant to the 1994 Bluefin Tuna Action Plan Resolution*, (1996); ICCAT, *Recommendation by ICCAT Regarding Panama Pursuant to the 1994 Bluefin Tuna Action Plan Resolution*, (1996); CCSBT, *Action Plan*, (2000).

RFMO which undertook steps towards trade restrictive measures. The bluefin tuna action plan resolution recommended non-discriminatory trade restrictive measures against countries involved in IUU fishing.[1012] In 1996 two recommendations required the Contracting Parties to take appropriate measures to prohibit the import of Atlantic bluefin tuna and its products from Belize, Honduras and Panama.[1013] In subsequent years similar recommendations followed for catches of tuna by Bolivian longline vessels (1998), for swordfish from Belize, Honduras and Equatorial Guinea (2000)[1014] as well as for bigeye tuna from Belize, Honduras, Cambodia, Equatorial Guinea and Saint Vincent and the Grenadines (2001).[1015] The success of such measures is difficult to assess. However, it can be recognized that trade bans like the one for bigeye tuna according to the measure of 2001 were suspended in 2003 for St. Vincent and the Grenadines due to continuing progress made in order to achieve full compliance with ICCAT measures.[1016] Another indicator for the success is the fact that several non-compliant countries like Panama and Honduras have become contracting parties in the meantime.

The action plan which has been adopted by the CCSBT in 2000 applies only to non-members. It entitles the Commission "to decide to impose trade-restrictive measures consistent with Members' international obligations on [southern bluefin tuna] products, in any form, from the non- Members".[1017] In 2003 the CCSBT notified Belize, Cambodia, Honduras, Equatorial Guinea and Indonesia that their vessels had been identified as conducting fishing activities in a manner which diminished the effectiveness of the conservation and management measures for southern bluefin tuna.[1018] These resolutions could have been used for the adoption of trade restrictive measures. In practice no trade sanctions have been taken against any of the non-compliant fishing countries. The main reasons included concerns about WTO consistency, consistency of approach

---

1012 Clarke, 2010, p.62.

1013 ICCAT, *Recommendation by ICCAT Regarding Belize and Honduras Pursuant to the 1994 Bluefin Tuna Action Plan Resolution,* (1996), Para.a; ICCAT, *Recommendation by ICCAT Regarding Panama Pursuant to the 1994 Bluefin Tuna Action Plan Resolution,* (1996), Para.a.

1014 Le Gallic, 2008, pp.861-862.

1015 ICCAT, *Recommendation by ICCAT regarding Bolivia pursuant to the 1998 Resolution concerning the unreported and unregulated catches of tuna by large-scale longline vessels in the Convention area,* (2000).

1016 ICCAT, *Recommendation by ICCAT concerning the trade sanction against St. Vincent and the Grenadines,* (2002).

1017 CCSBT, *Action Plan,* (2000), Para.6.

1018 CCSBT, *Resolutions pursuant to the 2000 Action Plan,* (2003), Para.1.

with other countries as well as uncertainty about whether non-compliance was intentional.[1019]

D. Summary

The goal of this chapter was to assess the current performance of the five tuna RFMOs in the context of the international legal framework for the management of international tuna fisheries. A further aim was the comparison of the tuna RFMOs with regard to the implementation of the requirements laid down in their respective constituent instruments.

The analysis of the evolution of catch and the stock status has shown that there are different types of tuna RFMOs. In the Convention areas of the two oldest RFMOs, IATTC and ICCAT comparatively high catch levels were reached already in the 1970s. The two youngest RFMOs, IOTC and WCPFC have shown continuously increasing catch rates since the 1980s until their highest catch levels were reached in recent years. The CCSBT which was established in 1994 is a special case. Despite its recent establishment it has to be considered that the highest catch levels for southern bluefin tuna were reached already in the 1960s. In the areas of competence of WCPFC and IOTC comparatively few stocks are overfished and little overfishing is occurring. In contrast, IATTC, ICCAT and CCSBT are struggling in keeping catches within sustainable limits. In particular, the stocks managed by ICCAT and CCSBT are often overfished and overfishing is occurring. It can be noted that, at present, the latter organizations have to rebuild several of their stocks while IOTC and WCPFC are facing the different challenge of maintaining their future catches at sustainable limits.

The general organizational structure of the tuna RFMOs is quite similar. All of them are composed by a Commission which is the executive body, subsidiary bodies to support the work of the Commission and an independent Secretariat with permanent staff. All bodies meet at least once a year and have a series of different functions. The main function of the Commissions is the adoption of binding and non-binding conservation and management measures on issues like catch and effort limits. The main difference between the scientific bodies is how the scientific information is produced. In ICCAT, IOTC and CCSBT scientific advices are the outcome of the effort of various working groups which are composed by national scientists from the parties. The ICCAT has further established specific panels for species groups which can directly propose recommendations to the Commission. In contrast, in the IOTC the working groups on species groups are not expected to directly advise the Commission. The special feature of the CCSBT is that national scientists are supported by an independent

---

1019 Tarasofsky, 2007, p.5, T. Markus, Wege zu einer nachhaltigen EU-Fischereiaußenhandelspolitik, *Europarecht (EuR) - Vol. 48* (2013), p.706.

expert panel which facilitates consensus or provides own views. In the IATTC information is provided by working groups but its Secretariats proper scientific staff has wider competences than the staff of every other tuna RFMO. Another different model is applied by the WCPFC where the scientific body gets its scientific data from external science providers.

All tuna RFMOs are composed by members or contracting parties and so called cooperating non-members or cooperating non-contracting parties. The terminology varies among the tuna RFMOs. All tuna RFMOs are generally open to new members but the organizations differ with regard to possible accession. In ICCAT and CCSBT the new member´s signature and a declaration on compliance is sufficient for accession. IOTC, WCPFC and IATTC require formal consultation and a qualified decision of the current members. A general problem of all RFMOs is the fact that existing RFMO members are usually not interested in providing new members with shares of the total catch. If this is taken into account then RFMOs are rather restrictive or closed.

In general, all constituent instruments of the tuna RFMOs endeavor consensus based decision making. In CCSBT and IATTC consensus is mandatory for all decisions. The other three tuna RFMOs provide the possibility to decide on certain decisions by a qualified majority. Exempted are the most important decisions like those on allocation of catch and effort. The constituent instruments of the three tuna RFMOs with voting procedures provide the possibility of objections or seeking of review. The WCPFC provides the most progressive approach to avoid objections.

Dispute settlement procedures are included in the instruments of WCPFC, IATTC, IOTC and CCSBT. No such procedures can be found in the Convention of ICCAT which is the only constituent instrument that predates the UNCLOS. The dispute settlement procedure of WCPFC comes closest to the provisions of the Fish Stocks Agreement which apply, mutatis mutandis, to any dispute between members of the Commission. IATTC and IOTC currently lack a compulsory dispute settlement mechanism. The only tuna RFMO with practical experience in dispute settlement is CCSBT. The southern bluefin tuna dispute has shown that the interpretation of the dispute settlement procedures in the constituent instruments of the tuna RFMOs can be powerful and may even lead to an exclusion of compulsory dispute settlement under UNCLOS.

Total catch and fishing mortality is regulated by different strategies. The first tuna RFMOs IATTC and ICCAT initially used output control. When overfishing due to increasing fishing capacity became a problem both incorporated input control and technical measures as well. In contrast, the younger IOTC and WCPFC were established when global overfishing already was a problem. Therefore they show a clear preference for input control and technical measures. Only the CCSBT differs from these patterns. Output control was already used long before the establishment of the CCSBT and it is still the dominating approach.

Most of the tuna RFMOs have developed criteria for the allocation of catch and effort similar to those provided by Article 11 of the Fish Stocks Agreement. However, it has to be recognized that in practice in all tuna RFMOs allocation is still mainly based on political negotiations. In the CCSBT the Convention provides allocation criteria but allocation is mainly based on an unpublished scheme from 1986. The ICCAT has adopted a non-binding document defining criteria but in practice those criteria have resulted to be too inclusive. A comprehensive list with guidelines for the development of allocation criteria is also provided by the Convention of the WCPFC. However, so far no formal criteria have been established nor has there been any formal allocation process. The Antigua Convention of the IATTC requires the development of allocation criteria but there are no guidelines for the establishment of such criteria. In practice the criteria are based on a resolution on fleet capacity and on the stock status and historical catches. The IOTC which has not developed any allocation criteria at all is currently discussing options for the allocation of fishing opportunities including the development of allocation criteria.

The application of the precautionary approach by tuna RFMOs is rather poor. Many of the stocks are overexploited or even depleted. There are few examples where precautionary measures have been taken. Among the constituent instruments of the tuna RFMOs only those recently adopted by IATTC and WCPFC are making explicit reference to the precautionary approach. However, formal reference points have only been developed by WCPFC and IOTC. WCPFC has recently adopted formal limit reference points for three main target species. The limit is the fishing mortality at a level no greater than F_{MSY}. In 2012 IOTC has adopted a resolution on the Implementation of the Precautionary Approach including provisional target and limit reference points. The target reference points are B_{MSY} and F_{MSY} and the limit reference points are between 40%-50% of B_{MSY} or between 30%-50% above F_{MSY} depending on the species. The IATTC is using F_{MSY} as an informal target reference point. CCSBT which is also using only informal target reference points adopted a Management Procedure in 2011 in order to ensure that the spawning stock biomass achieves an interim rebuilding target. The ICCAT has not established formal reference points neither but it is using F_{MSY} and B_{MSY} as informal target reference points.

All of the tuna RFMOs provide on their website a list of vessels that are authorized to fish. This is one of the tools that have been developed to strengthen flag State´s responsibility. IATTC, ICCAT, IOTC as well as WCPFC have further established a list of vessels that have been involved in IUU fishing. The lists of IUU vessels from ICCAT and IOTC include only non-member vessels while

IATTC and WCPFC are listing also vessels flying the flags of members and co-operating non-members.

Satellite based vessel monitoring systems (VMS) for vessels with certain lengths are mandatory in all tuna RFMOs. A difference of IOTC and WCPFC to other tuna RFMOs is the explicit statement that the VMS is applying only on the high seas and with regard to the WCPFC that the activation of the VMS is only mandatory in the southern and central part of the Convention Area. The CCSBT has no own VMS but it requires compliance with the systems of ICCAT, IOTC, WCPFC and CCAMLR.

The most comprehensive observer programmes have been developed by IATTC and WCPFC. They require 100 percent observer coverage of purse seine vessels and five percent for longline vessels. The regional observer scheme of the IOTC requires at least five percent coverage of for each gear type. The ICCAT, with a regional observer programme for bluefin tuna and national observers in the bigeye tuna fishery, is covering only a very small portion of all ICCAT fisheries. The programme of the CCSBT has a target observer coverage of 10 percent for catch and effort monitoring for each fishery. During at-sea transshipment observer coverage on the receiving carrier vessel is mandatory in all tuna RFMOs.

The most progressive port states measures have been adopted by the IOTC. They are based on the Port States Measures Agreement and include most of its substantive duties. The ICCAT has established a port inspection scheme and in addition there are port state measures in other recommendations. Despite the fact that most of the relevant recommendations predate the Agreement there are several measures which are conforming to its obligations. The legal framework on port state measures of WCPFC lacks a systematic and comprehensive approach. The obligations on the use of ports and the denial of entry are conform to the obligations of the Agreement but the insufficient obligations on inspections or the designation and capacity of ports are weak points. Also weak are the measures of IATTC and CCSBT. Both have neither a port inspection scheme nor a port inspection programme in place yet.

Tuna RFMOs are using a variety of possible trade related measures against vessels and states that have been involved in IUU fishing. Catch and trade documentation schemes have been established by ICCAT and CCSBT. The catch documentation schemes were intended to overcome the shortcomings of the trade based schemes. IOTC and IATTC have no catch documentation schemes but similar trade documentation schemes for bigeye tuna. WCPFC has not developed any scheme yet. ICCAT, IATTC IOTC and WCPFC prohibit importations of tuna and tuna-like species that have been caught by vessels that are on a IUU vessel list. Action plans for the adoption of specific trade restrictive measures have been developed so far only by ICCAT and CCSBT. These RFMOs required their members to imposed trade bans on certain non-compliant countries. The ICCAT is the only tuna RFMO which has adopted trade restrictions.

Chapter IV

Tuna fisheries management in the Western- and Central Pacific Ocean

A.   Introduction

The previous chapter has shown that the stock status of tuna and tuna-like species in the WCPO is comparatively good and that the WCPFC has developed a comprehensive legal framework for the management of these species. This chapter will, based on the assumption that the WCPFC is in a good position to manage its stocks sustainably in the long term, provide an in depth analysis of the current international legal framework for fisheries management in the WCPO. The underlying question to be answered is whether the comparatively good stock status in the WCPO has anything to do with the RFMO management. The WCPO is particularly interesting because the WCPFC has been only established in 2004 and its progressive legal framework draws on the most modern legal instruments for international fisheries management. Another reason for the focus on the WCPO is the fact that it is globally the most important catch area for tuna and tuna like species. In 2009, the catch of the principal market tuna species was almost 2.5 million mt, representing more than half of the global catch.[1020] Also important is the unique characteristic of the WCPO tuna fishery. It is a multi-species and multi-gear fishery with powerful and well organized developing Pacific Island countries and territories which rely heavily on the revenues from the tuna fishery.

The analysis comprises the development both before and since the establishment of the WCPFC in order to identify crucial provisions and to highlight any remaining challenges. For a better understanding of the current performance of the WCPFC, it was essential to analyze the sub-regional institutional framework in the Western and Central Pacific Ocean (WCPO) that had developed for the most part prior to the establishment of the WCPFC. This type of cooperation, mainly among small island developing countries, is unique in international fisheries management and represents a key pillar for sustainable tuna fisheries management in the WCPO. The main part of the analysis will look more closely at the legal framework of the WCPFC, and will evaluate the conservation and management measures (CMMs) that have been adopted based on the provisions of the WCPFC Convention. The principal question to be answered in this part is whether the present situation with comparatively good stock statuses can be maintained over the coming years, or if it is just a peak in performance?

1020   WCPFC, *Summary Report - Seventh Regular Session of the Scientific Committee*, (2011), p.3.

The answer to this depends on the quality of the decisions the WCPFC makes, both today and in the future, which in turn relies upon its legal framework.[1021]

B. Sub-regional cooperation in the WCPO between the Pacific Island countries and territories

The coastal States in the tropical region of the WCPO cover with their EEZs a huge area (Figure 4.1). For the management of tuna fisheries this area is very important because around 80% of the catch of the principal market tuna species skipjack tuna, yellowfin tuna, bigeye tuna, albacore tuna is made there.[1022] In particular the Pacific Island region is crucial for sustainable management of the fisheries. The region consists of 14 independent countries[1023] and 8 territories[1024] (hereinafter jointly called PICTs).[1025] Indonesia, the Philippines, Australia and New Zealand are no included in the group of PICTs. Around 57 percent of all WCPO catches are made In the EEZs of the PICTs.[1026] With the introduction of the EEZ in the 1970s and 1980s,[1027] most of them had gained jurisdiction over highly productive tuna grounds that promised great revenues.[1028] The overall EEZ area of the PICTs covers more than 30 million square kilometers, which is

---

1021 An important source of information for this chapter was the personal experience of the author during a three-month research stay at the Secretariat of the WCPFC in 2011. Information provided by the staff of the Secretariat, as well as conversations with delegates and external experts at two annual meetings, was essential to a fuller understanding of the work of the WCPFC in particular and of international tuna fisheries management in general.

1022 S. Harley & J. Hampton, 'Status of Tuna Stocks in the Western and Central Pacific Ocean and Scientific Challenges', *Navigating Pacific fisheries : legal and policy trends in the implementation of international fisheries instruments in the Western and Central Pacific region* (Wollongong: Australian National Centre for Ocean Resources and Security, 2009), p.188.

1023 The Cook Islands, the Federated States of Micronesia, Fiji, Kiribati, the Marshall Islands, Nauru, Niue, Palau, Papua New Guinea, Samoa, the Solomon Islands, Tonga, Tuvalu and Vanuatu.

1024 American Samoa, French Polynesia, Guam, New Caledonia, Northern Marianas, Pitcairn Islands, Tokelau, and Wallis and Futuna.

1025 R. Gillett & I. Cartwright, *The future of Pacific Island fisheries* (Noumea: SPC, 2010), p.1; R. Gillett, *A short history of industrial fishing in the Pacific islands*, RAP PUBLICATION 2007/22 (Bangkok: FAO Regional Office for Asia and the Pacific, 2007), p.1.

1026 Q. Hanich, et al., A collective approach to Pacific islands fisheries management: Moving beyond regional agreements, *Marine Policy - Vol. 34(1)* (2010), p.86.

1027 For more information about the introduction of the EEZ in the PICs, see R. Hannesson, The exclusive economic zone and economic development in the Pacific island countries, *Marine Policy - Vol. 32(6)* (2008).

1028 Gillett, 2010, p.4.

about 28 percent of the world's total EEZs.[1029] Unlike in other oceans, these small island developing countries and territories - typically sparsely inhabited islands with little land surface but vast EEZs - are important players in the tuna fisheries management. It is a logic consequence of this that sustainable fisheries management in the WCPO can only be possible when the PICTs are actively involved in the regional management process.

Figure 4.1   Maritime Claims in the Western and Central Pacific Ocean

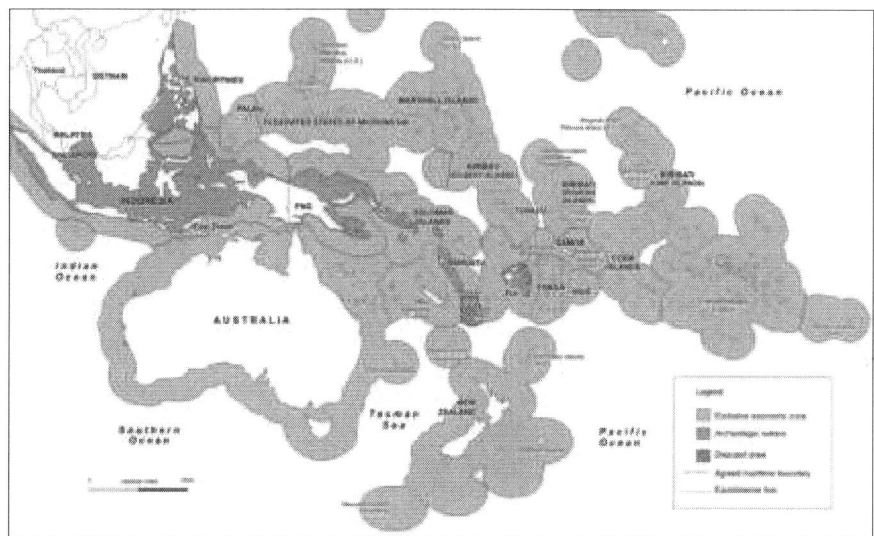

Source: Hanich et al. (2009)[1030]

The PICTs have established several subregional bodies and legal instruments in order to strengthen cooperation (Table 4.1 shows the participation). The Secretariat of the Pacific Community (SPC) is a sub-regional consultative and advisory body which offers a broad range of services to the marine and fisheries sector. Its Oceanic Fisheries Programme (OFP) is providing scientific and data management services to its members and to organizations like the WCPFC. The Pacific Island Forum Fisheries Agency (FFA) is a regional fisheries body that facilitates the development of key regional instruments on fisheries manage-

---

1029   R. Gillett, *Review of the state of world marine fishery resources*, FAO Fisheries Technical Paper - Vol. 457 (Rome: FAO, 2005), p.144.

1030   Q. Hanich, et al., 'Oceans of Opportunity: The Limits of Maritime Claims in the Western and Central Pacific Region', *Navigating Pacific fisheries : legal and policy trends in the implementation of international fisheries instruments in the Western and Central Pacific region* (Wollongong: Australian National Centre for Ocean Resources and Security, 2009), p.22.

ment and access regulation. Its governing body, the Forum Fisheries Committee (FFC) gives policy and administrative guidance and direction to the FFA and it provides a forum for Parties to consult together on matters of common concern in the field of fisheries. The Nauru Agreement is a subregional instrument that has developed specific terms and conditions for tuna purse seine fishing licenses in the EEZs of its Parties. It coordinates and harmonizes fisheries management and access conditions in order to place its Parties (the PNA) in a stronger strategic position when negotiating with foreign fleets on access to their EEZ resources. Further instruments are the Palau Arrangement which focuses on increased returns for the PNA through access fees and economic development as well as the Federated States of Micronesia Arrangement which had been established to increase the number of domestic and locally based vessels in the PNA waters. The Niue Treaty is an instrument on cooperative fisheries surveillance and law enforcement by the FFA members. It facilitates reciprocal enforcement procedures and the development of Subsidiary Agreements. The most recently developed sub-regional instrument is the Te Vaka Moana Arrangement (TVMA). It can be seen as the southern version of the Nauru Agreement.

All subregional bodies and instruments share a common objective of sustainable management of tuna and tuna-like species as well as economic development of the PICTs. Their importance has not lessened since the establishment of the WCPFC. An evaluation of the performance of the WCPFC must therefore include an analysis of this sub-regional legal framework. It is important to understand the special configuration that makes tuna fisheries in the WCPO so different from tuna fisheries found in other parts of the world.

Table 4.1   *Participation in sub-regional bodies and arrangements*

	SPC	FFA	NA	PA	FSMA	NT	TVMA
American Samoa	✓						
Australia	✓	✓				✓	
Cook Islands	✓	✓				✓	✓
Federated States of Micronesia	✓	✓	✓	✓	✓	✓	
Fiji	✓	✓				✓	
France	✓						
French Polynesia	✓						
Guam	✓						
Kiribati	✓	✓	✓	✓		✓	
Marshall Islands	✓	✓	✓	✓	✓	✓	
Nauru	✓	✓	✓	✓	✓	✓	
New Caledonia	✓						
New Zealand	✓	✓				✓	✓
Niue	✓	✓				✓	✓
Northern Mariana Islands	✓						
Palau	✓	✓	✓	✓	✓	✓	
Papua New Guinea	✓	✓	✓	✓	✓	✓	
Pitcairn Islands	✓						
Samoa	✓	✓				✓	✓
Solomon Islands	✓	✓	✓	✓	✓	✓	
Tokelau	✓	✓				✓	✓
Tonga	✓	✓				✓	✓
Tuvalu	✓	✓	✓	✓		✓	
United States of America	✓						
Vanuatu	✓	✓				✓	
Wallis and Futuna	✓						

Source: Unterweger, based on information provided by the websites of the SPC and FFA.

SPC = Secretariat of the Pacific Community; FFA = Pacific Island Forum Fisheries Agency; NA = Nauru Agreement; PA = Palau Agreement; FSMA = Federated States of Micronesia Arrangement; NT = Niue Treaty; TVMA = Te Vaka Moana Agreement

1. *Secretariat of the Pacific Community*

The first sub-regional body relevant to the management of tuna and tuna-like species in the WCPO was the South Pacific Commission, as the Secretariat of the Pacific Community (SPC)[1031] was originally called.[1032] It was founded on 6th February, 1947, by Australia, France, New Zealand, the Netherlands, the United Kingdom and the United States of America.[1033] The 'Agreement establishing the South Pacific Commission' (SPC Agreement) came into force on 29th July, 1948.[1034] Its purpose was to encourage and strengthen international cooperation in promoting the economic and social welfare and advancement of the peoples of the non-self-governing territories in the South Pacific region.[1035] Between 1965 and 1983, in the context of increasing independence of the Pacific Island states, all 22 PICTs became members with full voting rights.[1036] Today, the SPC is composed of these PICTs and four of the founding countries.[1037] The SPC Agreement does not provide a precise definition of the area of competence by lines of longitude and latitude but in the past the Commission was using a map showing an area that coincides with part of FAO Statistical Areas 71 and 77.[1038]

1031 Originally named the South Pacific Commission but at the 50th anniversary conference in 1997 the name was changed to reflect the organization's Pacific-wide membership.

1032 G. Lugten, *The role of international fishery organizations and other bodies in the conservation and management of living aquatic resources.*, FAO Fisheries Circular - No. 1054 (Rome: FAO, 2010), p.110.

1033 Participating governments: Australia, France, New Zealand, the Netherlands, the United Kingdom and the United States of America.

1034 Agreement establishing the South Pacific Commission (SPC), 1948.

1035 Ibid. Preamble.

1036 http://www.spc.int/en/about-spc/history.html.

1037 The remaining founding countries are Australia, France, New Zealand and the United States of America. The Netherlands and the United Kingdom have both relinquished their interests in the region.

1038 S.H. Marashi, *Summary Information on the Role of International Fishery and other Bodies with Regard to the Conservation and Management of Living Resources of the High Seas*, FAO Fisheries Circular - No. 908 (Rome: FAO, 1996).

All members have to meet annually to discuss work programme and budget issues.[1039] The Secretariat of the SPC which has been established according to Article XIII of the Agreement is located in Noumea, New Caledonia.[1040] As a 'consultative and advisory body'[1041] the SPC has no direct fisheries management competence[1042] but it is providing a broad range of technical and policy services to the marine and fisheries sector.[1043] The SPC is required to study, formulate and recommend measures for the development and coordination of services affecting, economic and social rights and welfare of its Pacific Island members, particularly in respect to issues like fisheries.[1044] A further competence of the SPC is to make recommendations with regard to the establishment and activities of auxiliary and subsidiary bodies.[1045] Six technical divisions, a statistical programme, a strategic engagement and planning facility, and a directorate of operations and management have been developed so far. The technical division relevant to the sustainable management of marine resources is the Fisheries, Aquaculture and Marine Ecosystems division (FAME).[1046] This division is active through the Coastal Fisheries Programme (CFP), which deals with coastal fisheries, nearshore fisheries and aquaculture, and the Oceanic Fisheries Programme (OFP), which mainly focuses on programmes for tuna and tuna-like species. The OFP was established by the 1980 South Pacific Conference (originally as the Tuna and Billfish Assessment Programme) in order to continue and expand the work initiated by its predecessor, the Skipjack Survey and Assessment Programme.[1047]

In 2009 the SPC has published a strategic plan for 2010-2013. According to this plan the overall goal of the OFP is that "fisheries exploiting the region's resources of tuna, billfish and related species are managed for economic and ecological sustainability using the best available scientific information."[1048] Three

---

1039   Agreement establishing the South Pacific Commission (SPC), Art.V.12.

1040   Ibid. Art. XIII.39.

1041   Ibid. Art.IV.6.

1042   Lugten, 2010, p.110.

1043   A. Wright, et al., The cooperative framework for ocean and coastal management in the Pacific Islands: Effectiveness, constraints and future direction, *Ocean & Coastal Management - Vol. 49(9-10)* (2006), p.753.

1044   Agreement establishing the South Pacific Commission (SPC), Art.IV.6(a).

1045   Ibid. Art.IV.6(g).

1046   A.D. Lewis, 'The South Pacific Commission', *Getting ahead of the curve: conserving the Pacific Ocean's tunas, swordfish, billfishes and sharks* (Leesburg: National Coalition for Marine Conservation, 2000).

1047   http://www.spc.int/oceanfish/.

1048   Secretariat of the Pacific Community, *Strategic plan 2010-2013* (Noumea: SPC, 2009), p.16.

objectives that have been formulated, are: to provide high-quality scientific information and advice for regional and national fisheries management authorities on the status of, and fishery impacts on, stocks targeted or otherwise impacted by regional oceanic fisheries; to collect and analyze accurate and comprehensive scientific data for regional and national fisheries management authorities on fisheries targeting the region's resources of tuna, billfish and other oceanic species; and to improve understanding of pelagic ecosystems in the western and central Pacific Ocean.[1049]

Today, the OFP provides scientific and data management services to its members as well as to organizations like the WCPFC, in order to guarantee that the collection and processing of scientific data on fishing activities in the region is carried out efficiently and without duplication of effort.[1050] In fact, most of the stock assessments presented at the Scientific Committee meeting of the WCPFC are produced by scientist from the OFP.[1051] Important in this regard are especially the region-wide tuna tagging programs in order to collect critical information for the assessment of the tuna and tuna-like species[1052] and the biannually published 'Western and Central Pacific Fisheries Commission (WCPFC) Tuna Fishery Yearbook' which presents annual catch estimates in the WCPFC Statistical Area since 1950.[1053]

In 2009, a data exchange agreement was signed between the WCPFC and the SPC.[1054] Basically, it was agreed that the WCPFC would provide the SPC with aggregated catch and effort data from all WCPFC members, and that the SPC would provide the WCPFC with aggregated catch and effort data from the fleets of SPC members. In the same year, in a Revised Memorandum of Understanding, both organizations further agreed to establish and maintain cooperation in respect of matters of common interest to the two organizations.[1055] The major goals of the memorandum are to encourage:

- reciprocal participation in the relevant meetings of each organization;

1049 Ibid. pp.16-17.

1050 C. Hunt, Management of the South Pacific tuna fishery, *Marine Policy - Vol. 21(2)* (1997), p.162.

1051 Observation made by the author during the Seventh Regular Session of the Scientific Committee.

1052 http://www.spc.int/OceanFish/en/major-projects/pttp.

1053 Secretariat of the Pacific Community, 2010.

1054 Data exchange agreement between the Western and Central Pacific Fisheries Commission (WCPFC) and the Secretariat of the Pacific Community (SPC), 2009.

1055 WCPFC SPC-OFP Revised Memorandum of Understanding 2009.

- the collaboration of national scientists in the scientific work undertaken by, or on behalf of, the Commission;

- the active and regular exchange of relevant meeting reports, information, project plans, documents, and publications regarding matters of mutual interest, up to the limits allowed by the information-sharing policies agreed by the members of each organisation; and

- consultation on a regular basis to enhance co-operation and minimize duplication.

2. Pacific Islands Forum Fisheries Agency

The South Pacific Forum Fisheries Agency Convention (FFA Convention)[1056] was signed on 10th July 1979 and entered into force on 9th August, 1979 after having been ratified by 17 PICTs.[1057] The Convention established the Pacific Island Forum Fisheries Agency (FFA)[1058] which formed a cornerstone in the later configuration of tuna fisheries management in the WCPO. The FFA members combined their resources with the prevision that it would be easier to manage and develop the newly proclaimed EEZs together in order to maximize the economic benefits.[1059] About 50% of the catch (1,200,000 mt) of the principal market tuna species is made in the waters of the FFA member countries.[1060] In the Preamble of the FFA Convention they recognize "their common interest in the conservation and optimum utilisation of the living marine resources of the South Pacific region and in particular of the highly migratory species." The Preamble further expresses the desire of the members to promote regional co-operation and co-ordination in respect of fisheries policies and to facilitate the collection, analysis, evaluation and dissemination of relevant statistical scientific and economic information.[1061] The area of competence of the FFA is the South Pacific Region.[1062] The FFA Convention does not provide a precise definition of this

---

1056 South Pacific Forum Fisheries Agency Convention, 1979.

1057 The 17 FFA members are: Australia, the Cook Islands, the Federated States of Micronesia, Fiji, Kiribati, the Marshall Islands, Nauru, New Zealand, Niue, Palau, Papua New Guinea, Samoa, the Solomon Islands, Tokelau, Tonga, Tuvalu and Vanuatu.

1058 The former name of the Pacific Islands Forum Fisheries Agency was the South Pacific Islands Forum Fisheries Agency.

1059 Q. A. Hanich, et al., Sovereignty and cooperation in regional Pacific tuna fisheries management: Politics, economics, conservation and the vessel day scheme, *Australian Journal of Maritime and Ocean Affairs - Vol. 2(1)* (2010), p.2.

1060 Harley & Hampton, 2009, p.188.

1061 South Pacific Forum Fisheries Agency Convention, Preamble.

1062 Ibid. Preamble.

area by lines of longitude and latitude but it is bounded approximately by Australia, New Zealand, Hawaii and the Philippines.[1063]

The FFA with its headquarters in Honiara, the Solomon Islands has neither a management mandate nor does it have the authority to enforce decisions of its governing body, the Forum Fisheries Committee (FFC).[1064] None of the two bodies is entitled to determine the TAC or to allocate the surplus catch to foreign countries.[1065] Article III of the FFA Convention lays down that "the Parties recognize that effective co-operation for the conservation and optimum utilization of the highly migratory species of the region will require the establishment of additional international machinery to provide for co-operation between all coastal states in the region and all states involved in the harvesting of such resources." More than 30 years later, the WCPFC became that 'additional international machinery' which was supposed to fill the gap.[1066] Another body with a management mandate for at least some of the FFA EEZs is formed by the Parties to the Nauru Agreement which was adopted in 1982.[1067]

Although without a management mandate the FFA and the FFC are very important for the cooperation between its Parties. The FFC which is comprised of one representative from each party has to meet at least once a year in order discuss strategies and to adopt decisions.[1068] Decisions usually require consensus but where this is not possible decisions can be taken by a two-thirds majority of the Parties present and voting.[1069] The Committee has: to provide detailed policy and administrative guidance and direction to the Agency; to provide a forum for Parties to consult together on matters of common concern in the field of fisheries; and to carry out such other functions as may be necessary to give ef-

---

1063 S. F. Herrick, et al., Access fees and economic benefits in the Western Pacific United States purse seine tuna fishery, *Marine Policy - Vol. 21(1)* (1997), p.84.

1064 T. Aqorau, *Cooperative Management of Shared Fish Stocks in the South Pacific*, FAO Fisheries Report No. 695, Supplement - Papers presented at the Norway- FAO expert consultation on the management of shared fish stocks (Rome: FAO, 2002).

1065 J. M. Van Dyke & S. Haftel, *Tuna management in the Pacific: an analysis of the South Pacific Forum Fisheries Agency* (Honolulu: 1981), p.18.

1066 S. Tarte, 'The Convention for the Conservation and Management of Highly Migratory Fish Stocks in the Western and Central Pacific Ocean: Implementation Challenges from a Historical Perspective', *Navigating Pacific fisheries: legal and policy trends in the implementation of international fisheries instruments in the Western and Central Pacific region* (Wollongong: Australian National Centre for Ocean Resources and Security, 2009), p.208.

1067 See the next subchapter.

1068 South Pacific Forum Fisheries Agency Convention Art.IV.1.

1069 Ibid. Art.IV.2.

fect to the Convention.[1070] A particular task for the Committee is to promote intra-regional co-ordination and co-operation regarding:

- harmonization of policies with respect to fisheries management;
- co-operation in respect of relations with distant water fishing countries;
- co-operation in surveillance and enforcement;
- co-operation in respect of onshore fish processing;
- co-operation in marketing; and
- co-operation in respect of access to the 200 mile zones of other Parties.[1071]

In the past the FFA has harmonized their policies in various areas. Important examples are the FFA Vessel Monitoring System (VMS); minimum terms and conditions of access for foreign fishing vessels; controlling and monitoring of transshipment; requirements for reporting and data collection; establishment of an observer program; and port state enforcement provisions.[1072] A particular success was the facilitation of the development of several key regional arrangements dealing with the access by foreign fishing fleets into EEZs.[1073] The most prominent ones are the Nauru Agreement, the Palau Arrangement, the FSM Arrangement, the Niue Treaty and the Te Vaka Moana Arrangement.[1074]

In 2005 the FFA has developed a strategic plan for 2005 to 2020.[1075] In this plan the FFA distinguishes between programs for fisheries management and fisheries development. The goals with regard to fisheries management are to: develop model management and legal frameworks and assist members to develop and implement arrangements that assure members; and to establish and maintain mutually effective and beneficial relationships with regional and international bodies, with a clear focus on the WCPFC.[1076] The goal with regard to fisheries development is to develop the capacity of members to create sustainable livelihoods for their people from the sustainable harvest, processing and marketing of their tuna resources.[1077]

1070 Ibid. Art.V.1.

1071 Ibid. Art.V.2.

1072 Aqorau, 2002.

1073 Hanich, et al., 2010, p.3.

1074 Further information will be provided in the analysis of the sub-regional agreements and arrangements within this sub-chapter.

1075 Fisheries Forum Agency, *Strategic Plan 2005-2020* (Honiara: FFA, 2005).

1076 Ibid. p.13.

1077 Ibid. p.14.

In 2009, in a memorandum of understanding, the Secretariat of the FFA and the Secretariat of the WCPFC recognized the need for a complementary relationship in order to promote the sustainable use, conservation and management of highly migratory fish stock in the WCPO.[1078] The two main agreements in the memorandum are to exchange information on activities and programmes of work on highly migratory fish stocks and associated and dependent species in the Pacific Islands region,[1079] and to hold a meeting between both Secretariats at least once a year in order to exchange information on activities of mutual interest, and to explore ways of minimizing duplication in their work.[1080] On the one hand, the memorandum recognized the role of the FFA, recalling its mission to support and enable its members to achieve sustainable fisheries and the highest levels of social and economic benefits in harmony with the broader environment.[1081] On the other hand, it also acknowledges the objectives of the WCPFC Convention in terms of conservation and management requirements for the whole Convention area, including the establishment of conservation and management measures.[1082] The memorandum also expresses the desire to put in place an arrangement to support the implementation of Article VII(e) of the FFA Convention and Article 22 of the WCPFC Convention, both of which deal with the cooperation between the two organizations.[1083]

3. Nauru Agreement

On 11th February 1982, shortly after the establishment of the FFA, the Nauru Agreement Concerning Cooperation on the Management of Fisheries of Common Interest (Nauru Agreement)[1084] was signed and entered into force on 4th December 1982.[1085] The members are a subgroup of eight FFA members, called 'Parties to the Nauru Agreement' (PNA). These are the Federated States of Micronesia, Kiribati, the Marshall Islands, Nauru, Palau, Papua New Guinea, the Solomon Islands and Tuvalu which control with their combined EEZs a huge area of about 14 million square kilometers, representing almost half of the total

---

1078   WCPFC-FFA Memorandum of Understanding, 2009, Preamble, Sentence 1.

1079   Ibid. Para.1.

1080   Ibid. Para.2.

1081   Ibid. Preamble, Sentence 3.

1082   Ibid. Preamble, Sentences 4-6.

1083   Ibid. Preamble, Sentence 10.

1084   Nauru Agreement Concerning Cooperation on the Management of Fisheries of Common Interest, 1982.

1085   D. J. Doulman, 'Fisheries co-operation: the case of the Nauru Group', *Tuna Issues and Perspectives in the Pacific Islands Region* (Honolulu: East-West Center, 1987), p.3.

EEZ area of all PICTs.[1086] With about 1,100,000 mt most part of the overall FFA catches is made within EEZs of the PNA.[1087] The purpose of the Nauru Agreement was to coordinate and harmonise fisheries management and access conditions in order to place the PNA in a stronger strategic position when negotiating with DWFNs on access to their EEZ resources.[1088] The PNA had recognized that DWFNs like Japan, Korea, Taiwan and the United States of America which were fishing in the in the South pacific for tuna since the 1930s[1089] could weaken their positions in negotiations when they play one State off against another.[1090] Since January 2010 the PNA office is located in Majuro, the Marshall Islands.

The Agreement applies within the EEZs, the so called 'fisheries zones', of the PNA.[1091] In these zones the PNA have a management mandate.[1092] Article I states that, in order to improve cooperation, the parties "shall seek, without any derogation of their respective sovereign rights, to coordinate and harmonise the management of fisheries with regard to common stocks within the Fisheries Zones, for the benefit of their peoples." A crucial element of the Agreement is the requirement to "seek to establish a coordinated approach to the fishing of the common stocks in the Fisheries Zones by foreign fishing vessels".[1093] In this regard, the PNA have to establish principles for the granting of priority to applications by fishing vessels of the Parties over applications by foreign fishing vessels.[1094] The Agreement recognized the need for a more integrated approach in terms of licensing. The Parties had to establish uniform terms and conditions for the licensing of foreign vessels requiring them to: apply for and possess a license or permit; have an observer on board; use standardized logbooks maintained on a day to-day basis; make timely reporting of the vessel's location; and

1086   D. J. Doulman, 'Development and expansion of the tuna purse seine fishery', *Tuna issues and perspectives in the Pacific islands region* (Honolulu: East-West Center, 1987), pp.257-277.

1087   http://www.pacifical.com/the_pna_countries.html.

1088   Hanich, et al., 2010, p.4.

1089   Y. Matsuda & Y. Ouchi, *Legal, political and economic constraints on Japanese strategies for distant water tuna and skipjack fisheries in Southeast Asian seas and the Western Central Pacific*, Memoirs of the Kagoshima University Research Centre for the South Pacific (Honolulu: East-West Environmental and Policy Institute, 1984).

1090   M. W. Lodge, Minimum terms and conditions of access: Responsible fisheries management measures in the South Pacific region, *Marine Policy - Vol. 16(4)* (1992), p.280.

1091   Nauru Agreement Concerning Cooperation on the Management of Fisheries of Common Interest, Arts 1 and 2.

1092   United Nations Convention on the Law of the Sea Art.56(1)(a).

1093   Nauru Agreement Concerning Cooperation on the Management of Fisheries of Common Interest Art.II.

1094   Ibid. Arts.II(a).

use a standardized identification.[1095] Other optional terms and conditions should include: payment of an access fee; complete catch and effort data reporting; reporting of additional information; licensing only for vessel over which the flag states has authority and other terms and conditions as necessary.[1096] The Parties should seek to develop standardized licensing procedures by adopting uniform measures and procedures relating to the licensing of foreign fishing vessels, and to explore the possibility of establishing a centralized system for the licensing.[1097] The envisaged benefits were better coordination and easier cooperation between the parties. In addition, the Parties were requested to cooperate and to coordinate the monitoring and surveillance of foreign vessel fishing activities by exchanging information and also exploring the possibility of joint surveillance.[1098]

In order to facilitate the implementation of the Agreement and to attain its objectives, parties were also required to make additional arrangements where necessary.[1099] To date, three Implementing Arrangements (IAs) have been agreed, all of them setting the terms and conditions of access to the fisheries zones of the parties. In 1982, the first IA established a regional fishing vessel register in order to guarantee compliance and to enable the blacklisting of vessels which were involved in IUU fishing.[1100] PNA countries were required to both participate in and comply with the Procedures for the Establishment and Operation of the South Pacific Forum Fisheries Agency Regional Register of Fishing Vessels, which had been adopted by the FFA in 1983.[1101] The second IA (1990) established the minimum terms and conditions for fishing in PNA waters.[1102] It addressed transshipment at sea, high seas catch reporting, the maintenance of log books and the use of observers.[1103] The most recent IA, orig-

---

1095 Ibid. Arts.II (b)(i-v).

1096 Ibid. Arts.II (c)(i-v).

1097 Ibid. Art.III.

1098 Ibid. Art.VI.

1099 Ibid. Art.IX.

1100 B. Kuemlangan, *National Legislative Options to Combat IUU Fishing*, FAO Fisheries Report - No. 666 (Rome: FAO, 2000).

1101 An Arrangement Implementing the Nauru Agreement Setting Forth Minimum Terms and Conditions of Access to the Fisheries Zones of the Parties, 1982 Art.I.

1102 A Second Arrangement Implementing the Nauru Agreement Setting Forth Additional Terms and Conditions of Access to the Fisheries Zones of the Parties, 1990.

1103 Ibid. Art.I.

inally from 2008 and amended in 2010, contained provisions for catch retention, FAD closure, closure of high seas areas and monitoring.[1104]

So far, there is no memorandum of understanding between the WCPFC and the PNA. During the meetings of the WCPFC, the PNA do not explicitly act as a block but rather as members of the FFA.[1105] However, the role the PNA play in setting the agenda for WCPFC meetings should not be underestimated. Some of the conservation and management measures adopted by the WCPFC have been based on the provisions of their Implementing Arrangements. The most obvious examples are the closure of the high seas pockets and the Vessel Day Scheme.[1106]

In December 2011, the PNA had a major victory regarding the management of tuna within their EEZs. The purse seine fishery on unassociated, free schools of skipjack tuna was certified as sustainable by the Marine Stewardship Council.[1107] With an annual catch volume of around 440,000 mt this fishery accounts for almost 20% of the overall catch of principal market tuna species in the WCPO.

4. *Palau Arrangement*

The Palau Arrangement for the Management of the Purse Seine Fishery in the Western and Central Pacific (Palau Arrangement) was adopted in April 1992 and came into force in November 1995.[1108] The members are the Parties to the Nauru Agreement (PNA).[1109] The purpose of the Arrangement was the conservation of tuna and tuna-like species and the increase of returns through access

---

1104 A third arrangement implementing the Nauru Agreement setting forth additional terms and conditions of access to the fisheries zones of the parties, 2008, Art.I.

1105 Observation made by the author during the Seventh Regular Session of the Scientific Committee and the Seventh Regular Session of the Technical and Compliance Committee.

1106 See the next subchapter.

1107 For further information see
http://www.msc.org/track-a-fishery/certified/pacific/pna_western_central_pacific_skipjack_tuna.

1108 Palau Arrangement for the Management of the Purse Seine Fishery in the Western and Central Pacific, 1992.

1109 See the previous subchapter.

fees and economic development.[1110] Prior to its establishment the total number of purse seine vessel catching yellowfin and bigeye tuna in the western Pacific had increased significantly from 14 in 1979 to 123 in 1988, raising concern regarding the economic and biological status of the stocks.[1111] The new Arrangement was seen as the "first serious attempt by the Pacific Island States to impose some form of limits to tuna fishing".[1112] It is subordinate to both the South Pacific Forum Fisheries Agency Convention and the Nauru Agreement.[1113] The Preamble states that the Parties are having regard to the objectives of these two treaties, particularly with regard to the promotion of regional co-operation and the co-ordination of fisheries policies.[1114]

The members meet annually for a Management Meeting where the stock status is reviewed and necessary measures for management and conservation are established.[1115] Decisions are made by consensus and are binding to the parties.[1116] All parties are obliged to enforce the decisions on their vessels and nationals.[1117] The Arrangement applies to all tuna and tuna-like species taken by purse seine vessels in the EEZs of the Parties.[1118] Fishing licenses can only be allocated to vessels with a good standing on the Regional Register.[1119] If this requirement is fulfilled, allocation has to follow a determined order with the

---

1110 T. Aqorau & A. Bergin, Ocean governance in the Western Pacific purse seine fishery - The Palau Arrangement, *Marine Policy - Vol. 21(2)* (1997), p.174; M. W. Lodge, The development of the Palau Arrangement for the management of the western Pacific purse seine fishery, *Marine Policy - Vol. 22(1)* (1998), p.2; C. Reid, et al., An analysis of fishing capacity in the western and central Pacific Ocean tuna fishery and management implications, *Marine Policy - Vol. 27(6)* (2003), p.450.

1111 Pacific Islands Forum Fisheries Agency, *The Western Pacific Purse Seine Fishery: A Summary of Concerns*, Forum Fisheries Agency Report - Number 90/27 (Honiara: Forum Fisheries Agency, 1990).

1112 T. Aqorau, Moving towards a rights-based fisheries management regime for the tuna fisheries in the Western and central pacific ocean, *International Journal of Marine and Coastal Law - Vol. 22(1)* (2007), p.131.

1113 Aqorau & Bergin, 1997, p.176.

1114 Palau Arrangement for the Management of the Purse Seine Fishery in the Western and Central Pacific, Preamble.

1115 Ibid. Art.3.1.

1116 Ibid. Art.4.

1117 Aqorau & Bergin, 1997, p.177

1118 Palau Arrangement for the Management of the Purse Seine Fishery in the Western and Central Pacific Art.2.1.

1119 Ibid. Art.5.1.

preference for domestic vessels over foreign vessels.[1120] The Arrangement provides criteria for the allocation of licenses distinguishing between fleets and individual vessels.[1121] The number of licenses that can be issued to fleets is provided by Annex 1.[1122] The overall number is limited to 205 purse seine vessels.[1123] However, this limit can be exceeded if a premium of at least 20% to the fees was paid for an additional license.[1124] The limit has to be reviewed by the management meeting and any alteration has to be approved by all parties[1125] With regard to individual vessels the vessel's record of compliance is the primary allocation criteria before the chronological order of applications submitted.[1126] Vessels with poor records have the lowest priority in allocation.[1127]

In 2000, the vessel limit was increasingly seen as ineffective in promoting the conservation and development interests because for new fleets it was difficult to enter the fishery.[1128] The Parties had the impression that the vessel limit was constraining their development aspirations.[1129] As a result, in October 2005, the Palau Arrangement was amended. The main element of the amended version is the so-called Vessel Day Scheme (VDS) which became effective on the 1st December, 2007. The VDS was a shift from an arrangement restricting vessel numbers to a programme limiting the total days which could be fished.[1130] The overall objective of the VDS is the enhanced management of purse seine fishing vessel effort in the waters of the Parties.[1131] By allocating fishing days among the PNA countries, the scheme aims to target increasing exploitation rates of the countries which are exempted from effort limitations due to their developing

---

1120   Ibid. Art.5.1(a-e).

1121   Ibid. Art.6.

1122   Ibid. Art.6.1.

1123   Ibid. Annex 1.

1124   Ibid. Art.6.2.

1125   Ibid. Art.6.3.

1126   Ibid. Art.6.4.

1127   Ibid. Art.6.5.

1128   Hanich, et al., 2009, p.41,

1129   T. Aqorau, Recent Developments in Pacific Tuna Fisheries: The Palau Arrangement and the Vessel Day Scheme, *International Journal of Marine & Coastal Law - Vol. 24(3)* (2009), p.563.

1130   S. Shanks, Introducing a transferable fishing day management regime for Pacific Island countries, *Marine Policy - Vol. 34(5)* (2010), p.988.

1131   Palau Arrangement for the Management of the Purse Seine Fishery in the Western and Central Pacific (as amended by VDS Working Group Meeting-Honiara, 7 & 13 October 2005), 2005, Art.2.1.

aspirations, especially of yellowfin, and to enable the Parties to economically control the fishery in their EEZs.[1132]

At a basic level, the VDS requires that all Parties seek to limit the level of fishing by purse seine vessels in their waters to the levels of Total Allowable Effort (TAE) previously agreed.[1133] Every three years, the TAE has to be reset and re-allocated in terms of fishing days.[1134] The central issue is the distribution of the fishing days among the Parties. The so-called Party Allowable Effort (PAE), which is "the total number of fishing days for a Management Year allocated to a Party,"[1135] is calculated in two parts: half is based on the distribution of the assessed relative biomass of skipjack and yellowfin within the EEZ of the respective Party over a 10 year period, and half is based on the average of the annual distribution of the number of vessel days fished in the waters of the Parties, based on the average over a seven year period.[1136] The Parties have to ensure that the number of fishing days by purse seine vessels in their waters does not exceeding their PAE in any one management year.[1137]

The PAE can only be transferred between Parties.[1138] A transfer between vessels of fishing companies is not mentioned in the amended Palau Arrangement. Transfer between Parties is possible if the transferred amount of days does not exceed 100 percent of the Parties' PAE and if it is not part of a PAE which has been already used.[1139] Parties can transfer up to 100 percent of their PAE within one management period, or up to 30 percent of its PAE from one management period to another.[1140] The VDS affects all vessels operating under a valid license issued under the Federated States of Micronesia Arrangement for Regional Fisheries Access (FSMA),[1141] as long as they are in the EEZ of the licensing

---

1132 Aqorau, 2009, p.558.

1133 Palau Arrangement for the Management of the Purse Seine Fishery in the Western and Central Pacific (as amended by VDS Working Group Meeting-Honiara, 7 & 13 October 2005) Art.2.2.

1134 Ibid. Art.2.3.

1135 Ibid. Art.1.1(x).

1136 Ibid. Art.5.1 and Schedule 1. With regard to the percentage see also Shanks, 2010, p.991.

1137 Palau Arrangement for the Management of the Purse Seine Fishery in the Western and Central Pacific (as amended by VDS Working Group Meeting-Honiara, 7 & 13 October 2005), Art.4.1.

1138 Ibid. Art.6.

1139 Ibid. Art.6.1.

1140 Ibid. Arts.7.1 and 7.2.

1141 See the next subchapter.

home Party.[1142] Exceptions, such as those for vessels operating under a valid license issued under the US Treaty,[1143] do weaken the Arrangement.[1144]

The Palau Arrangement was amended again in 2010, but this second amendment was less comprehensive than the first.[1145] The most important change was the shift of competencies from FFA to PNA. All Secretariat services and arrangements for meetings were transferred from the FFA to the PNA Office.[1146] This included adding a new responsibility of the Director of the Parties to the Nauru Agreement, who now, assists the Parties in the implementation and coordination of the provisions of the Arrangement, instead of the Director of the FFA.[1147] The shift shows the increasing importance of the PNA in terms of tuna fisheries management in the region.

5. FSM Arrangement

The Federated States of Micronesia Arrangement for Regional Fisheries Access (FSM Arrangement) was signed on 30th November, 1994, and entered into force on 23rd September, 1995.[1148] The Parties are six of the PNA: The Federated States of Micronesia, the Marshall Islands, Nauru, Palau, Papua New Guinea and the Solomon Islands. The Arrangement reflects the long-standing desire of the Parties to have their own domestic tuna industry.[1149] When it was established their domestic fishing fleets were, in comparison to foreign fleets, very small and

---

1142 Palau Arrangement for the Management of the Purse Seine Fishery in the Western and Central Pacific (as amended by VDS Working Group Meeting-Honiara, 7 & 13 October 2005), Art.3.1.

1143 The US Treaty is a multilateral treaty between 16 PICs and the United States that allows 40 purse-seiners to operate in the area. It shows the economic importance of tuna fishery for the PICs. Under the current US treaty that expires in 2013, the United States pays 18 million dollars a year. The new treaty that is currently being negotiated would increase the annual payment for 8,000 fishing days in the EEZs of the 16 PICs to 63 million dollars a year over the next 10 years.

1144 Palau Arrangement for the Management of the Purse Seine Fishery in the Western and Central Pacific (as amended by VDS Working Group Meeting-Honiara, 7 & 13 October 2005), Art.3.2(ii).

1145 Palau Arrangement for the Management of the Purse Seine Fishery in the Western and Central Pacific (Amended 11th September 2010), 2010.

1146 Ibid. Art.3.4.

1147 Ibid. Art.7.1.

1148 The Federated States of Micronesia Arrangement for Regional Fisheries Access, 1994

1149 T. Aqorau & A. Bergin, The Federated States Of Micronesia Arrangement for Regional Fisheries Access, *The International Journal of Marine and Coastal Law - Vol. 12* (1997), p.42.

their economic benefit from the tuna fishery was very low.[1150] The main objective of the Agreement was therefore "to establish a licensing regime under which fishing vessels of the Parties may gain access to the waters within the Arrangement Area on terms and conditions no less favourable than those granted by the Parties to foreign fishing vessels under bilateral and multilateral access arrangements."[1151] Basically, domestic and locally based vessels should get lower cost licenses and access to the waters of all PNA states.[1152] Like the Palau Arrangement, the FSM Arrangement is linked to both the Nauru Agreement and the FFA Convention. The Parties have to respect the objectives of both instruments, particularly with regard to the promotion of regional cooperation and the coordination of fisheries policies.[1153] With regard to the support of local development and the promotion of PNA vessels over foreign vessels the Arrangement represents a further development of the Nauru Agreement.[1154]

An important element of the Arrangement is the Register of Eligible Vessels which are fishing vessel of the Parties that have met specific criteria.[1155] This Register is to be maintained by the Director of the FFA who is the official Administrator.[1156] The general information which has to be provided for each vessel includes the name of the vessel, its international radio call sign, country of registration, regional register number, the name and address of the owner or owners, the name and address of operator and the name of the home party of the vessel.[1157] A party which is satisfied that a vessel meets the eligibility criteria can apply to enter the vessel on the Register by completing a form which is set out in Annex I to the Arrangement.[1158] The eligibility of a vessel is basically decided according to specific criteria which are evaluated using a set scoring system (Table 4.2).[1159] The criteria are the vessel flag, the number of nationals em-

---

1150   Ibid. p.42.

1151   The Federated States of Micronesia Arrangement for Regional Fisheries Access Art.2(c).

1152   Hanich, et al., 2010, p.5.

1153   The Federated States of Micronesia Arrangement for Regional Fisheries Access, Preamble.

1154   Nauru Agreement Concerning Cooperation on the Management of Fisheries of Common Interest Arts.II(a).

1155   The Federated States of Micronesia Arrangement for Regional Fisheries Access Art.1(e).

1156   Ibid. Art.7.

1157   Ibid. Art.3.1.

1158   Ibid. Art.3.2 and Annex I.

1159   Ibid. Annex III.

ployed, the local purchase and the onshore investment.[1160] A vessel must score a minimum of 25 points to be eligible.[1161]

Table 4.2   Points Evaluation System

Points	Equity	Vessel flag	Nationals employed	Local purchases (US$)	Onshore investment (US$)
10	100%	Party	> 50	> 700,000	> 5,000,000
8	> 50%		31-50	500,000-700,000	2,000,000-5,000,000
5	30-49%	State eligible to accede to this Arrangement	16-30	250,000-500,000	500,000-2,000,000
2	10-29%		5-15	50,000-250,000	100,000-500,000

Source: The Federated States of Micronesia Arrangement for Regional Fisheries Access, Annex III.

After the reception of a duly completed application the Administrator is required to enter the respective vessel on the Register of Eligible Fishing Vessels.[1162] A vessel which is registered can apply for a regional access license on behalf of its flag state.[1163] The final decision with regard to vessel eligibility is made by the Parties at the annual meeting of the Parties of the Nauru Agreement.[1164] The access for fishing requires further an access fee which has to be paid to the party that has the jurisdiction over the EEZ where the vessel has the regional access license for fishing.[1165] The access fee is calculated by the average regional catch per vessel multiplied by the average price of tuna multiplied by five percent.[1166] The fees are collected and distributed by the Administrator.[1167] Each Party has to get the sum of the regional catch within its EEZ multiplied by the regional price per tonne.[1168]

Parties are required to ensure that vessels flying their flags do not fish within the EEZ of other Parties unless these fishing activities have been authorized by

---

1160   Ibid. Annex III, Para.3, sent.1.

1161   Ibid. Annex III, Para.3, sent.2.

1162   Ibid. Art.3.3.

1163   Ibid. Art.6.3.

1164   Ibid. Art.8.1(a).

1165   T. Aqorau, *Analysis of the responses of the Pacific Island States to the fisheries provisions of the Law of the Sea Convention* (Wollongong: University of Wollongong, 1998), p.127.

1166   The Federated States of Micronesia Arrangement for Regional Fisheries Access Annex IV, Schedule 1.3.

1167   Ibid. Art.9.

1168   Ibid. Annex VI, Para.5.

the FSM agreement or another licensing agreement.[1169] In cases of non-compliance, the non-compliant party has to be dealt in accordance with the relevant laws and regulations of the Party in whose EEZ the incident occurred.[1170] All Parities must take reasonable measures to assist in the investigation if there is a request for help from any Party with respect to assistance in an investigation.[1171] Generally, all Parties have a duty to cooperate in the enforcement of the provisions of the FSM Arrangement and their national law and regulations.[1172] The joint surveillance has to be carried out in accordance with the provisions of the Niue Treaty.[1173]

6. Niue Treaty

The Niue Treaty on Cooperation in Fisheries Surveillance and Law Enforcement in the Pacific Region (Niue Treaty) was adopted by the FFA members on 9th July, 1992, and entered into force on 20th May, 1993.[1174] The Treaty is based on the principles of Article 73 of UNCLOS and Article 5 of the FFA Convention.[1175] Its main objective was "to promote maximum effectiveness in regional surveillance and enforcement through cooperation between countries, on a reciprocal or joint basis."[1176] The Treaty is seen as an outstanding example of the benefits of bilateral and regional cooperation between coastal states in order to combat IUU fishing.[1177]

The Parties have the general obligation to cooperate in the enforcement of their fisheries laws and regulations and to develop regionally agreed procedures for the conduct of fisheries surveillance and law enforcement.[1178] The treaty further contains provisions on the exchange of information,[1179] cooperation in prosecu-

---

1169   Ibid. Art.12.1.

1170   Ibid. Art.12.2.

1171   Ibid. Art.13.1.

1172   Ibid. Art.15.

1173   See the next sub-chapter.

1174   Niue Treaty on Cooperation in Fisheries Surveillance and Law Enforcement in the Pacific Region, 1992.

1175   Ibid. Preamble.

1176   F. Amoa, Introductory Note - The Niue Treaty on Cooperation in Fisheries Surveillance and Law Enforcement in the Pacific Region, *International legal materials - Vol. 32* (1993).

1177   Palma, et al., 2010, p.153.

1178   Niue Treaty on Cooperation in Fisheries Surveillance and Law Enforcement in the Pacific Region Art.III.

1179   Ibid. Art.V.

tions[1180] and cooperation in the enforcement of penalties.[1181] The crucial provisions of the Treaty are those on reciprocal enforcement procedures and Subsidiary Agreements.[1182] The term 'reciprocal' refers to the fact that a Party can be entitled to extend its fisheries surveillance and law enforcement activities to waters of another Party and vice versa.[1183] A 'Subsidiary Agreement' is an agreement or an arrangement entered into by any two or more Parties in accordance with the Niue Treaty.[1184]

The Niue Treaty was the first fisheries-related agreement in the world which put in place reciprocal enforcement procedures.[1185] A Party can "permit another Party to extend its fisheries surveillance and law enforcement activities to the territorial sea and archipelagic waters of that Party." The limitation to the territorial sea and archipelagic waters without addressing the EEZs has been seen as a potential weakness of the Treaty as most of the tuna catches are made within the EEZs.[1186] It was therefore recommended to expand the scope of Art VI to increase the practical value of the Treaty.[1187] Reciprocal enforcement procedures can be conducted "by way of provisions in a Subsidiary Agreement or otherwise."[1188] A Subsidiary Agreement can expand upon rights and obligations covered under the Treaty in their application between the Parties.[1189] Initially, the progress in finalizing subsidiary agreements was slow due to the high costs involved in running surveillance patrols.[1190] Since 2000, the interest of the FFA

1180 Ibid. Art.VII.

1181 Ibid. Art.VIII.

1182 Ibid. Art.VI.

1183 Ibid. Art.VI(1), sent.1.

1184 Ibid. Art.I(f).

1185 A. Bergin, Political and Legal Control over Marine Living Resources- Recent Developments in South Pacific Distant Water Fishing, *The international journal of marine and coastal law - Vol. 9(3)* (1994), pp.305-306.

1186 Aqorau, 1998, p.237.

1187 B.M. Tsamenyi, *Co-operation in Fisheries Law Enforcement in the South Pacific Region: An Analysis of the Niue Treaty* (Bangkok: SEAPOL Tri-Regional Conference on Current Issues in Ocean Law Policy and Management: Southeast Asia, North Pacific, and Southwest Pacific, 1994)

1188 Niue Treaty on Cooperation in Fisheries Surveillance and Law Enforcement in the Pacific Region Art.VI(1), sent.1.

1189 Ibid. Art.II(2).

1190 T. Aqorau, Illegal Fishing and Fisheries Law Enforcement in Small Island Developing States: The Pacific Islands Experience, *The International Journal of Marine and Coastal Law - Vol.15(1)* (2000), p.58.

members in surveillance and enforcement cooperation has increased.[1191] To date, six Subsidiary Agreements have been adopted between:

- Papua New Guinea and Australia;
- the Cook Islands and Samoa;
- the Cook Islands and Niue;
- Tonga and Tuvalu;
- the Republic of the Marshal Islands, the Federated States of Micronesia and Palau; and
- the Cook Islands, New Zealand, Niue, Samoa, Tonga and Tokelau[1192]

The latest Subsidiary Agreement is the Te Vaka Toa Arrangement Cook Islands, New Zealand, Niue, Samoa, Tonga and Tokelau.[1193] With regard to the critique on the limited area of application of the Niue Treaty which has been mentioned above it can be recognized that the Arrangement has expanded the scope. It applies "to the Fisheries Waters of each Participant and the high seas areas adjacent to each Participant's exclusive economic zone to the extent provided for in th[e] Arrangement."[1194]

A new approach of the FFA is the development of a multilateral subsidiary agreement with the possible participation of the United States and America and France.[1195] According to Article XIII(5) of the Niue Treaty such participation is possible if all parties agree.

7. Te Vaka Moana Arrangement

The Te Vaka Moana Arrangement (TVMA) is the most recently developed sub-regional body for the management of tuna and tuna-like species in the WCPO. In January 2010, the Arrangement was signed by the heads of fisheries administration of the Cook Islands, New Zealand, Niue, Samoa, Tokelau and Tonga.[1196] The TVMA can be seen as the southern version of the Nauru Agreement. All Parties are FFA members but no PNA. It can be observed that the participants are well aware of the performance of the PNA and that they are very in-

---

1191   Hanich, et al., 2010, p.6.

1192   Information based on the personal communication of the author with the FFA Secretariat.

1193   See the next sub-chapter.

1194   Te Vaka Toa Arrangement, 2010, Art.3.1.

1195   Hanich, et al., 2010, p.7.

1196   Te Vaka Moana Arrangement, 2010.

terested in developing a similar body in the south.[1197] One main driver for this new co-operation is certainly the proven economic benefit.

The main objectives of the Arrangement are improved cooperation between the participants, capacity development, enhanced sub-regional capability and information sharing.[1198] In terms of fisheries management, co-operation is intended to include the development of national and sub-regional approaches to secure collective strategic interests in shared fisheries and to improve their long-term sustainable economic performance.[1199] The possible areas of cooperation in fisheries development are vessel licensing, vessel access, processing, marketing, distribution and broader regional fisheries development research.[1200] The Arrangement also stipulates a meeting of the participants prior to any meeting of the WCPFC, and the sharing of information with organizations such as WCPFC and FFC.[1201] To date, no memorandum of understanding has been established with either of these organizations. The TVM participants went on to develop the Te Vaka Toa Arrangement (TVTA).[1202] This fisheries surveillance cooperation arrangement is a Subsidiary Agreement to the Niue Treaty, promoting co-operation between the participants regarding fisheries monitoring control and surveillance operations as well as enforcement activities.[1203] Due to the fact that the two arrangements have only been very recently established, it is not possible to analyze their implementation. Their institutional structures are still in development and there have been no formal interactions with the WCPFC.

C.   Regional cooperation - The WCPFC

1.   Negotiation process towards the WCPFC

 The first approach to develop an RFMO for the management of fisheries targeting tuna and tuna-like species in the WCPO was made by DWFNs in 1979 as a

---

1197   Observation made by the author during the Seventh Regular Session of the Scientific Committee and the Seventh Regular Session of the Technical and Compliance Committee.

1198   Te Vaka Moana Arrangement, Paras.1.a-c.

1199   Ibid. Para.5.1.

1200   Ibid. Para.6.1.

1201   Ibid. Paras.10.1.a and b.

1202   Te Vaka Toa Arrangement.

1203   Ibid. Para.1.

reaction to the establishment of the FFA.[1204] The DWFNs questioned the legitimacy and role of the FFA, pointing out that an RFMO based on Article 64 UNCLOS would be required in order to manage the tuna and tuna-like species throughout their whole migratory range.[1205] More specifically, they criticized the lack of complete data coverage for the whole region, the lack of jurisdiction to enable the proper enforcement of management measures on the high seas, and the lack of an appropriate framework for the proper conservation and control of fishing effort in the region as a whole.[1206] Another argument against the FFA was the fact that DWFNs could not join it.[1207]

It took until the middle of the 1990s when coastal states and DWFNs initiated the negotiation process for a fisheries management regime in the WCPO.[1208] The coastal states were the PICTs, New Zealand, Australia, Indonesia and the Philippines and the main DWFNs were Japan, the United States of America, Chinese Taipei, the Republic of Korea and China. The increased openness to the negotiation process was due to two reasons. The first was evolving international law, mainly in the form of the Fish Stocks Agreement. The negotiations towards the Fish Stocks Agreement had revealed for both groups the benefits of an RFMO.[1209] The second reason was the fact that fishing activities beyond the EEZs had been identified as having a negative impact on stocks within the EEZs. Particularly the PICTs realized that only a comprehensive RFMO could address this problem.[1210]

The formal negotiation process towards the WCPFC started in 1994 and included seven Multilateral High-Level Conferences on the Conservation and Management of Highly Migratory Fish Stocks in the Western and Central Pacific (MHLC).[1211] All relevant coastal states and DWFNs participated in this pro-

1204 A. K. Sydnes, Establishing a regional fisheries management organisation for the Western and Central Pacific tuna fisheries, *Ocean & Coastal Management - Vol. 44(11-12)* (2001), p.792.

1205 Aqorau, 1998, pp.98-104 Lodge, 1992, p.279.

1206 Lodge, 1998, p.8.

1207 Van Dyke & Haftel, 1981, pp.37-38.

1208 Cordonnery, 2002, pp.3-4.

1209 Sydnes, 2001, p.792.

1210 S. Tarte, Negotiating a Tuna Management Regime for the western and central Pacific: The MHLC process 1994, *Journal of Pacific History - Vol. 34(3)* (1999), pp.274-275.

1211 For a comprehensive summary of the negotiation process, see Sydnes, 2001.

cess.[1212] The first MHLC took place in Honiara, the Solomon Islands, in December 1994. It was very exploratory in nature because most of the participants were still waiting for the results of the UN Conference on Straddling and Highly Migratory Fish Stocks, which had begun in April 1993.[1213] As a result, the participants focused mainly on technical aspects and other key issues, such as stock status, the collection and exchange of data, transshipment, enforcement and commercial fishing operations.[1214]

The second MHLC, held in Majuro, the Marshall Islands, in June 1997, was shaped by two important facts. First, the Fish Stocks Agreement had been adopted two years earlier, and second, Satya Nandan, who had chaired the UN Conference on Straddling and Highly Migratory Fish Stocks, became chairman of the MHLC process.[1215] As a consequence, MHLC 2 was heavily influenced by the provisions of the Fish Stocks Agreement.[1216] The principal outcome of MHLC 2 was the Majuro Declaration.[1217] This Declaration demonstrated the positive and consensus-based approach of the whole conference and committed all parties to establishing a conservation and management regime within three years.[1218] For the upcoming negotiations, it was agreed that progressive principles from the Fish Stocks Agreement, such as the precautionary approach, management decisions based on best available science, ecosystem considerations and recognition of special requirements of small island developing states, should be included.[1219]

1212 Australia, Canada, China, the Cook Islands, the Federated States of Micronesia, Fiji, France, French Polynesia, Indonesia, Japan, Kiribati, the Republic of the Marshall Islands, Nauru, New Caledonia, New Zealand, Niue, Palau, Papua New Guinea, Philippines, the Republic of Korea, Samoa, Solomon Islands, Chinese Taipei, Tonga, Tuvalu, the United States of America, the United Kingdom of Great Britain and Northern Ireland, on behalf of Pitcairn, Henderson, Ducie and Oeno Islands, Vanuatu, Wallis and Futuna. For an overview of representation see Final Act of the Multilateral High-Level Conference on the Conservation and Management of Highly Migratory Fish Stocks in the Western and Central Pacific, 2000.

1213 Henriksen, et al., 2006, p.173.

1214 FFA, *Record of Proceedings of the Multilateral High-Level Conference on South Pacific Tuna Fisheries*, (1995).

1215 Henriksen, et al., 2006, p.173.

1216 Sydnes, 2001, pp.794-795.

1217 Majuro Declaration, 1997.

1218 Tarte, 1999, p.277.

1219 WCPFC, *MHLC 2 - Report of the Conference*, (1997).

Prior to the third MHLC in Tokyo in June 1998, growing tensions between the FFA countries and the DWFNs became apparent.[1220] These tensions had the potential to slow down the negotiations, and so, five weeks before the conference, the chairman decided to present a draft text for the potential WCPFC Convention, based mainly on the Fish Stocks Agreement.[1221] This move, which was intended to elaborate the structure and powers of the future RFMO, had not been expected by most of the participants and it promoted committing discussions.[1222]

For the subsequent conferences (MHLCs 4-7), all of which were held in Honolulu, Hawaii, the negotiations were conducted by smaller working groups in order to ensure more efficient progress.[1223] The working groups dealt with specific issues like quota allocation, observer programme or financial arrangements.[1224] During the MHLCs 4-7 the chairman often had to play an active role due to ongoing disagreement and conflicts.[1225]

Despite all these obstacles, on 4th September, 2000, the Convention for the Conservation and Management of Highly Migratory Fish Stocks in the Western and Central Pacific Ocean was adopted, making it the first international convention for regional fisheries management since the Fish Stocks Agreement.[1226] Due to opposition to some parts of the final draft, the Convention was adopted not by consensus but by a two-thirds majority, 19 states in favour, two states against (Japan and Korea), and three abstentions (China, France, and Tonga).[1227]

During the final conference of the MHLC process, the parties adopted a resolution for the establishment of Preparatory Conferences to establish the WCPFC.[1228] These Conferences were held in order to clear organizational issues

---

1220   Tarte, 1999, p.277-278.

1221   Draft Articles for a Convention on the Conservation and Management of Highly Migratory Fish Stocks in the Western and Central Pacific Ocean, 1998; WCPFC, *MHLC 3 - Report of the Conference*, (1998).

1222   Henriksen, et al., 2006, p.174.

1223   Ibid. p.175.

1224   Sydnes, 2001, p.796.

1225   Tarte, 1999, p.278.

1226   Convention on the Conservation and Management of Highly Migratory Fish Stocks in the Western and Central Pacific Ocean.

1227   Sydnes, 2001, p.801.

1228   WCPFC, *Resolution I establishing a Preparatory Conference for the Establishment of the Commission for the Conservation and Management of Highly Migratory Fish Stocks in the Western and Central Pacific Ocean*, (2001).

and to develop the basis for future regulations.[1229] The last Preparatory Conference was held in Pohnpei, the Federated States of Micronesia, from 6th to 7th December, 2004, only a few months after the Convention had entered into force.[1230]

2.   Convention area

The WCPFC convention area corresponds to the Western and Central Pacific Ocean (WCPO).[1231] Prior to the establishment of WCPFC, the term 'Western Central Pacific' referred to the comparatively small FAO statistical area 71.[1232] Today, the FAO definition is not sufficient to describe the enormous convention area, which also covers parts of the statistical areas 57, 61, 67, 77 and 81. Figure 3.7 shows a map of the Convention area provided by the FAO. It has to be considered that this map also includes the territorial sea and archipelagic waters which are not included in the Convention area. The illustrated northern limit is also only indicative as it is not defined in the Convention text.

The official definition of the WCPFC convention area is provided by Article 3.1 of the Convention. As mentioned before, the limits of the area are not equally precise for all boundaries. Latitudinal and longitudinal coordinates are only provided for the south and for the east. The southern boundary extends due south from the south coast of Australia, along the 141° meridian of east longitude to its intersection with the 55° parallel of south latitude; from there it moves due east along the 55° parallel of south latitude to its intersection with the 150° meridian of east longitude; then due south along the 150° meridian of east longitude to its intersection with the 60° parallel of south latitude; and finally, due east along the 60° parallel of south latitude to its intersection with the 130° meridian of west longitude.[1233] The eastern boundary extends due north from the southern limit along the 130° meridian of west longitude to its intersection with the 4° parallel of south latitude; then due west along the 4° parallel of south latitude to its intersection with the 150° meridian of west longitude;

1229 WCPFC, *WCPFC Preparatory Conference: List of Documents*, (2004).

1230 WCPFC, *Final Report of the Preparatory Conference for the Establishment of the Commission for the Conservation of Highly Migratory Fish Stocks in the Western and Central Pacific Ocean on all Matters within its Mandate pursuant to Paragraph 9 of Resolution I*, (2004).

1231 Convention on the Conservation and Management of Highly Migratory Fish Stocks in the Western and Central Pacific Ocean Preamble, Art.2.

1232 http://www.fao.org/fishery/area/Area71/en. The FAO has established 27 major fishing areas for statistical purposes.

1233 Convention on the Conservation and Management of Highly Migratory Fish Stocks in the Western and Central Pacific Ocean, Art.3.1.

and finally due north along the 150° meridian of west longitude.[1234] There is an overlap with the convention area of the IATTC.[1235] The overlapping area is bounded by 150°longitude west, 130° longitude west, 4° latitude south and 50° latitude south. With the exception of an overlapping area, the WCPFC convention area is coterminous with the area of the IATTC.

Figure 4.2   WCPFC Convention area (large version)

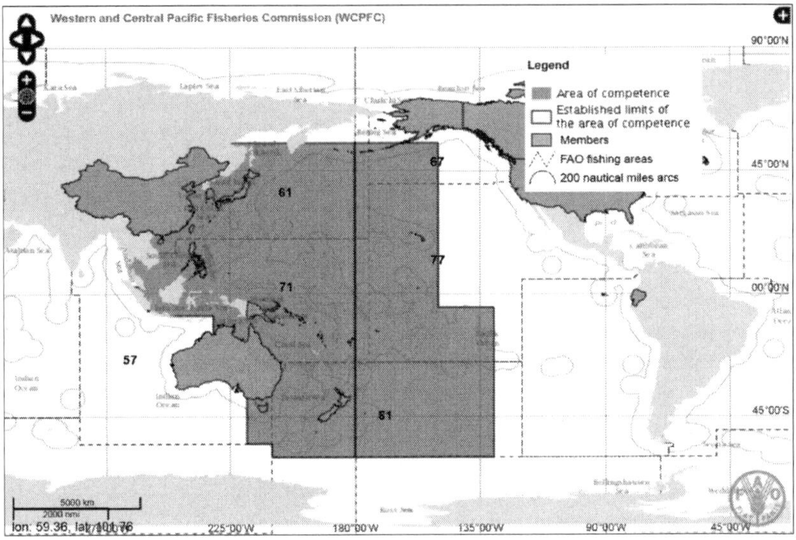

Source: FAO

No coordinates are provided for either the western or the northern boundaries. It is only stated that the Convention "shall be applied throughout the range of the stocks, or to specific areas within the Convention Area, as determined by the Commission."[1236] This lack of a precise definition has the potential to weaken the Commission's authority with regard to conservation and management in the respective areas. In terms of total catch weight, the tuna species which are most affected by the vague western boundary are skipjack, yellowfin and bigeye tuna.[1237] It is known that a great amount of these species, particularly

---

1234   Ibid. Art.3.1.

1235   The total tuna catch in this area varies between 18,000 and 20,000 mt per year. See also WCPFC, *WCPFC-IATTC Overlap Area*, Discussion Paper for WCPFC (Koror: WCPFC, 2011).

1236   Convention on the Conservation and Management of Highly Migratory Fish Stocks in the Western and Central Pacific Ocean, Art.3.3.

1237   Cordonnery, 2002, p.9.

juveniles, are caught in the archipelagic waters of Indonesia and the Philippines[1238] Both countries were not willing to include their archipelagic waters in the convention area.[1239] They referred to the provisions of Part IV of UNCLOS which states that all fisheries within archipelagic waters are subject to the sovereignty of the respective coastal state.[1240] Therefore archipelagic waters are, like territorial water, not included in the Convention area of the WCPFC.[1241] Other Archipelagic States in the WCPO are Papua New Guinea, the Solomon Islands, Vanuatu and Fiji (See Figure 4.2). A general problem with regard to archipelagic waters is that important areas are potentially exempted from the conservation and management measures (CMMs) of the WCPFC. This is reducing the effectiveness of measures because comprehensive management covering the whole migratory range of the stocks cannot be guaranteed.[1242] Another problematic area affected by the unclear western boundaries is the South China Sea.[1243] Significant tuna catches are made in this area but data provision remains poor. The reason for this is border disputes between China, Vietnam, Malaysia, the Philippines and Brunei which make it difficult to determine rights and duties.[1244]

No information is provided by the Convention to determine where the area ends in the north. During the MHLC process, attempts were made to determine a boundary at 50° north, but this was rejected by several delegations.[1245] As a consequence the boundary has been defined in terms of migration patterns of tuna and tuna-like species and through a subcommittee, the so called 'Northern Committee'.[1246] This Committee, which will be explicitly addressed in the subsection on governance, gives recommendations with regard to conservation and

---

1238 Langley, et al., 2011, Figure 6; Davies, et al., 2011, Figure 4.

1239 Henriksen, et al., 2006, p.172.

1240 United Nations Convention on the Law of the Sea, Art.49(2) and Art.2.

1241 Hanich, et al., 2009, p.26.

1242 Cordonnery, 2002, pp.8-9, J. Hampton, *Working Paper MHLC-2: The Convention Area,* presented at the 12th Meeting of the Standing Committee on Tuna and Billfish, (Papeete: Oceanic Fisheries Program, Secretariat of the Pacific Community, Noumea, New Caledonia, 1999), p.2.

1243 T. Aqorau, Tuna Fisheries Management in the Western and Central Pacific Ocean: A Critical Analysis of the Convention for the Conservation and Management of Highly Migratory Fish Stocks in the Western and Central Pacific Ocean and Its Implications for the Pacific, *International Journal of Marine & Coastal Law - Vol.16(3)* (2001), p.386.

1244 M. J. Valencia, Domestic politics fuels Northeast Asian maritime disputes *Asia Pacific Issues - No. 43* (2000), pp.4-5.

1245 Cordonnery, 2002, p.9.

1246 Henriksen, et al., 2006, p.172.

management measures for the stocks that occur 'mostly' in the area north of 20°
parallel of the north latitude.[1247]

3. Species covered by the Convention

In general, the Convention applies to all highly migratory fish stocks that occur
in the convention area.[1248] Highly migratory fish stocks include all fish stocks of
the species listed in Annex I of UNCLOS as well as all other species of fish, as
determined by the Commission.[1249] The Commission has to adopt conservation
and management measures (CMMs) in order "to ensure long-term sustainability [of these stocks] and promote the objective of their optimum utilization."[1250]
From the CMMs that have been adopted to date, it can be inferred that the
WCPFC is mainly focusing on the economically important principal market tuna species.[1251] Efforts have been mainly concentrated on the regulation of fisheries targeting these species. These so called 'target species' are also the species
for which stock assessments have been conducted most frequently. The target
species, for which CMMs have been adopted so far, are:

- yellowfin and bigeye tuna;[1252]

- albacore tuna;[1253]

- Pacific bluefin tuna;[1254]

- swordfish;[1255] and

1247 Convention on the Conservation and Management of Highly Migratory Fish Stocks in the
      Western and Central Pacific Ocean, Art.11.7.

1248 Ibid. Art.3.3.

1249 Ibid. Art.1(f).

1250 Ibid. Art.5(f).

1251 Sydnes, 2001, p.801.

1252 WCPFC, *Conservation and Management Measures for Bigeye and Yellowfin Tuna in the
      Western and Central Pacific Ocean*, (2005); WCPFC, *Conservation and Management Measures
      for Bigeye and Yellowfin Tuna in the Western and Central Pacific Ocean*, (2006); WCPFC,
      *Conservation and Magement Measures for Bigeye and Yellowfin Tuna in the Western and
      Central Pacific Ocean*, (2008); WCPFC, *Conservation and Management Measure for Temporary
      Extension of CMM 2008-01*, (2012).

1253 WCPFC, *Conservation and Management Measure for South Pacific Albacore*, (2005); WCPFC,
      Conservation and Management Measure for North Pacific Albacore, (2005); WCPFC,
      Conservation and Management Measure for South Pacific Albacore, (2010).

1254 WCPFC, *Conservation and Management Measure for Pacific Bluefin Tuna*, (2009); WCPFC,
      Conservation and Management Measure for Pacific Bluefin Tuna, (2010).

- striped marlin.[1256]

However, there is also a significant number of measures for non-target species. The Commission is required to "adopt measures to minimize waste, discards, catch by lost or abandoned gear, [...], catch of non-target species, both fish and non-fish species, [...] and impacts on associated or dependent species, in particular endangered species and promote the development and use of selective, environmentally safe and cost-effective fishing gear and techniques.[1257] The non-target species, for which CMMs have been adopted so far, are:

- sharks;[1258]
- sea turtles;[1259]
- seabirds;[1260] and
- cetaceans.[1261]

4. *Governance*

a) *Secretariat*

According to the Convention, the Commission had to establish a Secretariat as the head office of the WCPFC.[1262] It is located in Kolonia on Pohnpei, which is

---

1255 WCPFC, *Conservation and Management Measure for Swordfish in the Southwest Pacific*, (2006); WCPFC, *Conservation and Management of Swordfish*, (2008); WCPFC, *Conservation and Management of Swordfish*, (2009).

1256 WCPFC, *Conservation and Management Measure for Striped Marlin in the Southwest Pacific*, (2006); WCPFC, *Conservation and Management Measure for North Pacific Striped Marlin*, (2010).

1257 Convention on the Conservation and Management of Highly Migratory Fish Stocks in the Western and Central Pacific Ocean Art.5(f).

1258 WCPFC, *Conservation and Management Measure for Sharks in the Western and Central Pacific Ocean*, (2006); WCPFC, *Conservation and Management of Sharks*, (2008); WCPFC, *Conservation and Management for Sharks*, (2009); WCPFC, *Conservation and Management Measure for Sharks*, (2010); WCPFC, *Conservation and Management Measure for Oceanic Whitetip Shark*, (2012).

1259 WCPFC, *Conservation and Management of Sea Turtles*, (2008).

1260 WCPFC, *Conservation and Management Measure to Mitigate the Impact of Fishing For Highly Migratory Fish Stocks on Seabirds*, (2006); WCPFC, *Conservation and Management Measure to Mitigate the Impact of Fishing For Highly Migratory Fish Stocks on Seabirds*, (2007)

1261 WCPFC, *Conservation and Management Measure for Protection of Cetaceans from Purse Seine Fishing Operations*, (2012).

the main island of the Federated States of Micronesia. The Secretariat has several core functions, including:

- receiving and transmitting the Commission's official communications;[1263]

- facilitating the compilation and dissemination of data necessary to accomplish the objective of the Convention;[1264]

- preparing administrative and other reports for the Commission and the Scientific and Technical and Compliance Committees;[1265]

- administering agreed arrangements for monitoring, control and surveillance and the provision of scientific advice;[1266]

- publishing the decisions and promoting the activities of the Commission and its subsidiary bodies;[1267] and

- treasury, personnel and other administrative functions.[1268]

The head of the Secretariat is the Executive Director who is appointed for a term of four years and can be re-appointed for a second term.[1269] As the chief administrative officer of the Commission, he has to be present at all meetings of the Commission and its subsidiary bodies.[1270] He is also responsible for the appointment of any additional staff required by the Commission.[1271] Currently, this staff is composed of international experts and local employees.[1272] The Scientific Manager, together with an Assistant Science Manager, is responsible for all scientific issues. The Compliance and Monitoring section is led by the Compliance Manager. In this section, several staff members work on the Register of Fishing Vessels and the Vessel Monitoring System. The Regional Observer Programme is coordinated by the manager of the programme and his assistant. Both are mainly involved in the collection and processing of observer data. In

---

1262 Convention on the Conservation and Management of Highly Migratory Fish Stocks in the Western and Central Pacific Ocean Arts. 9.7 and 15.1.

1263 Ibid. Art.15.4(a).

1264 Ibid. Art.15.4(b).

1265 Ibid. Art.15.4(c).

1266 Ibid. Art.15.4(d).

1267 Ibid. Art.15.4(e).

1268 Ibid. Art.15.4(f).

1269 Ibid. Arts.9.7 and 15.2.

1270 Ibid. Art.15.3.

1271 Ibid. Arts.15.1 and 16.1.

1272 Observation made by the author during a research stay at the Secretariat in 2011.

addition, there is the Manager for Administration and Finance as well as other administrative staff, an IT Manager and local staff for maintenance.

b) *Commission*

aa) Composition

The Commission for the Conservation and Management of Highly Migratory Fish Stocks in the Western and Central Pacific Ocean is the governing body of the WCPFC.[1273] It is composed by 24 members, eight Participating Territories and one fishing entity. After its adoption, the Convention had been open for signature by the 25 countries and the territories, which had participated in the MHLC process.[1274] Article 34.4 states that "[e]ach Contracting Party shall be a member of the Commission established by this Convention." The seven participating territories are explicitly entitled to participate fully in the work of the Commission with right to attend all meetings and to speak at them.[1275] Taiwan (Chinese Taipei) has the right to participate as a 'fishing entity'.[1276]

Members:

- Australia, China, Canada, the Cook Islands, European Union, the Federated States of Micronesia, Fiji, France, Japan, Kiribati, the Republic of Korea, the Republic of Marshall Islands, Nauru, New Zealand, Niue, Palau, Papua New Guinea, the Philippines, Samoa, the Solomon Islands, Chinese Taipei, Tonga, Tuvalu, the United States of America, Vanuatu;

Participating Territories:

- American Samoa, the Commonwealth of the Northern Mariana Islands, French Polynesia, Guam, New Caledonia, Tokelau and Wallis and Fortuna:

Fishing entity:

- Taiwan (Chinese Taipei)

All participants of the MHLC process, except Indonesia, have signed the Convention. However, Indonesia as a participant of the MHLC process can accede

---

1273 Convention on the Conservation and Management of Highly Migratory Fish Stocks in the Western and Central Pacific Ocean Art.9.1.

1274 Ibid. Arts.34.1.

1275 Ibid. Arts.43.1 and 43.3.

1276 Ibid. Annex I.

to the Convention at any time in order to become a member.[1277] In practice, the delegates of Indonesia are participating in meetings of the WCPFC like as it would be a member.[1278] In contrast to other non-parties, they are sitting next to the members, participating territories and the fishing entity and participate actively in the negotiations.

Officially, Indonesia is a cooperating non-member (CNM). This term stands for another group of countries that participate in the fishery.[1279] They are not part of the Commission but they are allowed to fish in the Convention area, as long as they cooperate fully in the implementation of conservation and management measures (CMMs) adopted by the Commission.[1280] At meetings of the Commission they are only allowed to participate as observers.[1281]

Current CNMs are:

- Belize, the Democratic People's Republic of Korea, Ecuador, El Salvador, Indonesia, Mexico, Senegal, St Kitts and Nevis, Panama, Thailand and Vietnam.

All members, participating territories and the fishing entity were participants of the MHLC process. However, the Commission is not closed to new entrants. States with a real interest in the fisheries can become members if they are invited with the agreement of all contracting parties.[1282]

Since the Commission's meeting in 2005, the new abbreviation 'CCMs' has been used to refer to members,[1283] participating territories[1284] and cooperating non-members.[1285]

1277   Ibid. Arts.34.1.

1278   Observation made by the author during the Seventh Regular Session of the Scientific Committee and the Seventh Regular Session of the Technical and Compliance Committee.

1279   The role of CNMs will be explicitly addressed below.

1280   Convention on the Conservation and Management of Highly Migratory Fish Stocks in the Western and Central Pacific Ocean Art.32.4.

1281   Ibid. Art.32.5.

1282   Ibid. Art.35.2.

1283   Australia, China, Canada, the Cook Islands, the European Union, the Federated States of Micronesia, Fiji, France, Japan, Kiribati, Korea, the Republic of the Marshall Islands, Nauru, New Zealand, Niue, Palau, Papua New Guinea, Philippines, Samoa, the Solomon Islands, Chinese Taipei, Tonga, Tuvalu, the United States of America and Vanuatu.

1284   American Samoa, the Commonwealth of the Northern Mariana Islands, French Polynesia, Guam, New Caledonia, Tokelau and Wallis and Futuna.

bb) Tasks

Every year, usually in December, the Commission meeting is held.[1286] During this meeting, decisions about important issues are made in accordance with the requirements of the Convention. The specific functions of the Commission are laid down in Article 10. The main functions are to:

- determine the total allowable catch or total level of fishing effort within the Convention Area and adopt such conservation and management measures and recommendations as may be necessary to ensure the long-term sustainability of stocks;[1287]

- promote cooperation and coordination between members of the Commission to ensure compatibility between conservation and management measures for highly migratory fish stocks in areas under national jurisdiction and measures for the same stocks on the high seas;[1288]

- compile and disseminate accurate and complete statistical data to ensure that the best scientific information is available, while maintaining confidentiality, where required;[1289]

- establish appropriate cooperative mechanisms for effective monitoring, control, surveillance and enforcement, including a vessel monitoring system;[1290] and

- promote the peaceful settlement of disputes.[1291]

In its current form the Commission it is basically serving to coordinate management in order to achieve compatible measures for the EEZs and the high seas. The adoption of CMMs is one of the most important tasks at the annual meetings. In general, these measures apply either throughout the range of the stocks, or to specific areas within the Convention Area.[1292] They relate to catch quantities, levels of fishing effort, limitations on fishing capacity, areas and pe-

---

1285 Belize, Ecuador, El Salvador, Indonesia, Mexico, Senegal, Vietnam, Panama and Thailand.

1286 Convention on the Conservation and Management of Highly Migratory Fish Stocks in the Western and Central Pacific Ocean Art.9.3.

1287 Ibid. Art.10.1(a).

1288 Ibid. Art.10.1(b).

1289 Ibid. Art.10.1(e).

1290 Ibid. Art.10.1(i).

1291 Ibid. Art.10.1(e).

1292 Ibid. Art.3.3.

riods for fishing, minimum size of fish, fishing gear or technology and particular sub-regions or regions.[1293]

c)   *Subsidiary bodies of the Commission*

Four subsidiary bodies have been established in order to support the work of the Commission. They provide the Commission with scientific, technical and financial information. The most important bodies are the Scientific Committee and the Technical and Compliance Committee.[1294] The Commission has to consider the reports and recommendations of both committees on matters within their areas of expertise.[1295] Another important body is the Northern Committee. Decisions of the Commission with respect to stocks and species mainly located north of 20° north have to be based on the recommendations of this sub-committee.[1296] The fourth body, the Finance and Administration Committee, which meets along with the Commission, is proposing the budget but it is not directly involved in management decisions.[1297] This section will analyze the three subsidiary bodies that are crucial for conservation and management in order to show how they interact with both the Commission and other organizations.

aa)   Scientific Committee

Scientific information is crucial for the development of appropriate conservation and management measures, and that is why the Scientific Committee meeting is usually held in August, ahead of the annual Commission meeting. The objective of the Scientific Committee is "to ensure that the Commission obtains for its consideration the best scientific information available."[1298] The principal functions of the Committee are to:

- recommend research plans to the Commission;[1299]

- review the assessments, analyses, other work and recommendations prepared for the Commission by scientific experts;[1300]

1293   Ibid. Art.10.2.
1294   Ibid. Art.11.1.
1295   Ibid. Art.10.5.
1296   Ibid. Art.11.7.
1297   Ibid. Art.11.6.
1298   Ibid. Art 12.1.
1299   Ibid. Art.12.2(a).
1300   Ibid. Art.12.2(b).

- encourage and promote cooperation in scientific research;[1301]

- report its findings or conclusions on the status of target stocks or non-target or associated or dependent species in the Convention Area;[1302] and

- make reports and recommendations to the Commission as directed, or as it sees fit.[1303]

The Commission is entitled to engage the services of external scientific experts from regional organizations and other fisheries management, technical or scientific organizations in order to get the best information and advice.[1304] In practice, scientific data is primarily provided by the Secretariat of the Pacific Community (SPC) as the main holder of scientific information for the stocks south of 20° north, and the International Scientific Committee to study the tuna and tuna-like species of the North Pacific Ocean (ISC) for the stocks which mostly occur in the area north of 20° north.[1305] In order to improve and exchange scientific knowledge, the Convention requires that representatives of the SPC be invited to all meetings of the WCPFC.[1306] No explicit reference is made to the ISC but it is stated that other organizations or individuals with scientific expertise in matters related to the work of the Commission can be invited when necessary.[1307] The WCPFC has established memorandums of understanding with both organizations,[1308] and it also has a Data Exchange Agreement with the SPC.[1309] The provisions in the memorandum between WCPFC and ISC are less comprehensive than those in the memorandum with the SPC. The ISC has committed itself to provide scientific reports but it is not required to send representatives to the meetings of the Scientific Committee.[1310]

1301 Ibid. Art.12.2(c).

1302 Ibid. Art.12.2(e).

1303 Ibid. Art.12.2(g).

1304 Ibid. Art.13.1.

1305 The member countries are Canada, Chinese Taipei, Japan, the Republic of Korea, Mexico, the People's Republic of China and the United States of America. For further information, please see the section on the Northern Committee, below.

1306 Convention on the Conservation and Management of Highly Migratory Fish Stocks in the Western and Central Pacific Ocean, Art.12(4).

1307 Ibid. Art.12(4).

1308 WCPFC SPC-OFP Revised Memorandum of Understanding WCPFC ISC Memorandum of Understanding, 2007.

1309 Data exchange agreement between the Western and Central Pacific Fisheries Commission (WCPFC) and the Secretariat of the Pacific Community (SPC).

1310 WCPFC ISC Memorandum of Understanding.

The legal provisions are reflected by the practical performance of the two principal science providers. During the Scientific Committee meeting in 2011 for example, the representatives of the SPC played a crucial role.[1311] They provided comprehensive reports for all species that occurred in their area of competence, and during the discussions their information was essential in helping the delegates to decide on future management strategies. In comparison the information provided by the ISC was less comprehensive. Especially the stock assessment reports presented by delegates in their capacity as participants of the ISC Working Groups were less detailed and transparent than those of their SPC counterparts. Other science providers for the Scientific Committee Meeting in 2011 were individual CCMs, independent consultants and a representative of the IATTC. Their research, which dealt with both target and non-target species, had been conducted in agreement with the Commission.

The archipelagic waters of Indonesia and the Philippines as well as the EEZ of Vietnam which is located in the disputed South China Sea are not part of the Convention area.[1312] The three countries are therefore not required to provide data for these areas.[1313] However, comprehensive data of the whole Convention area is necessary in order to reduce uncertainty in the stock assessments conducted by the SPC. For that purpose, the West Pacific East Asia Oceanic Fisheries Management Project Steering Committee has been established in 2009.[1314] The Committee assists with effective implementation of the West Pacific East Asia Project which is addressing gaps in the data for the WCPO tuna fisheries.[1315] In the course of this project, Indonesia, Philippines and Vietnam are par-

---

1311 Observation made by the author during the Seventh Regular Session of the Scientific Committee.

1312 Henriksen, et al., 2006, p.172.

1313 Convention on the Conservation and Management of Highly Migratory Fish Stocks in the Western and Central Pacific Ocean, Art.23(a); Agreement for the Implementation of the Provisions of the United Nations Convention on the Law of the Sea of 10 December 1982 relating to the Conservation and Management of Straddling Fish Stocks and Highly Migratory Fish Stocks, Annex I,

1314 WCPFC, *Report of the third session of the WPEA OFP Project Steering Committee*, (2011).

1315 WCPFC, *West Pacific East Asia Oceanic Fisheries Management Project Steering Committee*, Summary Report (Port Vila: WCPFC, 2009). Art.1.1.

ticipating voluntarily in order to improve the quality of data collection and re-porting. The project is funded by several international and national donors.[1316]

The Secretariat of the WCPFC does not conduct scientific research itself, but it processes the data provided by the CCMs. Before the Scientific Committee meetings, CCMs must provide a so-called 'Part One Report' to the Secretari-at.[1317] The data provision is based on the on the obligation to "provide annually to the Commission statistical, biological and other data and information in ac-cordance with Annex I of the [Fish Stocks] Agreement and, in addition, such data and information as the Commission may require."[1318] The rules for data provision are laid down in a so called Regulation Paper.[1319] According to these rules, the CCMs have to provide estimates of annual catches, number of vessels active, operational catch and effort data, aggregated catch and effort data, as well as size composition data.[1320] Most of the countries do provide the data, but often late, which makes it difficult for the Secretariat to process in time for the Scientific Committee meeting. In addition, there are still uncertainties in the data, particularly with regard to operational catch and effort data. During the meeting in 2011, it was stated by some CCMs that more operational catch data is still needed in order to conduct more reliable stock assessments.[1321]

In general, political interests should not be allowed to influence the outcomes of the Scientific Committee meetings. However, in recent years, there has been increasing political influence on scientific issues.[1322] One example is on the stock

---

1316   The Global Environment Facility (GEF), Indonesia, Philippines and Vietnam (in-kind and cash contributions), the Australian International Development Assistance Agency (AusAID), the US National Marine Fisheries Service, the Government of Japan (though the WCPFC Japan Trust Fund), the Pacific Islands Forum Fisheries Agency (FFA), the Netherlands, and the WCPFC.

1317   The term ,Part One Report' is not legally defined but it is used by the Commission to distinguish between the scientific data which has to be provided by the CCMs prior to the Scientific Committee Meeting and the technical and compliance data which has to be provided prior to the Technical and Compliance Committee Meeting in the 'Part Two Report'.

1318   Convention on the Conservation and Management of Highly Migratory Fish Stocks in the Western and Central Pacific Ocean, Art.23.2(a).

1319   WCPFC, *Scientific Data to be Provided to the Commission - refined and adopted at the Fourth Regular Session of the Commission, Tumon, Guam, USA, 2-7 December,* (2007).

1320   Ibid. Paras.1-5.

1321   WCPFC, *Summary Report - Seventh Regular Session of the Scientific Committee,* (2011), pp.66 and 104.

1322   Observation made by the author during the Seventh Regular Session of the Scientific Committee.

status of bigeye tuna. Depending on the parameters used for the stock assessment, the degree of estimated overfishing varies. During the meeting in 2011, the CCMs argued for and against the use of certain parameters according to their own political interests rather than due to wider sustainability criteria.

bb)   Technical and Compliance Committee

The sustainable management of tuna and tuna-like species in the WCPO is only possible when all CCMs comply with their duties. Most of these duties are laid down in the CMMs. The levels of compliance with these measures is monitored and evaluated by the Technical and Compliance Committee. According to the Convention, the main functions of this Committee are to:

- provide the Commission with information, technical advice and recommendations relating to the implementation of, and compliance with, conservation and management measures;[1323]

- monitor and review compliance with conservation and management measures adopted by the Commission and make such recommendations to the Commission as may be necessary;[1324] and

- review the implementation of cooperative measures for monitoring, control, surveillance and enforcement adopted by the Commission and, again, make such recommendations to the Commission as may be necessary.[1325]

Before each Technical and Compliance Committee meeting, CCMs have to provide the so-called 'Part Two Report' to the Secretariat.[1326] The general requirement to provide technical and compliance data derives from the obligations of Article 23.2(a). The current data provision is based on a template for standardized data provision that has been developed by the Secretariat in 2012.[1327] The template requires technical and compliance data on: implementation of the conservation and management measures; monitoring and inspection activities;

1323   Convention on the Conservation and Management of Highly Migratory Fish Stocks in the Western and Central Pacific Ocean, Art.14.1(a).

1324   Ibid. Art.14.1(b).

1325   Ibid. Art.14.1(c).

1326   The term 'Part Two Report' is not legally defined but it is used by the Commission to distinguish between the technical and compliance data which has to be provided by the CCMs prior to the Technical and Compliance Committee Meeting and the scientific data which has to be provided prior to the Scientific Committee Meeting in the 'Part One Report'.

1327   WCPFC, *Revised template for the 2012 Annual Report (Part 2)*, (2012).

surveillance activities; investigation and prosecution activities; further monitoring surveillance and control measures as well as high seas transshipment activities.[1328] All reporting requirements are related to the respective paragraphs in the active CMMs.

During the annual Technical and Compliance Committee meeting in October - the last meeting before the annual meeting of the Commission - issues of non-compliance are discussed and then forwarded to the Commission. Due to the fact that the compliance issues deal with the responsibilities of flag states, the discussions at these meetings are already much more political than those of the Scientific Committee, as the CCMs attempt to prevent any possible compliance issues relating to vessels flying their flag from being forwarded to the Commission.[1329] Another important role of the Technical and Compliance Committee is to discuss the development of new CMMs and to prepare them in advance of the annual meeting of the Commission. Drafts for these measures are usually prepared prior to the Technical and Compliance Committee meeting by the members, in consultation with the Secretariat.

cc)   Northern Committee

As described above, due to the opposition of some delegations which pursued national interests, it was impossible to determine the northern borders of the Convention area.[1330] Japan in particular was against an extension of the Convention area beyond 20° north. This dispute called into question the adoption of the whole Convention. The establishment of the Northern Committee which usually meets in September, between the Scientific Committee meeting and the Technical and Compliance Committee meeting was, therefore, a compromise to ensure the Convention could be adopted.[1331]

1328   Ibid. Paras.2.1-2.6.

1329   Observation made by the author during the Seventh Regular Session of the Technical and Compliance Committee.

1330   Cordonnery, 2002, p.9.

1331   Henriksen, et al., 2006, p.173.

Figure 4.3   Area of competence of the Northern Committee

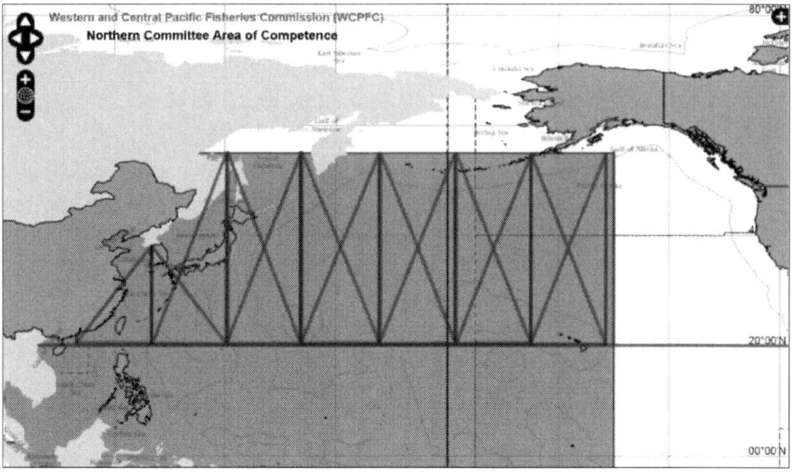

Source: FAO

The Northern Committee is composed by those member countries located in the area north of 20° north, together with other countries that fish in this area.[1332] Those are the USA, Canada, Japan, Korea, Taiwan, China and New Zealand. Members of the Commission which are not represented in the Northern Committee can send a representative to participate in the negotiations, but only as an observer.[1333] As a result of these restrictions, the Northern Committee is kept exclusive, and it is sometimes described as an RFMO within the RFMO.[1334] Despite this perception it has to be noted that the Northern Committee is not having independent decision making power and it is only acting as an advisory body to the Commission.[1335] The Convention lays down that the Northern Committee can only make "recommendations on the implementation of such conservation and management measures as may be adopted by the Commission for the area north of the 20° parallel of north latitude and on the formulation of such measures in respect of stocks which occur mostly in this area".[1336]

1332 Convention on the Conservation and Management of Highly Migratory Fish Stocks in the Western and Central Pacific Ocean, Art.11.7, sent.2.

1333 Ibid. Art.11.7.

1334 Observation made by the author during the Seventh Regular Session of the Scientific Committee and the Seventh Regular Session of the Technical and Compliance Committee.

1335 Cordonnery, 2002, p.9.

1336 Convention on the Conservation and Management of Highly Migratory Fish Stocks in the Western and Central Pacific Ocean, Art.11.7, sent.1.

At the time of the establishment of the WCPFC it was assumed that among other stocks this would include 20% of skipjack, 14% of bigeye, and 12% of yellowfin tuna.[1337] Figure 4.3 shows the area of competence of the Northern Committee. The map also includes the territorial sea and archipelagic waters which are not included in the Convention area. The illustrated northern limit is also only indicative as it is not defined in the Convention text.

The Committee has to adopt all recommendations to the Commission by consensus[1338] and the recommendations must be consistent with the general policies and measures adopted by the Commission.[1339] A weak point is uncertainties with regard to the cooperation between the Northern Committee and the ISC. In general, the legal basis for the ISC is rather poor. It is only based on a joint press release from the governments of the United States of America and of Japan.[1340] The main purposes of the ISC, according to that press release, are:

- to enhance scientific research and cooperation for conservation and rational utilization of the species of tuna and tuna-like fishes which inhabit the North Pacific Ocean during part or all of their life cycle; and

- to establish the scientific groundwork, if at some point in the future, it is decided to create a multilateral regime for the conservation and rational utilization of these species in this region.

It is not clear if the 'multilateral regime' referred to is actually the WCPFC or a new and separate tuna RFMO for the region north of 20° north. It will be interesting to see if there is an approach for such a separate RFMO in the future.

In the Convention there are no provisions that deal with the cooperation between the Northern Committee and the ISC. A Memorandum of Understanding (MOU) between WCPFC and ISC represents the only existing legal framework for cooperation between the Northern Committee and the ISC. It lays down that the Northern Committee can request from the ISC scientific information and advice regarding those stocks occurring mostly north of the 20° north.[1341] The ISC is entitled to provide the requested scientific information and advice to the Northern Committee but also to the Commission and the Scientific Committee.[1342] The memorandum reaffirms the purposes stated in the press release es-

---

1337 Cordonnery, 2002, p.9.

1338 Convention on the Conservation and Management of Highly Migratory Fish Stocks in the Western and Central Pacific Ocean, Art.11.7, sent.5.

1339 Ibid. Art.11.7, sent.7.

1340 http://isc.ac.affrc.go.jp/about_isc/press_release.html.

1341 WCPFC ISC Memorandum of Understanding Part I, Para.1.

1342 Ibid. Part I, Para.2.

tablishing the ISC and recognizes the competencies of the WCPFC in adopting measures for highly migratory fish stocks in the Convention Area. It also recognizes the role of the Northern Committee in making recommendations about the implementation of conservation and management measures for the area north of the 20° north and about the formulation of such measures in respect of stocks which occur mostly in this area. The memorandum also identifies the need to clarify the relationship between the ISC and the Northern Committee and the Scientific Committee. [1343] A remaining concern is the unclear competence of the Scientific Committee with regard to the so called 'northern stocks' (northern Pacific bluefin, northern albacore and the northern stock of swordfish).[1344]

5. Decision making

During the MHLC process the provisions on decision making were described as the 'make or break'.[1345] The Fish Stocks Agreement only had provided limited guidance by stating that states shall "agree on decision-making procedures which facilitate the adoption of conservation and management measures in a timely and effective manner."[1346] Due to the resulting uncertainty, decision making became an issue of 'hard bargaining'.[1347] It was discussed at length and in detail, especially by FFA member countries and DWFNs.[1348] The FFA members supported the majority vote, knowing that they would be in the majority in most negotiations.[1349] In contrast, the DWFNs were concerned about being outnumbered by FFA members, and so much less confident about the benefits of majority voting.[1350]

Despite these challenges the outcome of the MHLC process was comprehensive decision-making procedures. Consensus voting became the general decision-

---

1343   Ibid. Part I, Options for discussion. Para.1.

1344   Ibid. Part I, Options for discussion. Para.2.

1345   Sydnes, 2001, p.798.

1346   Agreement for the Implementation of the Provisions of the United Nations Convention on the Law of the Sea of 10 December 1982 relating to the Conservation and Management of Straddling Fish Stocks and Highly Migratory Fish Stocks

1347   Sydnes, 2001, p.798.

1348   Henriksen, et al., 2006, pp.182-183.

1349   Sydnes, 2001, p.799.

1350   Cordonnery, 2002, p.5.

making rule and it is mandatory for the most important decisions of the Commission.[1351] Those decisions include:

- the adoption and amendment of the rules of procedure for the conduct of meetings of the Commission and its subsidiary bodies;[1352]

- decisions with regard to the allocation of the total allowable catch or the total level of fishing effort, including decisions relating to the exclusion of vessel types;[1353]

- the decisions of the Northern Committee to the Commission;[1354]

- the adoption and amendment of financial regulations for the administration of the Commission and for the exercise of its functions;[1355]

- the adoption of the budget and its contribution scheme;[1356]

- the invitation of new members;[1357] and

- amendments to the Convention.[1358]

For other decisions the Commission can also opt to vote by majority.[1359] It was decided to use a voting system that involves chambered voting. This system excludes those decisions that require consensus, and makes a further distinction between questions of procedure and questions of substance.[1360] Decisions about questions of procedure have to be made by majority vote. Decisions about questions of substance require a three-quarters majority, provided that such a majority is made up of a three-quarters majority of both members and non-members of the FFA. It is not possible for a proposal to be defeated by two votes or fewer in either chamber. So far, majority voting has not yet been used by the Commission.[1361] When in practice, even after extensive debates, one or more parties

---

1351 Convention on the Conservation and Management of Highly Migratory Fish Stocks in the Western and Central Pacific Ocean, Art.20.1.

1352 Ibid. Art.9.8.

1353 Ibid. Art.10.2.

1354 Ibid. Art.11.7.

1355 Ibid. Art.17.2.

1356 Ibid. Arts.18.1 and 18.2.

1357 Ibid. Art.35.2.

1358 Ibid. Art.40.2.

1359 Ibid. Art.20.2, sent.1.

1360 Ibid. Art.20.2, sent.2.

1361 Observation made by the author during the Seventh Regular Session of the Scientific Committee and the Seventh Regular Session of the Technical and Compliance Committee.

could not agree on an issue, it was either decided not to adopt the measure or to incorporate certain exceptions.

Another relatively progressive element addresses opt-out behavior.[1362] In order to prevent parties from opting-out, those parties which voted against a decision or which were absent during the decision-making process can, within 30 days of the adoption of the decision, seek a review of the decision by a review panel. This must be on the grounds that the decision was inconsistent with the provisions of the WCPFC Convention, the Fish Stocks Agreement or UNCLOS,[1363] or that it unjustifiably discriminates against the respective member.[1364] If the review panel does not identify any problems, the decision becomes binding.[1365] In cases where the review panel recommends modifying, amending or revoking the decision, the Commission is required to modify or amend its decision at the next annual meeting.[1366]

6.   Settlement of disputes

The provisions for the peaceful settlement of disputes in the Convention of the WCPFC are identical to those of the Fish Stocks Agreement. The provisions of Part VIII of the Fish Stocks Agreement apply, mutatis mutandis, to any dispute between members of the Commission, whether or not they are also Parties to the Fish Stocks Agreement.[1367] States have to settle their disputes by negotiation, inquiry, mediation, conciliation, arbitration, judicial settlement, resort to regional agencies or arrangements, or any other peaceful means of their own choice.[1368] To date, no formal dispute settlement procedure has been needed in order to solve disputes between members of the WCPFC.[1369]

1362 Convention on the Conservation and Management of Highly Migratory Fish Stocks in the Western and Central Pacific Ocean, Art.20.6.

1363 Ibid. Art.20.6(a).

1364 Ibid. Art.20.6(b).

1365 Ibid. Art.20.8.

1366 Ibid. Art.20.9.

1367 Ibid. Art.31.

1368 Agreement for the Implementation of the Provisions of the United Nations Convention on the Law of the Sea of 10 December 1982 relating to the Conservation and Management of Straddling Fish Stocks and Highly Migratory Fish Stocks, Art.27.

1369 Observation made by the author during the Seventh Regular Session of the Technical and Compliance Committee.

7. Cooperating non-members

As mentioned earlier the WCPFC has 11 so called cooperating non-members (CNMs).[1370] The WCPFC is the tuna RFMO with by far the most CNMs.[1371] The CNMs are non-parties which "cooperate fully in the implementation of conservation and management measures adopted by the Commission with a view to ensuring that such measures are applied to all fishing activities in the Convention Area."[1372] According to the Fish Stocks Agreement CNM are entitled to participate in a fishery if they agree to apply the conservation and management measures established by the respective RFMO.[1373] This provision of the Fish Stocks Agreement which applies only to its parties is intended to facilitate the exclusion of 'free-riding' parties who have not taken responsibility for conservation or management.

Like the Fish Stocks Agreement the WCPFC Convention does, due to the *pacta tertiis rule*, not entitle the Commission to exclude non-complying states from fishing on the high seas if they are not a party to the Fish Stocks Agreement or not a member to the WCPFC. However the members are required to take the necessary measures to prevent any activities of vessels of non-parties which undermine the effectiveness of conservation and management measures.[1374] The Commission is required to "draw the attention of any State which is not a Party to this Convention to any activity undertaken by its nationals or vessels flying its flag which, in the opinion of the Commission, affects the implementation of the objective of this Convention."[1375]

In 2004, the first CMM dealing with CNMs was adopted. This measure basically states that non-parties with an interest in fishing in the Convention Area can request that the Commission grant them the status of a CNM.[1376] Applicants are required to provide their reasons for seeking CNM status[1377] and to comply

---

1370 Belize, the Democratic People's Republic of Korea, Ecuador, El Salvador, Indonesia, Mexico, Senegal, St Kitts and Nevis, Panama, Thailand and Vietnam.

1371 IATTC: 2; ICCAT: 5; CCSBT: 2; IOTC: 2.

1372 Convention on the Conservation and Management of Highly Migratory Fish Stocks in the Western and Central Pacific Ocean Art.32.4

1373 Agreement for the Implementation of the Provisions of the United Nations Convention on the Law of the Sea of 10 December 1982 relating to the Conservation and Management of Straddling Fish Stocks and Highly Migratory Fish Stocks, Art.8.3.

1374 Convention on the Conservation and Management of Highly Migratory Fish Stocks in the Western and Central Pacific Ocean Art.32.1.

1375 Ibid. Art.32.2.

1376 WCPFC, *Conservation and Management Measure - Cooperating Non-Members*, (2004), Para.1.

1377 Ibid. Para.2.

with all the requirements of the Convention.[1378] The Commission then has to assess their status annually, to ensure it can be renewed, as long as they are still in compliance with the objectives and requirements of the Convention.[1379]

The second measure for CNMs was adopted in 2009, and replaced the first one. Its provisions are in line with those of the previous measure, but they are more detailed and more specific. Under it, not only can vessels intending to fish in the Convention area apply for the status of a CNM, but those which are already fishing there can as well.[1380] This enables states with active vessels to become CNMs. A new requirement is the explicit commitment to accept high seas boarding and inspections.[1381] Other new provisions address the competences of the Technical and Compliance Committee and the Commission. Although the Commission still has the final say about whether a non-party is accorded CNM status, the new measure transfers more responsibility to the Technical and Compliance Committee. According to the new measure, the Technical and Compliance Committee must assess applications for CNM status and give recommendations and technical advice to the Commission.[1382] When a CNM fails to comply with any of the conservation and management measures, the Commission is required to take appropriate action like revoking of CNM status and/or sanctions and penalties against such CNMs.[1383]

8.   Cooperation with other tuna RFMOs

Tuna and tuna-like species cannot always be clearly assigned to a single Tuna RFMO. Some stocks migrate between the areas of two or more RFMOs, and there are overlapping areas where two RFMOs claim jurisdiction. For sustainable management in the WCPO, it is therefore crucial that the WCPFC cooperates with its neighboring tuna RFMOs. Accordingly, the Convention required the Commission to make suitable arrangements for consultation, cooperation and collaboration with other relevant intergovernmental organizations that pursued related objectives.[1384] The tuna RFMOs which are explicitly named in the Convention are the Inter American Tropical Tuna Commission (IATTC), the

---

1378   Ibid. Para.3.

1379   Ibid. Para.4.

1380   WCPFC, *Conservation and Management Measure - Cooperating Non-Members*, (2009).

1381   Ibid. Para.2(c).

1382   Ibid. Para.3.

1383   Ibid. Para.15.

1384   Convention on the Conservation and Management of Highly Migratory Fish Stocks in the Western and Central Pacific Ocean, Art.22.1.

Commission for the Conservation of Southern Bluefin Tuna (CCSBT) and the Indian Ocean Tuna Commission (IOTC).[1385]

The Inter American Tropical Tuna Commission (IATTC) has the longest shared border with the WCPFC. Most of the stocks of the principal market tuna species migrate at least partially between these two Convention areas, and so cooperation with regard to the management of these species was both important and necessary.[1386] The Antigua Convention, like the WCPFC Convention, contains provisions that stipulate cooperation with other RFMOs.[1387] To date, two memorandums of understanding have been signed. In a general memorandum of understanding, both organizations defined their areas of competence and the manner of cooperation.[1388] The second memorandum determined conditions for the exchange of data.[1389] A matter of concern is the overlapping area. Several options for the management of this area were discussed at the eighth annual meeting of the Commission.[1390] The options included the management of the area by a single organization, management by gear type, management as a special management area, the application of measures from both Commissions, or working towards a harmonization of measures.[1391] The Executive Director of the WCPFC was instructed to communicate with the Executive Director of the IATTC on remaining issues.[1392]

In the south, the WCPFC Convention area overlaps with the migratory range of the southern bluefin tuna. The Commission for the Conservation of Southern Bluefin Tuna (CCSBT) claims competence over southern bluefin tuna throughout its migratory range,[1393] but the WCPFC has the competence to regulate the species within its area of jurisdiction[1394]. A memorandum of understanding was therefore established where both organizations agreed that the CCSBT would

---

1385 Ibid. Art.22.2.

1386 WCPFC, *Tunas and billfishes in the Eastern Pacific Ocean in 2010*, (2011).

1387 Convention for the Strengthening of the Inter-American Tropical Tuna Commission established by the 1949 Convention between the United States of America and the Republic of Costa Rica "Antigua Convention", Art.XXIV.

1388 Memorandum of Understanding between WCPFC and IATTC, 2009.

1389 WCPFC IATTC Memorandum of Cooperation on the Exchange and Release of Data, 2009.

1390 WCPFC, *Summary Report - Eighth Regular Session of the Commission*, (2012).

1391 Ibid. p.50.

1392 Ibid. p.51.

1393 Convention for the Conservation of Southern Bluefin Tuna, Art.12.

1394 Convention on the Conservation and Management of Highly Migratory Fish Stocks in the Western and Central Pacific Ocean, Art.22.

be the competent body for the development and implementation of conservation and management measures for southern bluefin tuna.[1395] It was further agreed that both organizations have to provide detailed catch data, and the CCSBT also has to provide stock assessment data.[1396] The memorandum also requires exchange of all data, scientific information and information on fisheries management on an annual basis, cooperation in investigations and studies of mutual interest, permanent reciprocal observer status at meetings, and methods of recognizing mutual conservation and management measures.[1397]

In the western part of the Convention area, there is another overlap, this time with the Indian Ocean Tuna Commission (IOTC). Some stocks and species migrate through the geographical areas of both organizations. The IOTC Agreement requires cooperation, particularly with RFMOs that deal with tuna and tuna-like species.[1398] In a memorandum of understanding, both organizations agreed to cooperate in the exchange of data and information, in the collaboration on research efforts relating to stocks and species of mutual interest, and in conservation and management measures for stocks and species of mutual interest.[1399] The cooperation has to include the reciprocal participation of observers in relevant meetings, the sharing of information about stocks and species of mutual interest, the promotion of the harmonization and compatibility of conservation and management measures and the exchange of other relevant information.[1400]

9.   Allocation of catch and effort

One of the functions of the Commission stated in Article 10 is to "determine the total allowable catch or total level of fishing effort within the Convention Area for such highly migratory fish stocks as the Commission may decide."[1401] For that purpose the Commission is entitled to adopt measures relating to:

- the quantity of any species or stocks which may be caught;[1402]
- the level of fishing effort;[1403]

1395   Memorandum of Understanding between WCPFC and CCSBT, 2006, Para.a.

1396   Ibid. Paras.b and c.

1397   Ibid. Para.d.

1398   Agreement for the Establishment of the Indian Ocean Tuna Commission, Art.XV.

1399   Memorandum of Understanding between WCPFC and IOTC, 2009, Para.1.

1400   Ibid. Para.2.1.

1401   Convention on the Conservation and Management of Highly Migratory Fish Stocks in the Western and Central Pacific Ocean, Art.10.1(a).

1402   Ibid. Art.10.2(a).

- limitations of fishing capacity, including measures relating to fishing vessel numbers, types and sizes;[1404]

- the areas and periods in which fishing may occur;[1405]

- the size of fish of any species which may be taken;[1406]

- the fishing gear and technology which may be used;[1407] and

- particular subregions or regions.[1408]

No preference given for input or output control. For the most important target species bigeye and yellowfin tuna there are effort limits for purse seine fisheries[1409] and catch limits for longline fisheries.[1410] The respective provisions are accompanied by technical measures like regulations for Fish Aggregating Device (FAD) or area closures.[1411]

A crucial question is whether the Commission is entitled to determine the total allowable catch (TAC) or total level of fishing effort (TAE) for the whole Convention area, or only for the high seas. In principle, members have the duty to cooperate for the purpose of achieving compatible measures in respect of highly migratory stocks.[1412] However, Article 10.1 states at the beginning that the sovereign rights of coastal states for the purpose of exploring and exploiting, conserving and managing highly migratory fish stocks within areas of national jurisdiction should not be prejudiced.[1413] According to that the sovereign rights for the EEZ remain at the coastal states and the Commission "has no authority to allocate rights to fish within EEZs in any manner that undermines the sovereign rights of coastal states."[1414]

1403   Ibid. Art.10.2(b).

1404   Ibid. Art.10.2(c).

1405   Ibid. Art.10.2(d).

1406   Ibid. Art.10.2(e).

1407   Ibid. Art.10.2(f).

1408   Ibid. Art.10.2(g).

1409   WCPFC, *Conservation and Magement Measures for Bigeye and Yellowfin Tuna in the Western and Central Pacific Ocean*, (2008), Para.10.

1410   Ibid. Paras.31-38.

1411   Ibid. Paras.17.b, and 19; WCPFC, *Conservation and Management Measure for the Eastern High-Seas Pocket Special Management Area*, (2010).

1412   Convention on the Conservation and Management of Highly Migratory Fish Stocks in the Western and Central Pacific Ocean Art.8.1.

1413   Ibid. Art.10.1.

1414   Hanich, 2009, p.226.

Coastal states are only required to "ensure that the measures adopted and applied by it to highly migratory fish stocks within areas under [their] national jurisdiction do not undermine the effectiveness of measures adopted by the Commission."[1415] This requirement lacks the necessary precision because it is not clear what is meant by 'undermine the effectiveness'. As a result, after the establishment of the WCPFC, two possible allocation models for the WCPO fisheries have been identified.[1416] In one model, allocation is based on coastal state or national allocation where, in the exercise of their sovereign rights, coastal states determine how much fish can be taken from their EEZ. In this model, the Commission takes the role of coordinator to ensure global limits are not exceeded. A second option is that the Commission determines the TAC or TAE and allocates it to each member of the Commission and each co-operating non-contracting party.

Currently, it seems especially due to the role of the Parties to the Nauru Agreement (PNA), that the first model is applied. Since the establishment of the WCPFC, the PNA as the most powerful sub-group have followed within their EEZs a progressive approach on allocation of effort. With the Vessel Day Scheme (VDS) they control the whole EEZ fishing effort through the allocation of fishing days. The VDS shows that the PNA are aware of their rights within the EEZs, and not willing to transfer more powers to the Commission.[1417] The Commission basically coordinates and ensures that global limits are not exceeded.

The rights of the coastal states are further protected by the decision making procedure. All decisions with regard to the allocation of the total allowable catch or the total level of fishing effort require consensus.[1418] In summary, it can be therefore concluded that the Commission is entitled to allocate catch and effort on the high seas, while the coastal states have the right to allocate catch and effort in their EEZs, as long as in doing so they do not 'undermine' the effectiveness of the measures adopted by the Commission.

The Commission of the WCPFC is required by its Convention to "develop, where necessary, criteria for the allocation of the total allowable catch or the total level of fishing effort for highly migratory fish stocks in the Convention Area."[1419] The Convention also provides guidelines for the development of the-

---

1415 Convention on the Conservation and Management of Highly Migratory Fish Stocks in the Western and Central Pacific Ocean Art.8.3.

1416 Aqorau, 2001, pp.393-394.

1417 Aqorau, 2009, pp.559-563.

1418 Convention on the Conservation and Management of Highly Migratory Fish Stocks in the Western and Central Pacific Ocean Art.10.2.

1419 Ibid. Art.10.1(g).

se criteria.[1420] According to Article 10.3, the Commission has to take into account:

- the status of the stocks and the existing level of fishing effort in the fishery;
- the respective interests, past and present fishing patterns and fishing practices of participants in the fishery;
- the extent of the catch being utilized for domestic consumption;
- the historic catch in an area;
- the needs of small island developing states, territories and possessions in the Convention Area which depend overwhelmingly on the exploitation of marine living resources for their economies, food supplies and livelihoods;
- the respective contributions of participants to the conservation and management of the stocks, including their provision of accurate data and their contribution to the conduct of scientific research in the Convention Area;
- the record of compliance by the participants with conservation and management measures;
- the needs of coastal communities which are mainly dependent on fishing the stocks;
- the special circumstances of a state which is surrounded by the EEZs of other states and has a limited EEZ of its own;
- the geographical situation of small island developing states which are made up of non-contiguous groups of islands having distinct economic and cultural identities of their own, but which are separated by areas of high seas;
- the fishing interests and aspirations of coastal states, particularly small island developing states, territories and possessions, in whose areas of national jurisdiction the stocks also occur.[1421]

In spite of these detailed provisions, which are even more comprehensive than the respective provisions in Article 11 of the Fish Stocks Agreement, the WCPFC has not yet developed any formal allocation criteria.[1422] An implicit

---

1420 For further information on the criteria see: Cordonnery, 2002, p.8.

1421 Convention on the Conservation and Management of Highly Migratory Fish Stocks in the Western and Central Pacific Ocean, Art.10.3.

1422 Henriksen, et al., 2006, pp.180-181.

form of allocation which is mainly based on historical catches and the interests of small island developing states are the catch and effort limitations for the target species.[1423] An example is the important CMM for bigeye and yellowfin tuna.[1424] The parties have to reduce their effort and catches to the historical 2001-2004 levels.[1425] Exempted from this measure are small developing state members[1426] and participating territories.[1427] This exemption shows where conservation interests of the Commission collide with the development interests of the PICTs.

10. *Precautionary approach*

The provisions for the application of the precautionary approach in the Convention are closely coordinated with those of the Fish Stocks Agreement and have been described as representative of the modern character of the whole Convention.[1428] The principal requirements can be found in Articles 5 and 6 of the Convention. According to Article 6.1, in order to apply the precautionary approach, members must:

- apply the guidelines set out in Annex II of the Fish Stocks Agreement as an integral part of the WCPFC Convention, and determine, on the basis of the best scientific information available, stock-specific reference points and the action to be taken if they are exceeded;[1429]

- take into account, inter alia, uncertainties relating to the size and productivity of the stocks, reference points, stock condition in relation to such reference points, levels and distributions of fishing mortality and the impact of fishing activities on non-target and associated or dependent species, as well as existing and predicted oceanic, environmental and socio-economic conditions;[1430] and

---

1423 Parris & Lee, 2009, p.256.

1424 WCPFC, *Conservation and Magement Measures for Bigeye and Yellowfin Tuna in the Western and Central Pacific Ocean*, (2008).

1425 Ibid. Paras.10, 31, 38.

1426 The term small developing state members and participating territories includes all PICTs. For further information see http://www.un.org/special-rep/ohrlls/sid/list.htm.

1427 WCPFC, *Conservation and Magement Measures for Bigeye and Yellowfin Tuna in the Western and Central Pacific Ocean*, (2008), Para.10.

1428 Cordonnery, 2002, p.6.

1429 Convention on the Conservation and Management of Highly Migratory Fish Stocks in the Western and Central Pacific Ocean, Art.6.1(a).

1430 Ibid. Art.6.1(b).

- develop data collection and research programmes to assess the impact of fishing on non-target and associated or dependent species and their environment, and adopt plans where necessary to ensure the conservation of such species and to protect habitats of special concern.[1431]

The establishment of conservation, or limit reference points, and management, or target reference points according to Annex II of the Fish Stocks Agreement is crucial for a successful implementation of the precautionary approach. The Commission has therefore the possibility to entitle scientific experts "to develop and recommend to the Commission and the Scientific Committee stock-specific reference points for the species of principal interest to the Commission"[1432] and to "assess the status of stocks against the reference points established by the Commission."[1433] Based on the best scientific information available the Commission has to determine stock-specific reference points.[1434]

Despite the progressive provisions of the Convention text, so far, there was no final agreement on the development of reference points. Since 2009 approaches for identification of appropriate reference points are discussed on the basis of a working paper but without success.[1435] No formal reference points have been adopted for any species.[1436] The current practice in the WCPO is that stock assessment outputs are given in relation to the B_{MSY} and F_{MSY}.[1437] Therefore the MSY can be identified as an informal limit reference point but target reference points are so far not apparent at all.[1438]

In current discussions, particularly around the more precautionary target reference points, most of the parties argue that the scientific basis for the definition of specific limits is not yet sufficient.[1439] Though this is partially true, all participants must admit that in this highly political discussion many states are simply not willing to risk economic losses due to lower catch limits. However, the parties are in fact acting against the Convention when they refuse to support the establishment of reference points. The absence of adequate scientific infor-

---

1431   Ibid. Art.6.1(c).

1432   Ibid. Art.13.2(b).

1433   Ibid. Art.13.2(c).

1434   Ibid. Art.6.1(a).

1435   Davies & Basson, 2009.

1436   WWF & TRAFFIC, *WWF & Traffic Statement to WCPFC – 8th Regular Session* (Guam: WWF & TRAFFIC, 2012), p.4.

1437   Campbell, 2010, p.3.

1438   Mooney-Seus & Rosenberg, 2007, pp.132-141.

1439   Observation made by the author during the Seventh Regular Session of the Scientific Committee.

mation is not an allowable reason for postponing or failing to take conservation and management measures.[1440] If the information is uncertain, unreliable or inadequate, the Commission is required to apply more cautious measures. At the moment, due to the importance of the issue in the context of to difficulties to understand all biological and socioeconomic factors, it might be therefore useful for the WCPFC to adopt at least interim target reference points.[1441]

11. Conservation and management of target species

The main target species in the WCPFC Convention area are the principal market tuna species skipjack, yellowfin, albacore, bigeye and Pacific bluefin tuna, and the billfish species striped marlin and swordfish. Since the establishment of the WCPFC, the Commission has adopted conservation and management measures for all of these target species. The conservation and management measures are mainly based on Article 5 of the Convention, which requires the parties "to ensure long-term sustainability of highly migratory fish stocks in the Convention Area and [to] promote the objective of their optimum utilization."[1442] The Commission must use the best scientific evidence available and the measures have to be designed to maintain stocks at, or restore stocks to, levels capable of producing the maximum sustainable yield.[1443] In addition, the precautionary approach has to be applied,[1444] but, as already shown, progress in adopting precautionary measures according to Article 6 has been rather limited.

In the following sections, the conservation and management measures that have been adopted for the target species will be analyzed in order to identify the most critical issues. 'Overfishing' and 'overfished' are two terms which are frequently used. In accordance with the practice of the Scientific Committee the term 'overfishing' is used when current fishing mortality is above the fishing mortality that can produce the Maximum Sustainable Yield (MSY). The term 'overfished' is used when the current biomass is below the biomass that can produce MSY. The management is described as 'sustainable' when the current fishing mortality is below the fishing mortality that can produce MSY and when the current biomass is above the biomass that can produce MSY.

1440 Convention on the Conservation and Management of Highly Migratory Fish Stocks in the Western and Central Pacific Ocean, Art.6.2.

1441 WWF, 2012.

1442 Convention on the Conservation and Management of Highly Migratory Fish Stocks in the Western and Central Pacific Ocean, Art.5(a).

1443 Ibid. Art.5(b).

1444 Ibid. Art.5(c).

a)   Bigeye, yellowfin and skipjack tuna

Bigeye, yellowfin and skipjack tuna are the main target species in the WCPO tuna fisheries. In 2010 they represented with 2,285,324 mt approximately 95 percent of the overall tuna catch in the WCPO.[1445] Bigeye tuna is slightly more often caught by longline than by purse seine, yellowfin tuna is mainly caught by purse seine but also by longline while skipjack tuna is mainly caught by purse seine and to a minor part by pole and line.[1446] To date, three measures have been adopted for the conservation and management of bigeye and yellowfin tuna.[1447] A fourth measure has been adopted in 2012 including for the first time explicit provisions for the conservation and management of skipjack tuna.[1448]

At its first meeting in 2005, the Scientific Committee recognized that overfishing was likely occurring in the bigeye stock and was probably occurring in the yellowfin stock.[1449] Therefore, the principal objective of the first CMM 2005-01 was that the total level of purse seine and longline effort should not be increased beyond the level of 2004, or the average level of 2001-2004.[1450] A need for further action was also identified with regard to temporary closures[1451] and management plans for fish aggregating devices (FADs)[1452] in the purse seine

---

1445   WCPFC, *Summary Report - Seventh Regular Session of the Scientific Committee*, (2011), p.3.

1446   Ibid. p.4.

1447   WCPFC, *Conservation and Management Measures for Bigeye and Yellowfin Tuna in the Western and Central Pacific Ocean*, (2005); WCPFC, *Conservation and Management Measures for Bigeye and Yellowfin Tuna in the Western and Central Pacific Ocean*, (2006); WCPFC, *Conservation and Magement Measures for Bigeye and Yellowfin Tuna in the Western and Central Pacific Ocean*, (2008).

1448   WCPFC, *Conservation and Management Measure for Bigeye, Yellowfin and Skipjack Tuna in the Western and Central Pacific Ocean*, (2012).

1449   WCPFC, *Summary Report - First Regular Session of the Scientific Committee*, (2005), pp.33-34.

1450   WCPFC, *Conservation and Management Measures for Bigeye and Yellowfin Tuna in the Western and Central Pacific Ocean*, (2005), Paras,1, 8,10 and 17.

1451   Ibid. Para.11.

1452   P. Guillotreau, et al., Fishing tuna around Fish Aggregating Devices (FADs) vs free swimming schools: Skipper decision and other determining factors, *Fisheries Research - Vol. 109(2-3)* (2011); D. G. Itano & K. N. Holland, Movement and vulnerability of bigeye (Thunnus obesus) and yellowfin tuna (Thunnus albacares) in relation to FADs and natural aggregation points, *Aquatic Living Resources - Vol. 13(4)* (2000).

fishery targeting skipjack tuna which was seen as responsible for the increased mortality of juvenile bigeye and yellowfin tuna.[1453]

One year later, it was clear that the provisions of the first measure were not going to be enough to maintain the bigeye and yellowfin stocks at levels capable of producing the MSY. In response, the Commission endorsed scientific advice that a further reduction in fishing mortality (25 percent for bigeye and 10 percent for yellowfin) was necessary to keep their biomass at levels that could produce MSY.[1454] Despite this, no such provisions were included in the CMM 2006-01. The reference levels for fishing effort remained the same as in the previous measure, and no new provisions were included at all for longline fishery.[1455] The use of temporary closures as well as the management of FAD use was still only in development.[1456]

The first two measures failed to meet their objectives. In 2008, the Scientific Committee announced that there was still a very high probability of overfishing of bigeye tuna,[1457] and in 2007 the assessment of yellowfin showed that the possibility of overfishing was also relatively high.[1458] This meant that a new measure was needed. The following CMM 2008-01 was the most comprehensive measure for a target species to date.[1459] Due to the complexity of the fishery, it has specific provisions for the different gear types and the different legal zones.

The overall goal of CMM 2008-01 with regard to the purse seine fishery was a reduction of the fishing mortality of bigeye tuna by at least 30 percent over a three year period (2009 to 2011) and no increase in fishing mortality for yellowfin tuna.[1460] All CCMs, with the exception of the small island developing states, were required to ensure that the level of purse seine fishing effort, measured in days fished by their vessels in areas of the high seas, would not exceed 2004 levels or the average from 2001-2004.[1461] Explicit provisions for 2009 required PNA countries to implement the conservation and management meas-

---

1453 WCPFC, *Conservation and Management Measures for Bigeye and Yellowfin Tuna in the Western and Central Pacific Ocean*, (2005) Paras.1, 8,10,and 17.

1454 WCPFC, *Summary Report - Third Regular Session of the Commission*, (2006), p.7.

1455 WCPFC, *Conservation and Management Measures for Bigeye and Yellowfin Tuna in the Western and Central Pacific Ocean*, (2006), Paras.1 and 3.

1456 Ibid. Paras.5-7.

1457 WCPFC, *Summary Report - Fourth Regular Session of the Scientific Committee*, (2008), p.24.

1458 WCPFC, *Summary Report - Third Regular Session of the Scientific Committee*, (2007), p.22.

1459 WCPFC, *Conservation and Magement Measures for Bigeye and Yellowfin Tuna in the Western and Central Pacific Ocean*, (2008).

1460 Ibid. Para.8.

1461 Ibid. Para.10.

ure within their EEZ through domestic processes and legislation, including the Vessel Day Scheme, while non-PNA CCMs were required to implement compatible measures in their EEZs.[1462] Another new provision was the two month closure of purse seine fishing on FADs between 20° north and 20° south.[1463] It applied both within the EEZ and on the high seas and included mandatory observer coverage during the closure. The provisions for 2010 and 2011 were more detailed. The PNA countries were required to implement the Third Agreement Implementing the Nauru Agreement within their EEZs, with provisions for catch retention, FAD closure, closure of high seas areas and monitoring requirements that were compatible with those for the high seas.[1464] The FAD closure was extended to three months and observer coverage remained mandatory for this longer period.[1465] Some DWFNs proposed that, instead of the high seas FAD closure, members should be allowed to simply reduce bigeye catches, where they could demonstrate the capacity to implement appropriate measures.[1466] In practice, these countries were exempted from the requirement if they reduced their catches by at least 30 percent. Moreover, the Commission was instructed to consider the development of a Vessel Day Scheme for the high seas, compatible with one established by the PNA.[1467] The WCPFC is the only tuna RFMO with high seas areas that are surrounded by the EEZs of coastal states. Illegal, unreported and unregulated fishing operations from and within these high seas pockets had undermined the effectiveness of the measures in the past. CMM 2008-01 therefore closed these so-called 'Western High Seas Pockets' to fishing from 2010, and considered an additional closure of the other high seas pockets in the area between 20° north and 20° south.[1468] Other provisions for the management of purse seine fishery included management plans for FADs, research on juvenile tuna catch mitigation, catch retention and monitoring.[1469] Developing skipjack purse seine fisheries between 20° north and 20° south and with less than two percent yellowfin and bigeye by-catch,

---

1462 Ibid. Paras.11 and 12.

1463 Ibid. Paras.11 and 13.

1464 Ibid. Para.17. See also A third arrangement implementing the Nauru Agreement setting forth additional terms and conditions of access to the fisheries zones of the parties.

1465 WCPFC, *Conservation and Magement Measures for Bigeye and Yellowfin Tuna in the Western and Central Pacific Ocean*, (2008), Paras.17(b) and 19.

1466 Ibid. Paras.15 and 20.

1467 Ibid. Para.21.

1468 Ibid. Para.22.

1469 Ibid. Paras.23-29.

which have 100% observer coverage and legitimate development plans were exempted from this measure.[1470]

For longline fishery, CMM 2008-01 required a mandatory catch reduction. All members and participating territories with catches higher than 2,000 metric tonnes (mt) in 2004 had to reduce their total catch of bigeye by 2012 to 70 percent of their catch in 2004 or of the average annual catch from 2001-2004.[1471] Similarly, the yellowfin catch made by longliners was not allowed to increase beyond the average level from 2001-2004.[1472] These provisions did not apply to small island developing states or to members which had a total longline bigeye tuna catch limit of less than 5,000 mt and were landing exclusively fresh fish.[1473] Other exceptions included the maintaining of China's catch limit at 2004 levels and the confirmation that any reduction should not result in catches below 2,000 mt.[1474]

All commercial tuna fisheries, other than purse seine and long line targeting, were further required to ensure that their total capacity would not exceed the average level for the period 2001-2004 or 2004.[1475] Artisanal fisheries and those fisheries catching less than 2,000 mt of bigeye and yellowfin were exempted.

Although it is still too early to say if CMM 2008-01 will be successful in achieving its objectives, there are clear indicators that challenges still remain for the future. In 2011, the Scientific Committee showed that the purse seine effort had increased significantly between 20° north and 20° south,[1476] while longline catches of bigeye tuna had apparently decreased significantly.[1477] However, due to the increases in purse seine effort, the potential reduction of longline catches was not enough to stop overfishing. The Scientific Committee showed that levels of bigeye catch were unlikely to be sustainable in the long term and that overfishing was still occurring.[1478] It was concluded that the stock was at minimum approaching an overfished state and, depending on the level of recruitment, it could already be in an overfished state. Another matter of concern was that, due to catches of small juveniles, the MSY had been reduced to less than half of the levels prior to 1970. Therefore, the Scientific Committee recommend-

---

1470   Ibid. Para.30.

1471   Ibid. Paras.31and 32.

1472   Ibid. Para.33.

1473   Ibid. Paras. 34 and 35.

1474   Ibid. Paras. 36 and 37.

1475   Ibid. Para.39.

1476   WCPFC, *Summary Report - Seventh Regular Session of the Scientific Committee*, (2011), p.35.

1477   Ibid. p.3.

1478   Ibid. p.26.

ed a minimum reduction in fishing mortality of 39 percent of the 2004 level, or a 28 percent reduction of the average level from 2001-2004.[1479] For yellowfin tuna, it was noted that overfishing was not occurring and that the stock was not in an overfished state.[1480] It seems that the existing measures were enough to enable the WCPFC to manage yellowfin sustainably. However, regional differences indicate that exploitation rates in the western equatorial region are comparatively high and should not be increased.[1481] As for bigeye, if the mortality of juvenile yellowfin could be reduced, greater overall yields would be obtained due to an increased MSY. A further major challenge for the WCPFC is the number of exemptions which, in effect, weaken the measure. The general exemption of small island developing states from mandatory catch and effort reduction could become a problem.[1482] Further catch and effort increases by small island developing states have the potential to undermine the efforts of other members to limit the same. The other exemptions provide loopholes and make it difficult to evaluate compliance across all members.

A third conservation and management measure on bigeye and yellowfin tuna, supposed to strengthen the previous conservation and management measures, was scheduled for 2011.[1483] This measure should have included, for the first time, explicit provisions for the management of skipjack tuna. However, because members were not willing to accept additional restrictions on their fleets, it was not possible to develop and adopt a new conservation and management measure. Instead, it was agreed to extend CMM 2008-01 until 2013, and to make some amendments to it.[1484] These amendments could have a rather negative impact on the stocks.[1485] They included a reopening, albeit with conditions attached, of the Western High Seas Pockets for the Philippines, an increased number of total days under the Vessel Day Scheme in the EEZs of the PNA and increased bigeye catch limits for Chinese longline vessels.[1486]

1479   Ibid. p.35.

1480   Ibid. p.38.

1481   Ibid. p.47.

1482   WCPFC, *Conservation and Magement Measures for Bigeye and Yellowfin Tuna in the Western and Central Pacific Ocean*, (2008) Paras.10 and 32.

1483   WCPFC, *Summary Report - Eighth Regular Session of the Commission*, (2012), p.35.

1484   WCPFC, *Conservation and Management Measure for Temporary Extension of CMM 2008-01*, (2012), Para.1.

1485   H. Amanda, et al., *WCPFC8 disappoints on enhanced management of bigeye and yellowfin stocks*, FFA Fisheries Trade News - Volume 5 (Honiara: Forum Fisheries Agency, 2012), pp.1-4.

1486   WCPFC, *Conservation and Management Measure for Temporary Extension of CMM 2008-01*, (2012), Paras.2, 3 and 5.

In 2012 the fourth measure on bigeye and yellofin tuna with explicit provisions for the conservation and management of skipjack tuna could be adopted finally.[1487] The most important innovation of the measure was that formal limit reference points for any species in the WCPO could be established for the first time.[1488] The agreed limit for bigeye, yellowfin and skipjack tuna is the fishing mortality at a level no greater than the fishing mortality that produces the maximum suatainable yield.[1489] The Commission further envisaged the adption of target reference points for the near future.[1490] Another progresive element of the new measure was the explicit catch limits for longline catches of bigeye tuna.[1491] All CCMs catching bigeye with longlines agreed to adhere to specific catch limits.[1492]

b)   *South Pacific albacore tuna*

The South Pacific albacore tuna, mainly caught by longline vessels, is an important target species, particularly for the PICTs which are located in the South Pacific. The total catches which had been relatively stable since the 1960s increased significantly after 1995.[1493] When the WCPFC was established in 2004, catch levels still appeared to be sustainable even with some increases in fishing mortality and yields but there were critical biological uncertainties which required further conservation and management.[1494]

To date, two CMMs have been adopted for the management of South Pacific albacore. The main goal of the first CMM 2005-02 was to limit the increasing number of fishing vessels targeting albacore in the area south of 20° south.[1495] CCMs were required not to increase the number of their fishing vessels to either 2005 levels or to the average level of the period 2000-2004.[1496] No further guidance was provided on how this should be done. As with other measures, it was required that the legitimate rights and obligations of small island developing

1487   WCPFC, *Conservation and Management Measure for Bigeye, Yellowfin and Skipjack Tuna in the Western and Central Pacific Ocean*, (2012).

1488   Ibid. Para.1.

1489   Id. at. Paras. 2-4.

1490   WCPFC, *Conservation and Management Measure for Bigeye, Yellowfin and Skipjack Tuna in the Western and Central Pacific Ocean*, (2012), Para.1.

1491   Id. at. Para. 26.

1492   Id. at. Attachment F.

1493   WCPFC, *Summary Report - First Regular Session of the Scientific Committee*, (2005), p.34.

1494   WCPFC, *Conservation and Management Measure for South Pacific Albacore*, (2005), Preamble.

1495   Ibid. Para.1.

1496   Ibid. Para.1.

state and territory CCMs should not be prejudiced.[1497] These CCMs for which South Pacific albacore is an important component of the domestic tuna fishery within their EEZs "may wish to pursue a responsible level of development of their fisheries".[1498] In 2010, the first measure was replaced by CMM 2010-05, but in fact only one paragraph had been added, requiring the annual reporting of all albacore catches south of 20° south.[1499] Effort limits remained the same as in the previous measure.

In 2011, the Scientific Committee concluded that the stock was being managed sustainably because no overfishing was occurring and the stock was not over-fished.[1500] However, it was stated that any further increases in catch and effort could result in catch rate declines. A particular point of concern was the possible transfer of effort from the Indian Ocean and the North Pacific.[1501] This was mainly related to fishing activities on the high seas. Therefore, especially the FFA members had a strong interest in the revision of CMM 2010-05. Basically, they wanted to limit further increases in catch and effort which could affect the catches within their EEZs.[1502] Due to time constraints during the eighth annual Commission meeting, it was not possible to review the measure, but it became an agreed priority for the next annual meeting.[1503] At the time of writing no further measure has been adopted.

c)   North Pacific albacore tuna

North Pacific albacore tuna occurs in the convention areas of the WCPFC and the IATTC. It is therefore crucial that both organizations cooperate to ensure sustainable management throughout the whole migration range of the species. In 2005, at its first session, the Scientific Committee announced that overall productivity levels of North Pacific albacore were declining and they recommended a modest reduction in fishing mortality in order to maintain the spawning biomass above MSY.[1504]

1497   Ibid. Para.2.

1498   Ibid. Para.2.

1499   WCPFC, *Conservation and Management Measure for South Pacific Albacore*, (2010), Para.4.

1500   WCPFC, *Summary Report - Seventh Regular Session of the Scientific Committee*, (2011), p.65.

1501   Ibid. p.64.

1502   Observation made by the author during the Seventh Regular Session of the Scientific Committee.

1503   WCPFC, *Summary Report - Eighth Regular Session of the Commission*, (2012), pp.48-49.

1504   WCPFC, *Summary Report - First Regular Session of the Scientific Committee*, (2005), p.24.

In the same year, the IATTC adopted a binding resolution for albacore tuna,[1505] and the WCPFC followed suit with a corresponding conservation and management measure.[1506] The main goal of CMM 2005-03 was to ensure that the total level of fishing effort for North Pacific albacore did not increase beyond current levels.[1507] No explicit reference year was determined but in practice the reference period is 2002-2004.[1508] A crucial role was assigned to the Northern Committee, which has to monitor, together with other scientific bodies, the stock status, report back to the Commission and make recommendations.[1509] Further consultation with the IATTC was seen as essential in order to reach agreement on a consistent set of uniform measures.[1510]

The CMM of 2005 is the only measure for north Pacific albacore to date. In 2011, the Scientific Committee concluded that overfishing was not occurring and that the stock was not likely to be in an overfished condition.[1511] A representative of the International Scientific Committee for Tuna and Tuna-like Species in the North Pacific Ocean announced that sustainability was not threatened but there was uncertainty with regard to the estimates of biomass and fishing mortality, and so further research should be conducted in the near future.

d)   Pacific bluefin tuna

Pacific bluefin tuna mainly occurs in the Convention area north of 20° north. The total catch weight of this species is relatively small when compared to skipjack or yellowfin tuna, but it is the most valuable tuna species of the Pacific Ocean. In 2012, a single fish was sold for more than 700,000 $US.[1512] Until 2009, no conservation and management measure had been adopted for Pacific bluefin tuna. In this year, the Scientific Committee announced that spawning stock biomass had increased significantly since 2005, but it also predicted a short-term decline of the biomass if fishing mortality continued at the level of the years

---

1505   IATTC, *Resolution on Northern Albacore Tuna*, (2005).

1506   WCPFC, *Conservation and Management Measure for North Pacific Albacore*, (2005).

1507   Ibid. Para.1.

1508   WCPFC, *Summary Report - Seventh Regular Session of the Scientific Committee*, (2011), p.70.

1509   WCPFC, *Conservation and Management Measure for North Pacific Albacore*, (2005) Para.5.

1510   Ibid. Para.8.

1511   WCPFC, *Summary Report - Seventh Regular Session of the Scientific Committee*, (2011), p.72.

1512   In 2012, one Pacific bluefin tuna of 269 kg was sold for $736,000 at Tsukiji fish market in Tokyo. See also http://www.bbc.co.uk/news/world-asia-pacific-16421231.

running up to 2009.[1513] Unfortunately, it was not explicitly stated whether the stock was healthy and above the level required to sustain the current catches.

The overall management objective of CMM 2009-07, as the first measure, was to ensure that the current level of fishing mortality rate of Pacific bluefin tuna would not be increased.[1514] Therefore the CCMs had to take the necessary measures to ensure that in 2010 the total fishing effort of vessels targeting northern Pacific bluefin tuna was limited to 2002-2004 levels.[1515] Respective measures had to "take into account the need to reduce the effort on juveniles (age 0-3 years) to 2000-2004 levels."[1516] The latter provision addressed the problem of juvenile catches for aquaculture uses, but the wording 'take account' was still weak. CCMs were also required to strengthen the data collection system for better data quality and timeliness of data reporting.[1517] In addition, the Northern Committee had to conduct an annual review of the reports of the CCMs and of the International Scientific Committee for Tuna and Tuna-like Species in the North Pacific Ocean (ISC).[1518] Due to the migration range of Pacific bluefin tuna, it was further stated that the WCPFC should cooperate with the IATTC.[1519]

In 2010, at the tenth meeting of the ISC, an update to the 2009 analysis was presented.[1520] It was once again pointed out that it was crucial to reduce the fishing mortality of all age classes, but particularly of juveniles, to below 2002-2004 levels.[1521] The subsequent CMM 2010-04 basically restated the provisions of the earlier measure with only minor changes regarding conservation and management.[1522] For 2011 and 2012, the total fishing effort had to stay 'below' the 2002-2004 level, while in the first measure, the total fishing effort had not been allowed to be increased 'beyond' the same level. In addition, juveniles up to three years old were now explicitly included by this provision.[1523]

At the meeting of the Scientific Committee in 2011, a representative of the ISC presented information about the Pacific bluefin tuna, but only very briefly.[1524]

1513   WCPFC, *Summary Report - Fifth Regular Session of the Scientific Committee*, (2009), pp.38-40.

1514   WCPFC, *Conservation and Management Measure for Pacific Bluefin Tuna*, (2009), Para.1.

1515   Ibid. Para.2, sent.1.

1516   Ibid. Para.2, sent.2.

1517   Ibid. Para.3.

1518   Ibid. Para.5.

1519   Ibid. Para.6.

1520   ISC, *ISC10 Plenary Report*, (2010), pp.24-26.

1521   Ibid. p.26.

1522   WCPFC, *Conservation and Management Measure for Pacific Bluefin Tuna*, (2010).

1523   Ibid. Para.2.

1524   WCPFC, *Summary Report - Seventh Regular Session of the Scientific Committee*, (2011), p.73.

The Scientific Committee noted that no new stock assessments had been conducted prior to the meeting and that the stock status description and management recommendations from the previous meeting would remain unchanged.[1525] In particular, emphasis was given to the continuing importance of reducing and monitoring the fishing mortality of juvenile bluefin tuna in the 0-3 year age classes.[1526] Overall, the information that was made available on the stock status of Pacific bluefin tuna was rather poor. Due to the economic importance of the species, it would be desirable to have information comparable to that available for bigeye and yellowfin tuna for future meetings of the Scientific Committee.

e) *Swordfish*

Swordfish occurs in the convention area mainly south of 20° south and north of 20° north. Stock assessments for the southern stock are conducted by the Secretariat of the Pacific Community (SPC) and for the northern stock by the International Scientific Committee for Tuna and Tuna-like Species in the North Pacific Ocean (ISC). To date, conservation and management measures have been adopted only for the southern stocks.[1527]

In 2006, at its second meeting, the Scientific Committee concluded that both catch rates and the mean size of swordfish in the south-west Pacific had been declining.[1528] Further declines in biomass were predicted for the following years. An assessment indicated the possibility that the stock was already in an overfished state and also that overfishing was possibly occurring. Accordingly, the Scientific Committee recommended that there should be no further increases in fishing mortality. The resulting CMM 2006-03 required the limiting of the number of vessels targeting swordfish in the area south of 20° south to the number from any one year from 2000-2005 without shifting any excess fishing effort to the area north of 20° north.[1529] A further important element of the CMM was the requirement to cooperate in order to reduce uncertainties with regard to the stock status.[1530]

1525 Ibid. p.73.

1526 Ibid. p.74.

1527 WCPFC, *Conservation and Management Measure for Swordfish in the Southwest Pacific*, (2006); WCPFC, *Conservation and Management of Swordfish*, (2008); WCPFC, *Conservation and Management of Swordfish*, (2009).

1528 WCPFC, *Summary Report - Second Regular Session of the Scientific Committee*, (2006), p.34.

1529 WCPFC, *Conservation and Management Measure for Swordfish in the Southwest Pacific*, (2006), Para.1.

1530 Ibid. Para 3.

In 2008, stock assessment data showed that in the southwest Pacific no overfishing was occurring and that the stock was not in an overfished state.[1531] No stock status could be determined for the south central Pacific, and although there was no evidence for significant changes in fisheries it was noted that catches had increased to levels exceeding those in the south west Pacific. Despite some indications of an increase in stock abundance of swordfish in the south western Pacific region, the Scientific Committee recommended, as a precautionary measure, that there be no increase in either catch or effort.[1532] CMM 2008-05 replaced the earlier measure from 2006 and added that the total catch of swordfish in the area south of 20°S should be limited to the amount caught in any one year during the period 2000-2006.[1533] In addition, CCMs were required to nominate their maximum total catch, and to develop an independent catch verification review.[1534] A third conservation and management measure was adopted one year later.[1535] CMM 2009-03 replaced the earlier measure but did not add any new significant provisions.

During the Scientific Committee Meeting in 2011, it was stated that no new stock assessment had been conducted on the southern stocks,[1536] and so the assessment of 2009 and the corresponding recommendations were still current. The Committee criticized the fact that data provision was not sufficient to undertake a new assessment in 2012. The Spanish longline fleet came under particular criticism for not providing essential information. To date, no conservation and management measure had been adopted for the northern stock. At the Scientific Committee Meeting in 2011, a brief presentation was provided by the ISC.[1537] It was concluded that, according to previous assessments, the stock was being managed sustainably because no overfishing was occurring and the stock was not overfished. For 2013, the ISC was tasked with providing a new assessment of the northern stock.

f)   South Pacific striped marlin

Most catches of Pacific striped marlin are made within subequatorial and subtropical areas. One of the main catch areas is the southwest Pacific. The first and only stock assessment was presented during the second Scientific Committee

---

1531   WCPFC, *Summary Report - Fourth Regular Session of the Scientific Committee*, (2008), p.36.

1532   Ibid. p.38.

1533   WCPFC, *Conservation and Management of Swordfish*, (2008), Para.2.

1534   Ibid. Paras.3 and 5.

1535   WCPFC, *Conservation and Management of Swordfish*, (2009).

1536   WCPFC, *Summary Report - Seventh Regular Session of the Scientific Committee*, (2011), p.66.

1537   Ibid. p.74.

Meeting, in 2006.[1538] In spite of significant uncertainty with regard to important parameters, it was still concluded that the stock was close to overfishing and close to an overfished state.[1539] The Scientific Committee therefore recommended precautionary measures to stop the increase in fishing mortality.[1540] The resulting CMM 2006-04 basically requires the CCMs to limit the number of vessels fishing for south Pacific striped marlin south of 15°S, to the number from any one year from 2000-2004.[1541] CCMs had to provide also the number of the vessels that fished for south Pacific striped marlin from 2000-2004, and they had to nominate the maximum number of vessels permitted.[1542] They were further required to report the catch levels of the vessels that caught striped marlin not only as a target species but also as bycatch. An exception was provided for coastal state CCMs located south of 15° south which had already established a commercial moratorium on the landing of striped marlin caught within waters under their national jurisdiction.[1543]

By the Scientific Committee Meeting in 2011, no new stock assessment had been conducted and no new measures had been adopted. As a result, the stock status was uncertain, making it impossible to judge the effect of CMM 2006-04. However, the need for a new assessment was recognized by the Scientific Committee and a revised stock assessment was proposed for presentation at the Scientific Committee Meeting in 2012.[1544]

g)   North Pacific striped marlin

The Pacific striped marlin occurs partially north of 20° north. However, it is not totally clear which scientific body has the competence for this part of the stock.[1545] Pacific striped marlin is not explicitly defined as a 'northern stock', and it is therefore uncertain whether the ISC or the SPC is entitled to conduct the stock assessments and to give management recommendations to the Scientific Committee and the Commission.[1546] In general it might be argued that it is not important who is providing the scientific information but the issue becomes

---

1538   WCPFC, *Summary Report - Second Regular Session of the Scientific Committee*, (2006), p.212.

1539   Ibid. p.214

1540   Ibid. p.36.

1541   WCPFC, *Conservation and Management Measure for Striped Marlin in the Southwest Pacific*, (2006), Para.1.

1542   Ibid. Para.4.

1543   Ibid. Para.5.

1544   WCPFC, *Summary Report - Seventh Regular Session of the Scientific Committee*, (2011), p.69.

1545   See the sub-chapter on the functions and competences of the Northern Committee.

1546   WCPFC, *Summary Report - Third Regular Session of the Scientific Committee*, (2007), p.31.

important due to the fact that striped marlin in the North Pacific is one of the few stocks which seems to be in critical conditions. The first complete assessment, conducted in 2007 by the ISC, concluded that the fishing mortality rate of striped marlin in the north Pacific should be reduced.[1547] Unfortunately, the assessment provided no clear information on whether the stock was overfished or if overfishing was occurring. During the sixth Scientific Committee Meeting, three years later, it was stated that the north Pacific striped marlin was experiencing overfishing and the stock was considered to be depleted, so a measure for striped marlin had to be adopted urgently.[1548] CMM 2010-01 requires a phased catch reduction of 10 percent in 2011, 15 percent in 2012 and 20 percent in 2013 from the average catch level from 2000-2003.[1549]

It was planned that the measure be amended in 2011, based on a revised stock assessment.[1550] However, no new stock assessment had been conducted by 2011 and so no amendments could be made.[1551] The lack of a stock assessment for this overfished species forced the Scientific Committee to recommend that future stock assessments be made by the SPC as part of the Scientific Committee work programme if the ISC should again fail to provide stock assessment results for the meeting in 2012.[1552]

12.   Conservation and management of non-target species

Conservation and management measures have been also adopted for non-target or bycatch species, such as sharks, seabirds, sea turtles and cetaceans. These conservation and management measures are based on the requirement to adopt measures to minimize discards, the catches of non-target fish and non-fish species, and impacts on associated or dependent species, in particular endangered species, through the development and use of selective, environmentally safe and cost-effective fishing gear and techniques.[1553]

1547   Ibid. p.30.

1548   WCPFC, *Summary Report - Sixth Regular Session of the Scientific Committee*, (2010), p.65.

1549   WCPFC, *Conservation and Management Measure for North Pacific Striped Marlin*, (2010), Para.4.

1550   Ibid. Paras.8 and 9.

1551   WCPFC, *Summary Report - Seventh Regular Session of the Scientific Committee*, (2011), p.69.

1552   Ibid. p.69.

1553   Convention on the Conservation and Management of Highly Migratory Fish Stocks in the Western and Central Pacific Ocean, Art.5(e).

a) Sharks

Some of the most important non-target species are sharks. Although mainly caught as bycatch, they are often kept by fishermen, particularly due to the value of their fins. All over the world, fishermen catch sharks and remove their fins before throwing them back into the water. This so-called 'finning' and the fact that many shark species reproduce very slowly is causing particular concerns about their status.[1554] To date, four general measures have been adopted for the key shark species,[1555] all of which are divided in two parts. CMM 2010-07 is the most recent measure. In the first part it requires the implementation of the FAO International Plan of Action for the Conservation and Management of Sharks (IPOA Sharks), as well as an evaluation of the need to adopt or assess National Plans of Action.[1556] According to the IPOA Sharks the countries are requested, among others, to reduce fishing effort targeting sharks, to improve data collection and monitoring of shark fisheries or to obtain utilization and trade data on shark species.[1557] The second part of the CMM explicitly requires the full utilization of any retained sharks and limits the allowable amount of fins on board to a maximum of five percent of the total weight of the sharks.[1558] This is intended to prevent that sharks are exclusively caught because of their fins. Another objective of the measure was the release of incidentally caught sharks which are still alive. The CCMs which target tuna and tuna-like species were therefore required to take measures in order to encourage the release of live sharks.[1559]

In 2011, a measure was adopted explicitly for the oceanic whitetip shark.[1560] CMM 2011-04 was the first measure for a specific shark species, and was a reaction to the steeply declining catch rates and size trends of these shark species in the longline and purse-seine fisheries.[1561] The measure basically prohibits the

---

1554 B. Worm, et al., Global catches, exploitation rates, and rebuilding options for sharks, *Marine Policy - Vol. 40* (2013), pp.104-204; WCPFC, *Summary Report - Second Regular Session of the Scientific Committee*, (2006), p.41.

1555 WCPFC, *Conservation and Management Measure for Sharks in the Western and Central Pacific Ocean*, (2006); WCPFC, *Conservation and Management of Sharks*, (2008); WCPFC, *Conservation and Management for Sharks*, (2009); WCPFC, *Conservation and Management Measure for Sharks*, (2010).

1556 WCPFC, *Conservation and Management Measure for Sharks*, (2010), Paras.1-3.

1557 International Plan of Action for the Conservation and Management of Sharks, Appendix A, Para.II.C.

1558 WCPFC, *Conservation and Management Measure for Sharks*, (2010) Paras.6 and 7.

1559 Ibid. Para.10.

1560 WCPFC, *Conservation and Management Measure for Oceanic Whitetip Shark*, (2012).

1561 WCPFC, *Summary Report - Seventh Regular Session of the Scientific Committee*, (2011), p.93.

retaining, transshipping, storing or landing of any oceanic whitetip shark across the whole Convention area.[1562] All oceanic whitetip shark must be released causing as little harm as possible to the shark.[1563]

In 2011, the Scientific Committee also recommended mitigation measures for blue shark in the north Pacific.[1564] Unfortunately, no such measures were discussed and adopted during the annual meeting of the Commission. Despite this, the conservation and management of sharks appears to be, at the moment, an important issue at all WCPFC meetings and further measures can be expected soon.[1565]

b)   Seabirds

Seabird bycatch is mainly a problem in longline fishery. Two conservation and management measures for the mitigation of fishing impacts on seabirds have been adopted to date under the WCPFC.[1566] The most recent, CMM 2007-04, is divided in two parts. The first part requires the implementation of the FAO International Plan of Action for Reducing Incidental Catch of Seabirds in Longline Fisheries and to report on National Plans of Action.[1567] These plans should provide optional technical and operational mitigation measures[1568] in order to prevent access by seabirds to baited hooks. The possible technical measures include weighting the longline gear,[1569] setting the line below the water,[1570] or bird scaring curtains.[1571] The operational measures include night setting,[1572] ar-

---

1562 WCPFC, *Conservation and Management Measure for Oceanic Whitetip Shark*, (2012), Para.1.

1563 Ibid. Para.1.

1564 WCPFC, *Summary Report - Seventh Regular Session of the Scientific Committee*, (2011), p.94.

1565 Observation made by the author during the Seventh Regular Session of the Scientific Committee.

1566 WCPFC, *Conservation and Management Measure to Mitigate the Impact of Fishing For Highly Migratory Fish Stocks on Seabirds*, (2006); WCPFC, *Conservation and Management Measure to Mitigate the Impact of Fishing For Highly Migratory Fish Stocks on Seabirds*, (2007).

1567 WCPFC, *Conservation and Management Measure to Mitigate the Impact of Fishing For Highly Migratory Fish Stocks on Seabirds*, (2007), Paras.1 and 2.

1568 International Plan of Action for Reducing Incidental Catch of Seabirds in Longline Fisheries Para.16.

1569 International Plan of Action for Reducing Incidental Catch of Seabirds in Longline Fisheries, Technical note on some optional technical and operational measures for reducing the incidental catch of seabirds in longline fisheries Para II.1(a).

1570 Ibid. Para II.5.

1571 Ibid. Para II.2.

1572 Ibid. Para.III.1.

ea and seasonal closures[1573] or the release of live birds.[1574] The second part of the CMM explicitly determines that longline vessels have to use certain mitigation methods which are mainly intended to avoid birds being accidentally hooked or entangled in the line.[1575] In all areas south of 30° south and north of 23° north, at least two methods of avoidance have to be in use, and in the remaining areas at least one method must be in place.[1576] All longline vessels had to implement the measure by 2009.[1577]

During the Scientific Committee Meeting in 2011, different approaches to assessing interactions with seabirds were presented.[1578] Due to limitations in the data, it was recommended that further intersessional research be conducted and information be exchanged in order to evaluate the effectiveness of CMM 2007-04.[1579]

c)   Sea turtles

Soon after the establishment of the WCPFC, it was noticed that some fisheries in the WCPO had made unintentional captures of turtles.[1580] Four years later, CMM 2008-03 required the implementation of the FAO Guidelines to Reduce Sea Turtle Mortality in Fishing Operations, and aimed to ensure that all captured sea turtles were handled safely, in order to improve their survival.[1581] Generally, hard-shell sea turtles must be brought on board as soon as possible and, once they have recovered, they should be returned to the water using proper handling techniques, mitigating any injury.[1582] Accidental catches of sea turtles occur mainly in purse seine and longline fisheries. Special provisions for purse seine vessels therefore oblige CCMs to avoid the encirclement of sea turtles and to release them with dip nets should they become encircled or entangled by the net or by other fishing gear, such as fish aggregating devices.[1583] In

---

1573   Ibid. Para.III.3.

1574   Ibid. Para.III.5.

1575   WCPFC, *Conservation and Management Measure to Mitigate the Impact of Fishing For Highly Migratory Fish Stocks on Seabirds*, (2007) Part 1, Para.1 (Table 1).

1576   Ibid. Part 2, Paras.1 and 2.

1577   Ibid. Para.10.

1578   WCPFC, *Summary Report - Seventh Regular Session of the Scientific Committee*, (2011), pp.94-97.

1579   Ibid. p.97.

1580   WCPFC, *Summary Report - First Regular Session of the Scientific Committee*, (2005), pp.39-40.

1581   WCPFC, *Conservation and Management of Sea Turtles*, (2008), Para.1.

1582   Ibid. Para.4.

1583   Ibid. Para.5.

general, longline vessels are required to carry and use line cutters and de-hookers to handle sea turtles and to release them promptly using dip nets.[1584] More specific provisions apply to longline vessels that fish for swordfish using a shallow-set technique. They must make use of at least one of a number of methods to mitigate the capture of sea turtles.[1585] These methods include special hooks and bait. Unfortunately, there are still some weak elements in these provisions, among which is the fact that the CCMs can establish and enforce their own operational definitions of the terms shallow-set swordfish longline fisheries or large circle hooks.[1586] A standardized definition determined by the Commission is missing.

No issues explicitly related to turtles were discussed at the Scientific Committee Meeting in 2011. According to informal discussions during the meeting it appears that the measures were working and that incidental catches of turtles did not present a problem at the time.[1587]

d)   Cetaceans

Several tuna species school around large cetaceans and whale sharks. They use the bigger animals as shelter, or feed on the smaller fish that gather around them. Toothed cetaceans are also often found with tuna and tuna-like species because they feed on the same prey. As a consequence, both cetaceans and whale sharks are sometimes encircled by purse seine nets during fishing operations.[1588] This was recognized at the Scientific Committee meeting in 2011. The Committee recommended that mortality of whale sharks and cetaceans by fishing activities should be avoided and that best practice guidelines for the release of encircled whale sharks and cetaceans should be developed.[1589] During its next annual meeting, the Commission discussed these recommendations, but only a measure for cetaceans was ultimately adopted.[1590] This measure applies both on the high seas and within the EEZ, and it primarily prohibits the setting of purse seine nets on a school of tuna associated with a cetacean if the animal

---

1584   Ibid. Para.6.

1585   Ibid. Para.7.a.

1586   Ibid. Para.7.c.

1587   Observation made by the author during the Seventh Regular Session of the Scientific Committee.

1588   Gilman, 2011, p.593.

1589   WCPFC, *Summary Report - Seventh Regular Session of the Scientific Committee*, (2011), p.99.

1590   WCPFC, *Summary Report - Eighth Regular Session of the Commission*, (2012), pp.42-43.

has been sighted prior to the set.[1591] In cases of unintentional encirclement, all reasonable steps have to be taken to ensure the animal's safe release and the incident should be reported to the relevant flag state authority.[1592] In the future the Commission has further the possibility to adopt guidelines for the safe release of cetaceans which would have to be followed by the master of the vessel.[1593]

Unfortunately, due to the opposition of some countries, no measure could be adopted for whale sharks. However, it was concluded that the Scientific Committee will develop best practice guidelines for the release of encircled whale sharks and that the Commission will adopt a measure at its next annual meeting.[1594]

13.   Regulation for fishing gear/area

Other important measures that have been adopted by the Commission regulate the use of certain fishing gear or restrict fishing activities in specific areas. Gear-related measures are based on the requirement to promote the development and use of selective, environmentally safe and cost-effective fishing gear and techniques.[1595] Area-related measures are based on the requirement to pay special attention to the areas of high seas that are entirely surrounded by the EEZs of members of the Commission.[1596]

a)   Prohibition of large-scale drift nets on the high seas

Large-scale drift nets are non-selective fishing gear which has a negative environmental impact due to the bycatch of non-target fish species, cetaceans, seabirds and turtles.[1597] Drift nets with lengths of up to 60 km had been identified as a cause of harm through both the actual drift net fishing activity, but also by becoming traps after being lost by the user.[1598] In 1989 and 1991, two UN Gen-

---

1591   WCPFC, *Conservation and Management Measure for Protection of Cetaceans from Purse Seine Fishing Operations*, (2012), Para.1.

1592   Ibid. Para.2.

1593   Ibid. Para.3.

1594   WCPFC, *Summary Report - Eighth Regular Session of the Commission*, (2012), p.43.

1595   Convention on the Conservation and Management of Highly Migratory Fish Stocks in the Western and Central Pacific Ocean, Art.5(e).

1596   Ibid. Art.8.4.

1597   A. H. Richards, Problems of drift-net fisheries in the South Pacific, *Marine Pollution Bulletin - Vol. 29(1-3)* (1994), p.107.

1598   Ibid. p.107.

eral Assembly Resolutions provided that a global moratorium on all large scale pelagic drift-net fishing had to be established by 1992.[1599] These resolutions gained, despite its non-binding legal nature, strong mandatory effect.[1600]

In the south Pacific, drift nets had been banned already since 1989 by the Wellington Convention.[1601] The adoption of CMM 2008-04 through the WCPFC was a response to the remaining drift nets active in the WCPO.[1602] The measure not only prohibited the use of large-scale drift nets longer than 2.5 km in the high seas but also even the carriage of gear which would allow the use of large-scale drift nets on the high seas.[1603] The prohibition does not apply within the EEZ of a coastal state if the vessel owner can demonstrate that he has permission to use large-scale drift nets.[1604]

During recent meetings of the WCPFC, drift nets have not been explicitly discussed. At the Scientific Committee Meeting in 2011, delegates and scientists generally stated that drift nets did not currently present a problem in the WCPO.[1605]

b)   High seas FAD closures and catch retention

CMM 2008-01 had established measures for Fish Aggregating Device (FAD) closure and catch retention by purse seine vessels in the high seas between 20° south and 20° north.[1606] However, due to a lack of specifications, some provisions had not been applied consistently.[1607] As a result, CMM 2009-02 was es-

1599   United Nations General Assembly, *UN General Assembly Resolution 44/225: Large-Scale Pelagic Drift-Net Fishing and Its Impact on the Living Marine Resources of the World's Oceans and Seas*, (1989), Para.4(a); United Nations General Assembly, *UN General Assembly Resolution 46/215: Large-Scale Pelagic Drift-Net Fishing and Its Impact on the Living Marine Resources of the World's Oceans and Seas*, (1991), Para.3(c).

1600   Kaye, 2001, p.193.

1601   Convention for the Prohibition of Fishing with Long Driftnets in the South Pacific, 1989. The Wellington Convention is open for accession by the members of the FFA and the other States and Territories in the Convention area.

1602   WCPFC, *Summary Report - Fifth Regular Session of the Commission*, (2008), pp.33-34.

1603   WCPFC, *Conservation and Management Measure to Prohibit the use of Large Scale Driftnets on the High Seas in the Convention Area*, (2008), Paras.1 and 3.

1604   Ibid. Para.4.

1605   Observation made by the author during the Seventh Regular Session of the Scientific Committee.

1606   WCPFC, *Conservation and Magement Measures for Bigeye and Yellowfin Tuna in the Western and Central Pacific Ocean*, (2008), Paras.11,13, 15, 17, 19, 20, 23, 24, and 27.

1607   WCPFC, *Summary Report - Sixth Regular Session of the Commission*, (2009), pp.34-35.

tablished with two main objectives. The first was to ensure the consistent and robust application of both FAD closures and catch retention, and the second was to apply high standards to the application of FAD closures and catch retention.[1608]

In CMM 2008-01 there was no specification on the distance a fishing vessel has to maintain to a FAD during the closure and the definition of the term FAD did not include all possible FADs. FADs were only defined as "any man-made device, or natural floating object, whether anchored or not, that is capable of aggregating fish."[1609] CMM 2009-02 prohibited purse seine fishing within one nautical mile of an FAD during the closure, as well as prohibiting the use of vessels to aggregate fish, or to move aggregated fish.[1610] Regarding catch retention, the measure specified that a release of the fish is only allowed before the net is fully pursed and when less than half of the net has been retrieved.[1611] CMM 2008-01 had determined that fish can be discarded if it is unfit for human consumption.[1612] Due to the fact that it was not clear what was exactly meant by the term 'unfit for human consumption', a comprehensive series of definitions was provided by CMM 2009-02 detailing what was included or excluded by term.[1613] Included and therefore allowed to be discarded is fish that: is meshed or crushed in the purse seine net; is damaged due to shark or whale depredation; or has died and spoiled in the net where a gear failure has prevented both the normal retrieval of the net and catch and efforts to release the fish alive.[1614] Excluded and therefore prohibited to be discarded is fish that: is considered undesirable in terms of size, marketability, or species composition; or is spoiled or contaminated as the result of an act or omission of the crew of the fishing vessel.[1615] Furthermore, it was laid down that discards are only allowed once an observer has estimated the species composition of the fish to be discarded.[1616]

1608 WCPFC, *Conservation and Management Measure on the Application of High Seas FAD Closures and Catch Retention*, (2009), Para.1.

1609 WCPFC, *Conservation and Magement Measures for Bigeye and Yellowfin Tuna in the Western and Central Pacific Ocean*, (2008), Preamble.

1610 WCPFC, *Conservation and Management Measure on the Application of High Seas FAD Closures and Catch Retention*, (2009) Paras.4 and 5.

1611 Ibid. Para.8.

1612 WCPFC, *Conservation and Magement Measures for Bigeye and Yellowfin Tuna in the Western and Central Pacific Ocean*, (2008), Para.27.b.

1613 WCPFC, *Conservation and Management Measure on the Application of High Seas FAD Closures and Catch Retention*, (2009), Para.9.

1614 Ibid. Para.9.a.

1615 Ibid. Para.9.b.

1616 Ibid. Para.11.

c) *Eastern High Seas Pocket Special Management Area*

As already stated above, the WCPFC is the only tuna RFMO with high seas areas that are surrounded by the EEZs of coastal states. These areas are called high seas pockets. By 2010, the Western High Seas Pockets had been closed to fishing since CMM 2008-01.[1617] This measure encouraged the Commission to consider the closure of other high seas pockets in the area between 20° north and 20° south. Located in the most easterly part of the Convention area the so-called Eastern High Seas Pocket is surrounded by the EEZs of the Cook Islands, French Polynesia, and Kiribati. CMM 2010-02 deals explicitly with this pocket and it aims to reduce the illegal, unreported and unregulated fishing that had been caused by the previous lack of specific provisions.[1618] The Commission declared the pocket as a special management area. CCMs are required to inform the Commission about the activities of their vessels at least six hours before they enter this area and not later than six hours before they leave it.[1619] They also have to report sightings of any fishing vessel to the Secretariat and provide continuous near-real time Vessel Monitoring System data to the adjacent coastal states.[1620] Transshipment is only allowed in accordance with other relevant conservation and management measures and it had to be declared to the Commission.[1621] In order to monitor all fishing activities in the area, the Secretariat was instructed to maintain a live list of all active vessels on the WCPFC website.[1622]

Due to the fact that this measure was only implemented very recently, it is still too early to see if it has had any positive effects on the stocks of tuna and tuna-like species in the area. However, during the Technical and Compliance Committee Meeting in 2011, some CCMs criticized the sporadic notification of entry to and exit from their EEZ.[1623] They proposed an EEZ entry and exit notification scheme but, due to opposition from other members, no revisions have yet been made to the measure.

1617 WCPFC, *Conservation and Magement Measures for Bigeye and Yellowfin Tuna in the Western and Central Pacific Ocean*, (2008), Para.22.

1618 WCPFC, *Conservation and Management Measure for the Eastern High-Seas Pocket Special Management Area*, (2010).

1619 Ibid. Para.2.

1620 Ibid. Paras.3 and 4.

1621 Ibid. Para.6.

1622 Ibid. Para.5.

1623 WCPFC, *Summary Report - Seventh Regular Session of the Technical Compliance Committee*, (2011), pp.47-48.

d)   Prohibition of fishing on data buoys

In all oceans, data buoys are usually needed to gather information for weather and marine forecasts, for different types of research and in order to support search and rescue efforts. In 2009, it was recognized by the WCPFC that fishing for tuna and tuna-like species around data buoys had to be regulated because vessels which were using the data buoys as Fish Aggregating Devices had caused significant damage to the buoys.[1624] The following CMM 2009-05 basically prohibited any fishing operations within one nautical mile of data buoys in the high seas.[1625] CCMs were further required to oblige their fishing vessels to remove accidentally entangled data buoys with as little damage as possible.[1626] Since the adoption of the measure, the issue of fishing around data buoys had not been brought up at the WCPFC meetings anymore.

14. Monitoring, compliance and enforcement

Comprehensive provisions on monitoring, compliance and enforcement are crucial for the sustainable management of tuna and tuna-like species. The WCPFC Convention already includes some important provisions.[1627] In general all flag states are required to exercise effective control over the vessels flying their flag.[1628] In particular they have to take measures to ensure that their vessels comply with the measures adopted by the Commission and that these vessels are not engaged in unauthorized fishing within areas under national jurisdiction.[1629] Both the flag state and the Commission are required to maintain a record of the vessels authorized to fish in the Convention area.[1630] The flag states are further responsible that all of their vessels are using vessel monitoring systems which report to the flag state and the Commission.[1631] The Convention required the establishment of boarding and inspection procedures for of fishing

---

1624   WCPFC, *Summary Report - Fifth Regular Session of the Technical and Compliance Committee* (2009), p.31; WCPFC, *Summary Report - Sixth Regular Session of the Commission*, (2009), p.36.

1625   WCPFC, *Conservation and Management Measure Prohibiting Fishing on Data Buoys*, (2009), Para.1.

1626   Ibid. Para.5.

1627   Convention on the Conservation and Management of Highly Migratory Fish Stocks in the Western and Central Pacific Ocean, Part VI-VII.

1628   Ibid. Art.24.

1629   Ibid. Art.24.1.

1630   Ibid. Arts.24.4-6.

1631   Ibid. Art.24.8.

vessels on the high seas[1632] as well as the development of a regional observer programme.[1633] Other important provisions are provided on the duties of port states[1634] Members of the Commission can adopt port state measures to prohibit landings and transhipments of catch that has been made in contravention of conservation and management measures adopted by the Commission.[1635] Transshipment refers to operations where fishing vessel transfer their catch to a carrier or bunker vessel which collects and transports the catch of several fishing vessels in order to save fuel costs and fishing time. The Convention contains explicit provisions on transhipment.[1636]

In addition to these provisions the Commission was required to establish appropriate cooperative mechanisms for effective monitoring, control, surveillance and enforcement, including a vessel monitoring system.[1637] Therefore, several measures that specify the provisions of the Convention have been adopted in recent years. More comprehensive measures are still needed with regard to port states control and trade related regulation.[1638]

a)   *Marking and identification of fishing vessels*

All monitoring, compliance and enforcement operations rely on the unequivocal identification of fishing vessels. The Convention of the WCPFC laid down that vessels must be marked and identified with clear, distinct and uncovered markings in accordance with either the FAO Standard Specifications for the Marking and Identification of Fishing Vessels,[1639] or an alternative standard developed by the Commission.[1640] The FAO Standard Specifications are based on the International Telecommunication Union (ITU) Radio Call Signs (IRCS) system which is an established international system for the determination of the identity and nationality of vessels.[1641]

1632   Ibid. Art.26.

1633   Ibid. Art.28.

1634   Ibid. Art.27.

1635   Ibid. Art.27.3.

1636   Ibid. Art.29.

1637   Ibid. Art.10(i).

1638   See chapter III.

1639   FAO, *The Standard Specifications for the Marking and Identification of Fishing Vessels* (Rome: FAO, 1989).

1640   Convention on the Conservation and Management of Highly Migratory Fish Stocks in the Western and Central Pacific Ocean Annex III, Para.3.

1641   Palma, et al., 2010, p.115.

After the establishment of the WCPFC, it was necessary to develop more specific requirements for marking and identification, so CMM 2004-03 was adopted, applying to all fishing vessels of members that are authorized to fish on the high seas of the Convention area.[1642] The basic requirement is that these vessels be marked for identification with a WCPFC Identification Number (WIN). The WIN can be either the International Telecommunication Union Radio Call Sign or the characters allocated by the International Telecommunication Union.[1643] It must be the only vessel identification mark consisting of letters and numbers on the hull or superstructure of the vessel, apart from the vessel's name, identification mark and port of registry.[1644] Vessels without a WIN are not authorized to fish in the Convention area beyond areas of national jurisdiction.[1645] Offences under national law, such as not marking or wrongfully marking a vessel, have to be considered as a possible reason for refusing the authorization through the flag state.[1646]

b)   Record of fishing vessels and authorization to fish

All vessels that conduct fisheries-related activities within the Convention area of the WCPFC have to be registered with the Commission. The Convention therefore requires each member to maintain a Members Record of Fishing Vessels, listing all vessels flying its flag and fishing in the Convention area beyond its area of national jurisdiction.[1647] For each vessel, the country must provide, among other details, the name of the vessel, its owner and master, any previous flag used, the international call radio sign, length and gross register tonnage.[1648] All national records have to be collected by the Commission in the WCPFC Record of Fishing Vessels and the information of the record has to be circulated periodically.[1649]

The first measure which specified these provisions was adopted at the inaugural session of the Commission. In general, CMM 2004-01 states that only vessels which are able to effectively exercise their responsibilities under international

---

1642 WCPFC, *Conservation and Management Measure - Specifications for the Marking and Identification of Fishing Vessels*, (2004), Para.1.1.2.

1643 Ibid. Paras.2.1.1(a) and (b).

1644 Ibid. Para.2.1.3(a).

1645 Ibid. Para.2.1.3(b).

1646 Ibid. Paras.2.1.3 (c) and (d).

1647 Convention on the Conservation and Management of Highly Migratory Fish Stocks in the Western and Central Pacific Ocean, Art.24.4.

1648 Ibid. Art.24.5 and Annex IV.

1649 Ibid. Art.24.7.

law should be authorized to fish.[1650] With regard to the Members Record of Fishing Vessels, the measure basically restated the requirements of the Convention with some minor additions, such as a time frame for reporting.[1651] The process for the collection of national records through the Commission needed more detail. This included requiring that the WCPFC Record of Fishing Vessels had to be made publically available on a website and circulated in an annual summary.[1652] The explicit provision that vessels, which were not included in the WCPFC Record of Fishing Vessels could not be authorized to fish for, retain on board, transship or land tuna and tuna-like species on the high seas of the Convention area was crucial.[1653] The member states had to ensure that any activities of such vessels were prohibited.

In 2009, a second measure was adopted, replacing the earlier one. The most comprehensive changes were new provisions for the regulation of transshipping and bunkering.[1654] Due to the fact that such activities had often been linked to IUU fishing, it was decided to develop the WCPFC Interim Register of non-Member Carrier or Bunker Vessels.[1655] For the establishment of this register, a schedule running from 2010 to 2013 and beyond was agreed upon.[1656] The WCPFC Record of Fishing Vessels and the WCPFC Interim Register of non-Member Carrier or Bunker Vessels are closely related to the WCPFC IUU Vessel List.[1657] Vessels that do not fully comply with the requirements of the Convention and the conservation and management measures have to be placed on that list.[1658] Should that occur, the flag or responsible state is required to revoke the authorization for the vessel to fish beyond the national jurisdiction of its flag state.[1659] Currently the WCPFC publishes the Record of Fishing Vessels, the Interim Register of Non-Member Fish Carrier and Bunker Vessels and the WCPFC IUU Vessel List on its webpage.

1650 WCPFC, *Conservation and Management Measure - Record of Fishing Vessels and Authorization to Fish*, (2004), Para.1(a).

1651 Ibid. Paras.5-7.

1652 Ibid. Paras.9 and10.

1653 Ibid. Para.12.

1654 WCPFC, *Conservation and Management Measure - WCPFC Record of Fishing Vessels and Authorization to Fish*, (2009), Para.2.

1655 Ibid. Paras.24-25.

1656 Ibid. Paras.26-46.

1657 For further information on the WCPFC IUU Vessel List see the next sub-chapter.

1658 WCPFC, *Conservation and Management Measure - WCPFC Record of Fishing Vessels and Authorization to Fish*, (2009), Paras.16 and 34.

1659 Ibid. Para.23.

c)   Establishment of a list of IUU-vessels

The identification and listing of vessels that had conducted illegal unreported and unregulated (IUU) fishing in the Convention area was a critical task for the Commission in order to prevent such activities from continuing. The WCPFC Convention itself did not contain explicit provisions on IUU fishing, but the Commission has since adopted two measures which are based on the International Plan of Action to prevent, deter and eliminate illegal, unreported and unregulated fishing.[1660]

The first measure was CMM 2007-03, which required the Commission to identify vessels engaged in IUU fishing and to establish a list of these vessels (IUU vessel list or black list).[1661] IUU fishing includes fishing from a vessel which is not on the WCPFC record of authorized vessels (white list),[1662] fishing in the EEZ of a coastal state without a permit or in contravention of its laws or regulations, fishing without or with incorrect recording and reporting, or fishing in a closed area or during a closed season.[1663] All CCMs must inform the Commission about vessels they believe to be involved in IUU fishing. They must provide a list of the presumed IUU vessels to the Executive Director and notify the relevant flag state of each vessel about its inclusion on this list.[1664] The Executive Director has then to draw up a draft IUU vessel list and request those CCMs and non-CCMs with listed vessels to inform the vessel owner of its inclusion and the possible consequences.[1665] In the meantime, the listed vessels must be closely monitored by the CCMs, and the CCMs and non-CCMs with listed vessels have to communicate with the Executive Director.[1666] The Technical and Compliance Committee must then adopt a Provisional IUU Vessel List and make recommendations to the Commission with regard to the possible removal of vessels from the list.[1667] The list can be modified when accused countries are

---

1660   International Plan of Action to Prevent, Deter, and Eliminate Illegal, Unreported and Unregulated Fishing (IUU).

1661   WCPFC, *Conservation and Management Measure to Establish a List of Vessels presumed to Have Carried out Illegal, Unreported and Unregulated Fishing Activities in the WCPO*, (2007), Para.1; See chapter III.

1662   See chapter III.

1663   WCPFC, *Conservation and Management Measure to Establish a List of Vessels presumed to Have Carried out Illegal, Unreported and Unregulated Fishing Activities in the WCPO*, (2007) Paras.3.a,b,c and e.

1664   Ibid. Paras.4 and 5.

1665   Ibid. Paras,6 and 7.

1666   Ibid. Paras.8 and 9.

1667   Ibid. Para.14.

able to prove that they have adopted measures to ensure that the vessel will comply with all WCPFC measures.[1668] At its annual meetings, the Commission reviews the Provisional IUU Vessel List and adopts a new WCPFC IUU Vessel List.[1669]

CMM 2010-06 is identical to the first measure except for an additional annex which defines the criteria for the inclusion of other vessels that are under the control of the owner of any vessel that is already on the list.[1670] The basic requirement is that such vessels are included on the draft IUU Vessel List and, where appropriate, on the WCPFC IUU Vessel List.[1671] A vessel may be removed from the list if the relevant flag state provides proof that the vessel no longer has a common owner on record or that significant progress has been made to resolve the matter, and that the CCM that originally submitted the vessel for listing is satisfied with the progress.[1672]

During the Technical and Compliance Committee Meeting in 2011, the Executive Director advised the participants of the meeting that the Secretariat had not received any new nominations for vessels to be placed on the Draft IUU Provisional Vessel List for 2011.[1673] Five vessels were already on the WCPFC IUU Vessel List. The Committee recommended that the Commission remove one of these vessels because the vessel's ownership had changed and there was no link between the current and former owner. In addition, the vessel had been detained in port for over 10 months and it had been fined by the flag state. This vessel was removed from the list for 2012 but the other vessels remained listed.[1674]

d)   *Commission Vessel Monitoring System*

Since 1993, a Vessel Monitoring System (VMS) had been used successfully by the FFA members within their EEZs. After the establishment of the WCPFC, it

---

1668   Ibid. Para.25(a).

1669   Ibid. Para.20.

1670   Ibid. Para.3.j, WCPFC, *Conservation and Management Measure to establish a List of Vessels presumed to have carried out Illegal, Unreported and Unregulated fishing activities in the WCPO*, (2010), Para.3.j.

1671   WCPFC, *Conservation and Management Measure to establish a List of Vessels presumed to have carried out Illegal, Unreported and Unregulated fishing activities in the WCPO*, (2010), Annex A, Paras.8 and 17.

1672   Ibid. Annex A, Para.18.

1673   WCPFC, *Summary Report - Seventh Regular Session of the Technical Compliance Committee*, (2011), pp.15-17.

1674   WCPFC, *Summary Report - Eighth Regular Session of the Commission*, (2012), p.24.

was recognized that a VMS was also needed for the high seas area.[1675] The Convention therefore required that all vessels fishing on the high seas had to use near real-time satellite position-fixing transmitters and the Commission had to develop standards, specifications, and procedures for the establishment of the VMS.[1676] Coastal states could not be obliged to include their EEZs within the area covered by the VMS but they had the right to request a voluntary inclusion.[1677] So far only New Zealand, Cook Islands Niue, the United States of America, Australia and the Federated States of Micronesia have made use of that right.[1678] For members fishing in the EEZs of another coastal state member, it is mandatory that their vessels operate in accordance with the VMS requirements of the respective coastal state.[1679] In general, all members are required to cooperate in order to ensure compatibility between national VMSs and the high seas VMS.[1680]

Three measures dealing with the implementation of the VMS have been adopted to date. All of them restated the basic requirements of the Convention, but added terms, dates and specifications for the area of application. CMM 2006-06 determined that, beginning in 2008, the WCPFC VMS had to be activated by all vessels in the Convention area south of 20° south, and east of 175° east in the Convention area north of 20° north.[1681] The remaining area north of 20° north and west of 175° east was to be included at a later date, which had still to be determined by the Commission.[1682] All information had to be collected and documented by the Secretariat, which should develop and administer the VMS under the guidance of the Commission.[1683] In practice, it is mainly the Compliance Manager of the WCPFC and its staff who administers the VMS.[1684] The cornerstone of the VMS is the Automatic Location Communicator (ALC), previ-

---

1675 Convention on the Conservation and Management of Highly Migratory Fish Stocks in the Western and Central Pacific Ocean, Art.10(i).

1676 Ibid. Art.24.8, sents.1 and 2.

1677 Ibid. Art.24.8, sent.6.

1678 http://www.wcpfc.int/vessel-monitoring-system.

1679 Convention on the Conservation and Management of Highly Migratory Fish Stocks in the Western and Central Pacific Ocean Art.24.9.

1680 Ibid. Art.24.10.

1681 WCPFC, *Conservation and Management Measure - Commission Vessel Monitoring System*, (2006), Paras.2 and 6(b).

1682 Ibid. Para.3.

1683 Ibid. Para.7(a).

1684 Observation made by the author during a research stay at the Secretariat of the WCPFC.

ously called a near real-time satellite position-fixing transmitter.[1685] It informs the Secretariat about the position of vessels fishing in the relevant areas and has to be used in line with minimum standards[1686] which determine which data has to be transmitted and how the ALC has to be used.[1687]

The first measure had been revised and replaced twice. CMM 2007-02 did not require a general activation of the VMS in the area north of 20° north and west of 175° east, but it added that those vessels fishing in the areas south of 20° south and east of 175° east had to keep their ALC activated and continue reporting to the Commission even when moving into the northern area.[1688] In 2008, the Commission adopted Standards, Specifications and Procedures for the VMS which established the terms of implementation of the system.[1689] This document included methods to ensure ALCs complied with: WCPFC standards; inspection protocols; rules on polling, vessel reporting; measures to prevent tampering; and the obligations and roles of fishing vessels, CCMs, the FFA Secretariat and the Commission Secretariat. The third measure was CMM 2011-02, but, unlike the earlier measures, it only removed one paragraph of fairly minor importance.[1690]

Today, the VMS is an important pillar for the monitoring of fisheries-related activities in the WCPO. Unfortunately, no schedule for the inclusion of the area north of 20° north and west of 175° east has been developed yet. However, during the postponed Commission Meeting of 2011, the chair of the Northern Committee stated that this issue would be discussed at its next meeting.[1691] Another critical issue is the coexistence of the WCPFC VMS and the FFA VMS. The FFA administers the VMS within the EEZs of its parties and the WCPFC administers the VMS mainly on the high seas. As shown before, only few coastal states allowed the WCPFC to use the VMS data within their EEZs. That leaves loopholes for potential IUU fishing as illegal activities within the EEZs cannot be seen by the Compliance Manager and its staff. The WCPFC is further de-

---

1685 WCPFC, *Conservation and Management Measure - Commission Vessel Monitoring System*, (2006), Para.4(a).

1686 Ibid. Paras.7(d) and (e).

1687 Ibid. Attachment L, Annex 1.

1688 WCPFC, *Conservation and Management Measure - Commission Vessel Monitoring System*, (2007), Para.4.

1689 WCPFC, *Standards, specifications and procedures (SSP) for the fishing vessel monitoring system (VMS) of the Western and Central Pacific Fisheries Commission (WCPFC)*, (2008).

1690 WCPFC, *Conservation and Management Measure - Commission Vessel Monitoring System*, (2012) WCPFC, *Conservation and Management Measure - Commission Vessel Monitoring System*, (2007), Attachment L, Annex 1, Para.13.

1691 WCPFC, *Summary Report - Eighth Regular Session of the Commission*, (2012), p.22.

pendant on the FFA with regard to the VMS because, according to a Service Level Agreement between both organizations, the FFA provides the VMS data for the high seas to the WCPFC.[1692] During the Technical and Compliance Committee Meeting in 2011, the members evaluated how the joint WCPFC/FFA VMS could be improved. The development of clear objectives for the WCPFC VMS was identified as a high priority, as was cost reduction.[1693] The annual Commission meeting recognized these comments and concluded that an inter-sessional working group should convene on the recommendations prior to the following Technical and Compliance Committee Meeting, in 2012.[1694]

e)   Regional Observer Programme

The presence of on-board observers during fishing operations facilitates the monitoring of all fisheries-related activities. Comprehensive provisions for the development of a Regional Observer Programme (ROP) for the WCPO had already been included in the Convention.[1695] They explicitly instructed the Commission to develop a programme for the collection of verified catch data, other scientific data, and any additional information related to the fishery, and to monitor the implementation of the measures.[1696] A particular requirement of the Convention was the development of procedures and guidelines for the confidentiality of data, the dissemination of data and the definition of the rights and responsibilities of all participants.[1697]

To date, only one measure has been adopted for the establishment of the ROP.[1698] The comprehensive CMM 2007-01 scheduled the phased implementation of the programme, to begin in 2008 and to be reviewed in 2012.[1699] The implementation required that existing sub-regional and national programmes had to be used initially and that an international working group had to be allowed to develop a framework and other important elements.[1700] By the end of 2008,

---

1692   WCPFC, *Service Level Agreement between WCPFC and FFA*, (2011), Para.3.

1693   WCPFC, *Summary Report - Seventh Regular Session of the Technical Compliance Committee*, (2011), pp.26-28 and Attachment F.

1694   WCPFC, *Summary Report - Eighth Regular Session of the Commission*, (2012), p.31.

1695   Convention on the Conservation and Management of Highly Migratory Fish Stocks in the Western and Central Pacific Ocean, Part VII.

1696   Ibid. Art.28.1.

1697   Ibid. Arts.28.7(a),(b).

1698   WCPFC, *Conservation and Management Measure for the Regional Observer Programme*, (2007).

1699   Ibid. Para.2.

1700   Ibid. Attachment K, Annex C, Paras.1 and 2.

all sub-regional and national programmes had to be regarded as a part of the ROP, and from then until June 2012, coverage of at least five percent of the effort in each fishery was mandatory.[1701] Exemptions were included for those vessels fishing exclusively in the area north of 20° north, small vessels, as well as troll and pole and line vessels used for fishing skipjack and albacore.[1702] Additional provisions were then established by CMM 2008-01, requiring 100 percent observer coverage on purse seine vessels fishing in the area bound by 20° north and 20° south.[1703]

The ROP applies to three types of vessels: vessels fishing exclusively on the high seas; vessels fishing on the high seas and in the EEZ of one or more coastal states; and vessels fishing in the EEZs of two or more coastal states.[1704] CCMs with such vessels were required to ensure that the vessels were prepared to accept on-board observers, to meet the required level of observer coverage, to source the observers for their vessels and to inform their captains about the observer's duties.[1705] The Commission, on behalf of its subsidiary bodies, has to monitor and supervise the implementation of the ROP, develop its priorities and objectives, and assess the results.[1706] The Secretariat has to coordinate all ROP activities and authorize observer providers.[1707] In practice, the ROP is co-ordinated by the manager of the programme and his assistant.[1708] Two further annexes were attached to the measure in order to avoid conflict and misunderstandings.[1709] These annexes basically determined the rights and responsibilities of observers as well as those of vessel operators, captains and crew.

During the seventh meeting of the Technical and Compliance Committee, the WCPFC ROP Coordinator reported that more than 600 observers were involved in the ROP and that further training would be needed to meet future needs. For the area between 20° north to 20° south, it was recognized that, as required by CMM 2008-01, 100 percent observer coverage on purse seiners could be

---

1701   Ibid. Attachment K, Annex C, Paras.4 and 6.

1702   Ibid. Attachment K, Annex C, Paras.9 and 10.

1703   WCPFC, *Conservation and Magement Measures for Bigeye and Yellowfin Tuna in the Western and Central Pacific Ocean*, (2008), Para.28.

1704   WCPFC, *Conservation and Management Measure for the Regional Observer Programme*, (2007), Para.5.

1705   Ibid. Paras.7-10.

1706   Ibid. Para.11.

1707   Ibid. Para.12.

1708   Observation made by the author during a research stay at the Secretariat of the WCPFC.

1709   WCPFC, *Conservation and Management Measure for the Regional Observer Programme*, (2007) Attachment K, Annex A and Annex B.

achieved, but it was also stated that the implementation of the five percent observer coverage on longline vessels by June 2012 would be a challenge.[1710] The
ROP Coordinator also reported on cases where vessel masters or crew members
as well as ROP observers had complained about each other's conduct.[1711] Such
complaints have the potential to undermine the success of the ROP. Due to the
fact that these complaints could only be addressed at a national level, the
Committee encouraged the CCMs to improve their national systems for dealing
with complaints.[1712] Other problems identified were a lack of definitions for key
terms, unprocessed data, uncertainties with regard to observer coverage on
transshipment operations, and issues around the cross-endorsement of observers by WCPFC and IATTC.[1713] The Committee therefore recommended the for-
mation of a Technical Advisory Group in order to make progress on some of
these issues. This recommendation was accepted by the Commission at its annual meeting.[1714]

f)   Regulation of transshipment

Transshipment at sea is a common practice in many fisheries where processing
plants are far from the fishing grounds. Usually the catch is transshipped from
the fishing vessel to a carrier vessel which collects and transports the catch of
several fishing vessels in order to save fuel costs and fishing time.[1715] According
to the WCPFC Convention members were required to encourage their fishing
vessels, where practicable, to conduct transshipment in a designated port.[1716] In
waters under national jurisdiction all transshipment operations had to be con-

---

1710   WCPFC, *Summary Report - Seventh Regular Session of the Technical Compliance Committee*,
       (2011), p.32.

1711   Ibid. p.36.
       The main complaints of vessel masters or crew members concerned alcohol, observer
       requests to return to port, the lack of adequate funding support for observers when at
       sea. Observers complained particularly about hindrance, obstruction, restricted access,
       and intimidation while conducting their duties, influence to "not report" infractions, and
       travel and funding while at sea.

1712   Ibid. p.37.

1713   Ibid. pp.32-33.

1714   WCPFC, *Summary Report - Eighth Regular Session of the Commission*, (2012), p.36.

1715   M. Gianni & W. Simpson, 'Flags of Convenience, Transhipment, Re-Supply and At-Sea
       Infrastructure in Relation to IUU Fishing', *Fish Piracy: Combating Illegal, Unreported and
       Unregulated Fishing* (Paris: OECD Publishing, 2004), p.89.

1716   Convention on the Conservation and Management of Highly Migratory Fish Stocks in the
       Western and Central Pacific Ocean, Art.29.1.

ducted in accordance with the national laws of the coastal state.[1717] Transshipment on the high seas was not explicitly prohibited by the Convention, but it
was only allowed in accordance with certain requirements that ensure that the
quantity and species transshipped are verified.[1718] Only transshipment by purse
seine vessels was explicitly prohibited.[1719] The provisions with regard to the
procedures for obtaining and verifying transshipment data were still imprecise
and had to be developed by the Commission.[1720]

One measure for the regulation of transshipment has been adopted to date.
CMM 2009-06 specified the provisions of the Convention and aimed to further
regulate transshipment in order to avoid unregulated and unreported transshipment. The measure applies to all transshipment operations in the Convention area, except those where the fish was taken and transshipped wholly in
archipelagic waters or territorial seas.[1721] The measure could not achieve a total
ban on transshipment at sea, but an important specification was the requirement that observers from the Regional Observer Programme had to observe all
transshipment operations at sea.[1722] The measure also contained specific provisions for different fishing gears. Except for certain cases, transshipping at sea by
purse seine vessels remained prohibited in the whole Convention area.[1723] The
exemptions were small purse seine boats from Papua New Guinea and the Philippines, and New Zealand flagged domestic purse seine vessels.[1724] The provisions with regard to all other fishing gears - including longline, troll and pole
and line - distinguish between EEZs and the high seas. In the EEZ, transshipment is allowed, but it has to be conducted in accordance with relevant domestic laws and procedures.[1725] For the high seas, it is generally required that no
transshipment be conducted, but a vague exemption allows transshipment for
vessels that can prove that it is impracticable for them to operate without being
able to transship.[1726]

1717   Ibid. Art.29.2.

1718   Ibid. Art.29.4.

1719   Ibid. Art.29.5.

1720   Ibid. Art.29.3.

1721   WCPFC, *Conservation and Management Measure on Regulation of Transshipment*, (2009),
       Paras.2 and 3.

1722   Ibid. Paras.13-17.

1723   Ibid. Para.25.

1724   Ibid. Paras.25.a and b.

1725   Ibid. Para.33.

1726   Ibid. Para.34.

After the adoption of the conservation and management measure, transshipment received limited discussion, but one problem was mentioned by the Co-ordinator of the Regional Observer Programme during the Technical and Compliance Committee in 2011.[1727] He noted several issues with regard to observer coverage of transshipment operations that required resolution. The main problem was that the Secretariat had struggled to identify whether or not carriers were planning to transship at sea. That problem needed to be solved in the near future because in such cases an observer would be required on-board. The Committee did not make any explicit recommendations, but the matter was included in the considerations for further procedures and guidelines regarding the operation of the Regional Observer Programme.[1728]

g)   *Monitoring of landings of purse seiners at ports*

In 2009, the Commission identified irregularities in catch reports of bigeye tuna. It was assumed that the actual catches from the purse seine fishery had been significantly higher than the catches reported by the flag States.[1729] A potential method to improve the compliance with the reporting requirements was port monitoring. CMM 2008-01 already determined that landing, transshipment and commercial transaction of tuna and tuna products from fishing activities made in contravention of the CMM had to be prohibited by the CCMs.[1730] Landing and transshipping ports of CCMs were required to monitor tuna or tuna products in order to determine the amount of catch by species.[1731] These provisions provided a sound basis for port monitoring, but a substantial portion of purse seine catches were being landed in Thai ports before being processed in canneries, and, unfortunately, Thailand was not a CCM at that time. To address this, a measure had to be established in order to improve cooperation between the WCPFC and Thailand.

CMM 2009-10 referred to the provisions of CMM 2008-01 and basically required the members to work together to establish arrangements with a Non-CCM in order to enable the collection of species and size composition data from canner-

1727   WCPFC, *Summary Report - Seventh Regular Session of the Technical Compliance Committee*, (2011), p.32.

1728   Ibid. p.36.

1729   WCPFC, *Summary Report - Sixth Regular Session of the Commission*, (2009), p.38.

1730   WCPFC, *Conservation and Magement Measures for Bigeye and Yellowfin Tuna in the Western and Central Pacific Ocean*, (2008), Para.42.

1731   Ibid. Para.43.

ies in the Non-CCM.[1732] No specific countries were named in the measure but, from previous discussions, it was obvious that it was directed at Thailand. One year after the adoption of the measure, Thailand became a Cooperating Non-Member (CNM) of the WCPFC, having been approached by the Commission, particularly because of this monitoring issue.[1733] It was agreed that Thailand would provide data from its canneries in order to assist in the work of the Commission. The CNM status for Thailand only related to the acquisition and exchange of fishery information and data, and not to participatory rights for fishing in the high seas of the Convention area. During the postponed Commission meeting of 2011, the CNM status for Thailand was approved, with the requirement that Thailand continued to provide canneries data, while now, participatory rights applied to carrier and bunker vessels as well.[1734]

h)   Boarding and inspection procedures

Prior to the Fish Stocks Agreement, enforcement on the high seas was the exclusive duty of the flag States, but it became necessary to develop provisions for reciprocal boarding and inspection in order to prevent IUU fishing. The provisions in the WCPFC Convention are closely related to those of the Fish Stocks Agreement.[1735] In general, vessels used for boarding and inspection have to be clearly identifiable, authorized inspectors must comply with the procedures for high seas boarding and inspection, and members are required to ensure that their fishing vessels accept boarding by the inspectors.[1736] The Convention did not determine specific boarding and inspection procedures, so the members had to establish such procedures within two years of the Convention entering into force.[1737]

The Commission followed this requirement with CMM 2006-08. The comprehensive procedures are divided into requirements for inspectors or inspection vessels and requirements for the master of the fishing vessel and its flag State. Authorized inspection vessels and inspectors must be clearly identifiable, and

---

1732 WCPFC, *Conservation and Management Measure to Monitor Landings of Purse Seiners at Ports so as to Ensure Reliable Catch Data by Species*, (2009), Para.1.

1733 WCPFC, *Summary Report - Seventh Regular Session of the Commission*, (2010), p.13.

1734 WCPFC, *Summary Report - Eighth Regular Session of the Commission*, (2012), p.9.

1735 Agreement for the Implementation of the Provisions of the United Nations Convention on the Law of the Sea of 10 December 1982 relating to the Conservation and Management of Straddling Fish Stocks and Highly Migratory Fish Stocks, Arts.19, 20,21, and 23.

1736 Convention on the Conservation and Management of Highly Migratory Fish Stocks in the Western and Central Pacific Ocean, Arts.26.1 and 26.3.

1737 Ibid. Art.26.2.

must communicate with the fishing vessel prior to boarding and in a language that the vessel master can understand.[1738] The inspectors have the authority to inspect the vessel, its license, gear, equipment, records, facilities, fish and fish products and any relevant documents necessary to verify compliance with the measure.[1739] During the procedure, the inspector has to respect the internationally-accepted principles of good seamanship, not interfere with lawful operations of the fishing vessel and avoid actions that might affect the quality of the catch or pose any other kind of harassment.[1740] If no evidence of a serious violation is found, the inspection has to be completed within four hours.[1741] Inspectors are required to collect and clearly document any evidence that indicates a violation and to provide an interim report to the master as well as a full report to the flag State of the fishing vessels.[1742]

The master of the fishing vessel has to ensure that the inspectors are able to work safely. He has to follow the principles of good seamanship, accept and facilitate prompt and safe boarding as well as safe disembarkation.[1743] During the inspection, he must cooperate and assist the inspectors, allow them to communicate and provide them with reasonable facilities.[1744] A refusal to allow a boarding and inspection requires an explanation from the vessel master.[1745] The flag State of the fishing vessel has to direct the vessel master to allow the operation if there is no acceptable reason for delay.[1746] If the master does not comply with the respective direction, the flag State has to suspend the vessel's authority

---

1738 WCPFC, *Conservation and Management Measure - Western and Central Pacific Fisheries Commission Boarding and Inspection Procedures*, (2006), Paras.18-21.

1739 Ibid. Para.22.

1740 Ibid. Para.23.

1741 Ibid. Para.24.c.

1742 Ibid. Paras.24.d,e and g.

1743 Ibid. Paras.25.a,b and g.

1744 Ibid. Paras.25.c, e and f.

1745 Ibid. Para.26.

1746 Ibid. Para.27.

to fish. An inspector who identifies a serious violation[1747] must inform the flag State of the fishing vessel both directly and through the Commission.[1748] The flag State has to regard any interference by its fishing vessels with an authorized inspector or inspection vessel on the high seas in the same manner as they would any such interference occurring within its EEZ.[1749] All boardings and inspections carried out have to be reported annually to the Commission.[1750] The members are required to state in their Annual Report the actions that they have taken in response to alleged violations, including any proceedings begun or sanctions applied.[1751]

After the adoption of CMM 2006-08, there were no significant discussions about high seas boarding and inspection during Commission meetings. From the statements made by CCMs during the Technical and Compliance Committee Meeting in 2011, it seems that there is general satisfaction with the High Seas Boarding and Inspection Programme.[1752] All boarding and inspection operations had been conducted successfully and no serious issues of non-compliance with WCPFC regulations were reported. One year earlier, the Technical and Compliance Committee had noted the progress of the programme and the Secretariat had only been tasked to prepare updated language cards with all conservation and management measures in order to ensure mutual understanding between the inspectors and vessel masters.[1753]

1747 Ibid. Para.37. Serious violations include: fishing without a license, permit or authorization; failure to maintain sufficient records of catch and catch-related data; significant misreporting; fishing in a closed area or season; intentional taking or retention of prohibited species; significant violation of catch limits or quotas; using prohibited fishing gear; falsifying or intentionally concealing the markings, identity or registration of a fishing vessel; concealing, tampering with or disposing of evidence relating to investigation of a violation; multiple violations; refusal to accept a boarding and inspection; assaulting, resisting, intimidating, sexually harassing, interfering with, or unduly obstructing or delaying an authorized inspector; intentionally tampering with or disabling the vessel monitoring system; and other violations as may be determined by the Commission.

1748 Ibid. Para.32.

1749 Ibid. Para.39.

1750 Ibid. Para.40.

1751 Ibid. Para.41.

1752 WCPFC, *Summary Report - Seventh Regular Session of the Technical Compliance Committee*, (2011), pp.51-52.

1753 WCPFC, *Summary Report - Sixth Regular Session of the Technical and Compliance Committee* (2010), p.30.

i)   *Vessels without nationality*

According to Article 92 of UNCLOS, vessels without nationality are vessels not flying the flag of any state or vessels flying the flag of two or more states. Such vessels represent a problem for the sustainable management of tuna and tuna-like species because potential issues on non-compliance cannot be assigned to any flag State. RFMOs rely on clear identification of vessels in order to prevent IUU fishing. The Convention of the WCPFC did not establish any specific provisions to deal with vessels without nationality. In contrast, CMM 2007-03 laid down that any vessels fishing in the Convention area without nationality would be presumed to have carried out IUU fishing.[1754]

In 2009, a measure explicitly dealing with vessels without nationality, mainly based on the recommendations of the International Plan of Action to prevent, deter, and eliminate Illegal Unregulated and Unreported fishing, was developed. CMM 2009-09 determined that vessels fishing on the high seas without nationality would be presumed to be operating in contravention of the objectives of the Convention and its measures.[1755] Such vessels had to be considered as undermining the respective provisions and thus constituting a serious violation of the Convention.[1756] CCMs were required to take all necessary measures, including enacting domestic legislation if appropriate, to prevent fishing by vessels without nationality, and all sightings of such vessels that might be fishing in the high seas had to be reported to the Secretariat.[1757]

Since the adoption of CMM 2009-09, there had been no explicit discussions about the measure during the annual meetings. However, in 2010, one of the four vessels on the WCPFC IUU vessel list for 2011 had been listed because a CCM had reported the vessel for fishing in the Convention area without nationality.[1758] In its justification of the listing, the Commission exclusively referred to CMM 2007-03. Due to this incident the vessel is not currently allowed to be involved in any fishing activities in the Convention area.

1754   WCPFC, *Conservation and Management Measure to Establish a List of Vessels presumed to Have Carried out Illegal, Unreported and Unregulated Fishing Activities in the WCPO*, (2007), Para.3.h.

1755   WCPFC, *Conservation and Management Measure for Vessels without Nationality*, (2009), Para.1.

1756   Ibid. Para.2.

1757   Ibid. Paras.4 and 5.

1758   WCPFC, *Summary Report - Seventh Regular Session of the Commission*, (2010), p.32.

j)   Charter Notification Scheme

Charter vessels are commonly used by small island developing states in order to develop their tuna fisheries. Chartering can be either "an arrangement in which the owner of a vessel is compensated by another party for some form of use of the vessel, [or] an arrangement in which the owner of a vessel compensates another party for services related to access provided by that party."[1759] In the WCPO usually, local companies from coastal states are chartering vessels from developed DWFNs. Important chartering countries are the Solomon Islands and Fiji.[1760] Vessels are mainly provided by China and Chinese Taipei.[1761]

Prior to the adoption of the first CMM on charter notifications, the Commission was concerned that charter arrangements could promote IUU fishing activities or undermine the CMMs.[1762] The Convention did not provide any requirements for charter arrangements. A Charter Arrangements Scheme, including notification provisions, was therefore already proposed at the second meeting of the Commission.[1763] During subsequent meetings, the CCMs were unable to progress further with it, but CMM 2008-01 explicitly required the Commission to consider the implementation of such a scheme at its annual meeting in 2009.[1764]

CMM 2009-08 was adopted at the following meeting, and it established basic procedures for charter arrangements. The provisions apply to members and participating territories that charter, lease or enter into other mechanisms with foreign flagged vessels for the purpose of conducting fishing operations in the Convention area as an integral part of their domestic fleet.[1765] Those parties have to notify the Commission of any vessel to be identified as chartered by submitting information including the name of the fishing vessel, its WCPFC Identification Number, the name and address of the owner and charterer, the duration of the charter arrangement and the flag state of the vessel.[1766] In addi-

---

1759   R Gillett, *Catch Attribution in the Western and Central Pacific Fisheries Commission*, Paper prepared by Gillett, Preston and Associates for the Seventh Regular Session of the Technical and Compliance Committee Meeting (Pohnpei: WCPFC, 2011), p.29.

1760   WCPFC, *Charter Notifications for 2011*, (2011).

1761   Ibid.

1762   WCPFC, *Conservation and Management Measure - Charter Notification Scheme*, (2009), Preamble.

1763   WCPFC, *Conservation and Management Measures for Bigeye and Yellowfin Tuna in the Western and Central Pacific Ocean*, (2005), Para.5.

1764   WCPFC, *Conservation and Magement Measures for Bigeye and Yellowfin Tuna in the Western and Central Pacific Ocean*, (2008), Para.2.

1765   WCPFC, *Conservation and Management Measure - Charter Notification Scheme*, (2009), Para.1.

1766   Ibid. Para.2.

tion, the Executive Director and the flag state have to be notified of any new chartered vessels, any change in the information or the termination of the charter of any vessel.[1767] Only vessels listed on the WCPFC Record of Fishing Vessels or the WCPFC Interim Register of Non-CCM Carriers and Bunkers, are eligible for charter, and none that appear on the WCPFC IUU vessel list or the IUU List of another RFMO.[1768] All information on charted vessels had to be made available to the CCMs by the Executive Director.[1769]

During the postponed annual Commission meeting of 2011, a list of chartered vessels was provided by the Secretariat.[1770] One member noticed there was poor compliance with the measure because the list of chartered boats for 2011 did not correspond with information in Part One Reports nor with statements made during the meeting.[1771] Although no further comments were made, it will be important to see how the Commission deals with this issue in future meetings. At the meeting in 2011, CMM 2009-08 had been reviewed and extended.[1772] Several members had been working on a proposal for a revision, but the Commission could not agree on any changes.

Another important issue with regard to chartered vessels is the attribution of catches. At the Technical and Compliance Committee meeting in 2011, a catch attribution study was presented.[1773] The crucial question was whether the catch of a chartered vessel should be attributed to the chartering state or to the flag state.[1774] This and other questions on catch attribution were delegated for discussion to the Technical and Compliance Committee meeting in 2012.[1775]

k)   *Compliance Monitoring Scheme*

In order to ensure sustainable fisheries it is essential to monitor and review the levels of compliance with the CMMs. In the WCPFC, the Technical and Compliance Committee is charged with that task and it is further required to make rel-

---

1767   Ibid. Para.3.

1768   Ibid. Para.4.

1769   Ibid. Paras.5 and 6.

1770   WCPFC, *Charter Notifications for 2011*, (2011).

1771   WCPFC, *Summary Report - Eighth Regular Session of the Commission*, (2012), p.46.

1772   Ibid. pp.46-47.

1773   WCPFC, *Summary Report - Seventh Regular Session of the Technical Compliance Committee*, (2011), pp.41-42.

1774   Gillett, 2011.

1775   WCPFC, *Summary Report - Eighth Regular Session of the Commission*, (2012), p.47.

evant recommendations to the Commission.[1776] Unfortunately, the Convention does not explicitly describe how to monitor and review the measures, so in the past this had been a difficult task, particularly due to the lack of a clear structure. As a consequence, in 2008 the process for the development of a Compliance Monitoring Scheme was initiated, and two years later the first measure was adopted.[1777]

CMM 2010-03 developed a scheme that was basically intended to ensure the implementation of and compliance with obligations arising under the Convention and the associated measures.[1778] Based on the information provided in the annual Part Two Reports, the Executive Director was required to prepare individual Draft Compliance Monitoring Reports for each CCM.[1779] Due to the immense workload of this task, in practice the reports were prepared on behalf of the Executive Director by the Compliance Manager, their staff and several interns.[1780] According to the measure, the CCMs had to receive the reports five days after the receipt of their Part Two Report or at least 25 days in advance of the annual Technical and Compliance Committee meeting.[1781] Although only weakly worded, it was stated that each CCM 'may reply' with comments on their report, if possible at least 12 days in advance of the Meeting.[1782] The Executive Director is then supposed to include the comments and information provided by the CCMs, but given the comparatively small staff of the Secretariat and the other tasks that have to be completed prior to the meeting, this timeframe has proven to be rather short.[1783] The resulting reports form the basis for a Provisional Compliance Monitoring Report which has to be developed by the Technical and Compliance Committee.[1784] The final Compliance Monitoring Report has to be adopted by the Commission, including compliance status and

---

1776 Convention on the Conservation and Management of Highly Migratory Fish Stocks in the Western and Central Pacific Ocean, Art.14.1(b).

1777 WCPFC, *Summary Report - Fifth Regular Session of the Commission*, (2008), pp.30-31.

1778 WCPFC, *Conservation and Management Measure for Compliance Monitoring Scheme*, (2010), Para.1.

1779 Ibid. Para.9.

1780 Observation made by the author during a research stay at the Secretariat of the WCPFC.

1781 WCPFC, *Conservation and Management Measure for Compliance Monitoring Scheme*, (2010) Para.10.

1782 Ibid. Para.11.

1783 Observation made by the author during the Seventh Regular Session of the Technical and Compliance Committee.

1784 WCPFC, *Conservation and Management Measure for Compliance Monitoring Scheme*, (2010), Para.14.

recommendations for any corrective actions.[1785] In the next Part Two Report, the CCMs were required to include any actions taken to address the previous year's non-compliance.[1786] The Commission is entitled to take a graduated response to non-compliance, while specific responses were to be developed during the following meetings.[1787]

By 2011, the Compliance Monitoring Scheme had operated for a one-year trial. At the postponed annual Commission meeting of 2011, the first Final Compliance Monitoring Report was adopted.[1788] This report provided the results of the assessment for 27 CCMs. Without naming specific countries, a table indicated to which extent the requirements of the measures had been implemented.[1789] The requirements were grouped by: catch and effort limits; catch and effort reporting; spatial and temporal closures and gear restrictions; observer and Vessel Monitoring System requirements; and scientific data provision, reporting and handling. The level of compliance was categorized as follows: implemented; potential implementation issue and explanation provided; potential implementation issue and more information needed; implementation needed; or not applicable to the respective CCM.[1790]

Overall, the report showed that most of the requirements had been implemented by the CCMs, but for several measures more information was needed or further steps towards implementation were required.[1791] The most critical measures were CMM 2007-01 (Regional Observer Programme); CMM 2007-04 (seabirds); CMM 2008-03 (sea turtles); CMM 2008-04 (drift nets); CMM 2009-02 (FAD closures and catch retention); CMM 2009-04 (sharks; in particular the 5% fin to weight ratio); Standards, Specifications and Procedures for the fishing Vessel Monitoring System; as well as scientific data to be provided to the Commission (particularly estimates of annual catches for sharks, estimates of discards, and size composition data).[1792] The main challenges with regard to the further development of the scheme were the limited time and information available, and the lack of clarity about exactly how or to which CCMs particular aspects of measures apply.[1793] It was suggested that additional information be

---

1785 Ibid. Para.19.

1786 Ibid. Para.21.

1787 Ibid. Paras.22 and 23.

1788 WCPFC, *Summary Report - Eighth Regular Session of the Commission*, (2012), Annex N.

1789 Ibid. Annex N, Attachment 1.

1790 Ibid. Annex N, Para.6.

1791 Ibid. Annex N, Para.10.

1792 Ibid. Annex N, Para.11.

1793 Ibid. Annex N, Para.8.

used for more comprehensive evaluation and to develop clear guidelines and procedures for the Secretariat, CCMs, the Technical and Compliance Committee and the Commission. A special need was also identified with regard to clear and objective formats and the criteria for the review of the individual reports.

During the postponed annual Commission meeting of 2011, the scheme was supported by most of the members and it was agreed to extend a revised version of the first measure for another year. CMM 2011-06 included new provisions which clarified that the Draft and Provisional Compliance Monitoring Reports had to be constituted as non-public domain data, and the final Compliance Monitoring Reports as public domain data.[1794] The measure also imposed earlier deadlines in order to enable all participants to produce the best possible results.[1795] In addition, it was laid down that inter-governmental sub-regional agencies were allowed to participate in working groups as observers in order to provide advice and assistance to developing states.[1796] Contrary to the requirements of CMM 2010-03, no new provisions were included with regard to the graduated response to non-compliance. The definition of such responses remains a task for the Commission in the future but no specific dates were determined for this.[1797]

D. Summary and challenges

The question of whether the WCPFC will be able to maintain the present situation with comparatively good stock statuses over the coming years required a detailed analysis of the legal framework for fisheries management in the WCPO. Overall, the analysis has shown that there are the progressive legal structures on a regional and sub-regional level. Despite these structures, it became clear that conservation and management of tuna and tuna-like species in the WCPO continues to pose a multifaceted challenge due to the variety of countries and gear types involved in the fishery. Although the WCPFC is the competent body to coordinate conservation and management, it must be understood that future challenges can only be tackled through the cooperation of all countries involved in the fishery.

For the WCPFC, an early challenge was developing its own legal framework while also taking into account the existing sub-regional structures which had been established by Pacific Island Countries. Eight years after the establishment

---

1794 WCPFC, *Conservation and Management Measure for Compliance Monitoring Scheme*, (2012), Para.6.

1795 Ibid. Paras.10-13.

1796 Ibid. Para.8(i).

1797 Ibid. Para.23.

of the WCPFC, it can be seen that significant progress has been made in this area. With the SPC and the FFA there are Memoranda of Understanding which specified and formalized their cooperation with the WCPFC. The Secretariat of the Pacific Community as the oldest sub-regional body provides comprehensive scientific information like stock assessments which are crucial for the adoption of appropriate conservation and management measures. The Pacific Island Forum Fisheries Agency (FFA) harmonized the policies of Pacific Island Countries and Territories (PICTs) and facilitated the development of several key regional arrangements dealing with the access by foreign fishing fleets into EEZs. The WCPFC benefited from the existing structures established by the FFA. Examples are the regional observer program and the Vessel Monitoring System of the FFA which both had been integrated in the framework of the WCPFC. However, there are also remaining challenges with regard to crucial issues like allocation and the management of the overall capacity. Due to the increasing importance of some of these sub-regional organizations with regard to the management of the EEZ fisheries, effective coordination of the cooperation will remain a crucial task for the WCPFC into the future.

A second challenge is to overcome the dichotomy between the aims of Distant Water Fishery Nations (DWFNs) and those of the Pacific Island Countries and Territories (PICTs). Essentially, the DWFNs want to maintain their current percentage of the total catch, while the PICTs have fisheries development as a major goal. If both aims were fulfilled, this would inevitably result in an increase in the total fishing effort. The problem is that the effort in most of the fisheries in the WCPO is already at a high level and so, if fishing is to be kept within sustainable limits, there is no room for any further increases. Particularly the situation of the fisheries catching skipjack, yellowfin and bigeye tuna in the Convention area between 20° north and 20° south shows that any increase in effort by the PICTs would require an equivalent decrease of effort by the DWFNs. Overall, both sides must compromise and find a solution that does not result in further total effort increases. The task for the WCPFC is to adopt conservation and management measures which acknowledge and respect the development aspirations of the PICTs, while carefully balancing economic interests with wider conservation needs. Especially with regard to bigeye tuna stock which is suffering overfishing there is uncertainty if the WCPFC can cope with this task.

The integration of both the western and northern parts of the Convention area is a third challenge. In the western part, Indonesia, Philippines, and Vietnam have already made, in the course of the West Pacific East Asia Project, good progress in improving voluntarily the quality of their data collection and reporting. The efforts made by the WCPFC in cooperation with these countries have the potential to significantly reduce uncertainty of stock assessments conducted by SPC. With regard to the area north of 20° north it can be criticized that, the Northern Committee is acting like a sub-RFMO. As a consequence, some important requirements such as the mandatory use of the Vessel Monitoring System (VMS) or the observer programme are not applied in this area yet. The success in including the northern area will show if the WCPFC is able to close remaining gaps and to ensure comprehensive management throughout the whole Convention area.

A fourth challenge is to deal with the conflicting interests between specific fisheries that use different gear types to catch the same species. The most significant case is the conflict between purse seine and long line fisheries and their joint impact on the very nearly overfished bigeye tuna stock. One part of the problem is that in the purse seine fishery, where mainly skipjack and yellowfin tuna are targeted, there is a significant bycatch of juvenile bigeye tuna, particularly when Fish Aggregating Devices are used. These unintended catches are both ecologically and economically unsustainable in the long term. The other part of the problem is the high catch levels in the long line fishery which is catching mainly adult bigeye tuna. Although it is difficult to identify which gear type is ultimately responsible for the bad stock status of bigeye tuna, it must be recognized that, at least in the recent past, any reductions of longline catches have been offset by a corresponding increase in purse seine effort. Successfully managing this species sustainably will therefore depend on the ability of the WCPFC to reconcile the interests of both fisheries.

Ensuring the consistent implementation of the Convention through the Commission represents a fifth challenge. Although numerous conservation and management measures have been adopted, there are still important issues which need to be addressed. One example is the progressive provisions of the Convention on the precautionary approach, which are unique among tuna RFMOs. Unfortunately, there is a gap between what the Convention requires and actual legal practice. An essential element is the adoption of precautionary target reference points, but, to date, target reference points have not been established for any species. Another example is the development of criteria used for the allocation of the Total Allowable Catch or Total Allowable Effort. The Convention provides explicit guidance on how to develop such criteria, but, again, no significant progress has been made. Further areas with the potential for improvement are trade related and port state measures. If the Commission takes the provisions of the Convention seriously it has to implement all of its requirements. Sustainable management needs successful implementation to make the progressive provisions work.

A sixth challenge is to avoid that exemptions undercut conservation and management efforts. Several measures include exemptions from certain requirements, which serve to weaken the measures, acting as legal 'loopholes' for unsustainable fisheries. Examples of these exemptions include: the special regulations for the area north of 20° north; the possibility for certain countries to bypass the Fish Aggregating Device closure; the special status given to China with regard to catch reduction of bigeye tuna in the long line fishery; and the general exemption of small island developing states from effort reduction. Although it is recognized that exemptions are often required to ensure that a measure is adopted in any form, they should be used as rarely as possible, and all existing exemptions should be carefully reviewed and reconsidered.

Securing the compliance of all participants that are involved in the fishery represents a seventh challenge. Due to a lack of relevant provisions, it was very difficult for the Commission to comprehensively evaluate compliance in the first years after the establishment of the WCPFC. The recently established Compliance Monitoring Scheme is, however, a positive development in this area. This scheme has the potential to improve compliance with the existing requirements. The possibility of establishing clear criteria to control the compliance of every single member represents an important step forward. Only time will tell if this instrument can deliver what is expected of it. One problem that still needs to be addressed is the Commission's limited ability to issue penalties for non-compliance.

In summary, the WCPFC has succeeded in establishing a robust framework for the conservation and management of tuna and tuna-like species in the WCPO. However, there are several challenges which must be addressed over the coming years. The success of the WCPFC in meeting these challenges will ultimately depend on its ability to integrate the interests of all participants through the development of a well-balanced system that determines how the burden of conservation is to be shared.

Chapter V

Final conclusions

The overall objective of this work was to support a broader understanding of the international legal framework for the management of fisheries targeting tuna and tuna-like species and to examine whether or not this framework enables tuna RFMOs, and in particular the WCPFC, to manage their fisheries sustainably.

The analysis of the international legal framework has shown that there are several instruments which include provisions that address tuna fisheries management. The development of these instruments is characterized by increasing clarity and specificity over time, while, in parallel, newer and more progressive provisions were introduced to tackle the errors of the past. The most important international instruments are UNCLOS and the Fish Stocks Agreement. Together with other relevant instruments, both binding and non-binding, these treaties form a comprehensive legal framework that covers all important aspects of the management of tuna fisheries. This international legal framework has empowered policy makers to manage fisheries sustainably and it also provides the basis for the development of more specific regulations at a regional level. Unfortunately, simply having an international legal framework has not been enough to solve the problem of overfishing. The central challenge remaining is the effective implementation of the existing instruments through both fishing states and tuna RFMOs.

The Southern Bluefin Case and the Swordfish Case have demonstrated that international jurisdiction dealing with fisheries targeting tuna and tuna-like species is highly complex. In this complex environment, the dispute settlement procedures under the UNCLOS but also under the GATT have proven to be useful instruments to guide negotiations between the conflicting parties. Although in both cases, in the end, the disputes were solved through negotiations formally located outside the framework of UNCLOS, it is remarkable that the dispute settlement procedures were applied consistently and turned out important to clarify the competences of the institutions involved. In general, it can be seen that all conflicting parties fully cooperated with the dispute settlement bodies, even though they questioned the legitimacy of their establishment. This is a positive finding with regard to possible future cases involving tuna and tuna-like species.

The analysis of the current performance of the five tuna RFMOs revealed that their legal frameworks are quite similar. They have similar organizational structures, they are all based on comprehensive constituent instruments which are supplemented by numerous conservation and management measures, and they all have developed similar tools to ensure compliance. Unfortunately, they also share deficits like the insufficient application of formal and transparent allocation criteria. Despite the similarities, there are differences in some areas as well. Examples are decision making procedures, accession of new members, port states measures and trade related measures, or the application of the precautionary approach. With regard to the important precautionary approach, it could be shown that only IATTC and WCPFC provide explicit provisions in their constituent instruments and to date only IOTC and WCPFC have adopted formal reference points. However, time will tell if these provisions are able to make a real difference with regard to sustainable management. Many of the variances which have been identified in the legal frameworks reflect the regional and historical characteristics of the fisheries. It is therefore difficult to determine which tuna RFMO is providing the most suitable overall legal framework for sustainable management of fisheries targeting tuna and tuna-like species. Given the already existing similarities as well as the efforts of the tuna RFMOs to enhance cooperation, it can be further expected that the frameworks will become increasingly similar in the future. This applies, for example, to the adoption of port state measures and trade related measures. Although there are currently significant differences across the tuna RFMOs, it is very likely that there will be a progressive alignment of relevant measures in the near future as happened with the measures which established the more traditional compliance tools.

A general problem for all tuna RFMOs, and other RFMOs, is an insufficient separation between science and politics. Risk assessment is not separated sufficiently from risk management. It is common practice in international fisheries management that scientific recommendations are negotiated against the backdrop of political interests, even before the annual Commission meetings begin. In the WCPFC, the SPC, which is a comparatively independent consultative and advisory body without competence for fisheries management, provides the results of its assessments to the Scientific Committee. This Committee is supposed to transform the information from these assessments into scientific recommendations for the Commission, which should then adopt conservation and management measures in line with these recommendations at its annual meeting. However, as the Committee is composed by scientific delegates from the RFMO member states, the recommendations are often the result of negotiations that have already been politically driven and shaped.

In most cases, the effect of this political influence at the scientific level is that the resulting 'scientific' recommendations do not deliver the best possible outcome from a wholly scientific point of view. With regard to catch and effort limits, the recommendations are often above the limits that would be recommended by independent scientists. At the Commission meetings, these recommendations are usually negotiated again with the result that catch and effort limits are further increased. While a certain amount political influence at the Commission meetings is understandable, there is no doubt that much more effort should be made to keep at least the scientific assessments and recommendations purely scientific and protected from any political influence.

But how could this political influence be avoided or, at the very least, reduced? Ideally, the risk assessment should be conducted exclusively by independent scientists, and only the risk management should be the responsibility of delegates from the WCPFC member countries. This clear separation would also increase transparency, allowing the public to clearly compare the scientific recommendations on the one side with the conservation and management measures that are eventually adopted based on political decisions on the other. To this end, scientists should be enabled to work independently. That means that instead of working for a specific country, they should work for an independent institution. This could be achieved by establishing an independent scientific body or by strengthening the independence of the existing bodies. With regard to the WCPFC, there should be a mechanism to strengthen the authority of the SPC in order to prevent political interests from weakening or diluting the results of the assessments. This could be done by amending the existing Convention of the WCPFC or by adopting new rules of procedure which entitle the SPC to give recommendations directly to the Commission.

In practice, national interests would make it difficult to reach agreement among the fishing countries in order to develop a legal framework that could implement this approach. Another problem to be addressed would be the independent funding of such a body. But even if it were possible to get the funding and to either amend an existing treaty or develop a new one, there would be still the problem of finding international scientists who are truly independent. It would be absolute necessary that such scientists were paid by an international organization in order to guarantee they were financially independent from the political interests of their states of origin. In addition, it has to be considered that scientists are reliant to a significant degree on the cooperation of the fishing countries. Without the cooperation of these countries regarding the provision of catch data or the facilitation of onboard observations it would be very difficult to conduct stock assessments.

So in summary, on the one hand the international community should find a way to create or strengthen a scientific body which is really independent from national interests, and on the other hand there should be a willingness from the fishing countries to cooperate with these scientists to enable them to conduct their scientific assessments. However, even if this could be made to work successfully, there would always be the danger that some countries might try to infiltrate the scientific bodies in order to guide the assessments in a direction that suits their national interests. This could result in political influence being moved to another level, instead of being avoided. These factors have to be considered if the risk assessments are to be kept purely scientific and without political influence. In the end, the key challenge will be to keep the scientists as far as possible from political discussions but close enough to crucial scientific data that is produced and collected by the fishing countries.

The detailed analysis of fisheries management in the WCPO has shown that the legal framework governing tuna fisheries in this region does, without a doubt, provide the WCPFC with the tools to manage tuna and tuna-like species sustainably. The WCPFC, as the most recently established tuna RFMO, has incorporated some of the most progressive provisions from the international treaties in its Convention, and it has adopted numerous conservation and management measures based on the requirements of the Convention.

Particularly remarkable in the WCPFC is the role of the developing Pacific Island Countries. There is no other tuna RFMO where economically weak developing countries like the Pacific Island Countries have such an influence on conservation and management of tuna and tuna-like species. The main reason for this influence is their unique cooperation which has made them strong political players. The Pacific Island Countries have recognized that they can increase their power if they cooperate with regard to the use of their sovereign rights over the exploitation of tuna and tuna-like species in their EEZs. Once it became possible to establish EEZs, the Pacific Island Countries saw potential of a cooperative approach and worked to put it into practice. Today, in the decision making processes of the WCPFC, they always act as a bloc with a common agenda. This united front makes it difficult for economically stronger countries to pressurize individual Pacific Island Countries in order to push through their own political agenda.

It is hoped that other developing countries across the world learn from this extraordinary cooperation and work together to increase their bargaining power, especially when it comes to the use of natural resources in areas where they have the sovereign rights to exploit them. The example of the Pacific Island Countries could also be applied to the other tuna RFMOs, as they too are all, with the exception of the CCSBT, composed by a significant number of developing coastal states. The respective countries could review their current cooperation and amend existing treaties where possible or develop new ones in a similar way to the legal framework established by the Pacific Island Countries.

Other areas where cooperation could help developing states to achieve a better outcome could be the exploitation of other marine resources such as other types of fish, mining for mineral resources or the use of marine genetic resources. In all these areas, it is difficult for a single developing state to negotiate with economically strong countries but if they acted as bloc with a common agenda based on their legal rights they could advance their own interests together.

Transparency is another area where other international organizations could learn from the WCPFC. NGOs are allowed to participate in all WCPFC meetings including the negotiation sessions and they are also allowed to speak during the sessions. In other organizations, like the ICCAT, NGO participation is much more restricted. The opportunity to participate can provide several benefits to both NGOs and the public. The main benefit at the WCPFC is that the NGOs are able to observe the whole negotiation process, and can therefore clearly identify which countries are responsible for specific recommendations or decisions. This is a big advantage over the information that could be extracted from studying the summary reports of the meetings, because in these summary reports, recommendations or decisions are usually not related to specific countries. Another significant benefit at the WCPFC is the direct contact with the policymakers which enables NGOs to exchange opinions and arguments. The possibility of speaking during the sessions of the meetings in front of the delegates further allows the NGOs to address their messages to all participants of the meetings and to raise awareness with regard to important topics like the overexploitation of shark species. With this high degree of transparency, the WCPFC could serve as an example for other tuna RFMOs as well as for other international organizations.

A special characteristic of the WCPFC is that the meetings are sometimes held at remote venues like the small Pacific Island Countries. Unfortunately the high travel costs sometimes make it difficult for smaller NGOs to attend the meetings. This is a downside but it is outweighed by the importance given to the Pacific Island States by holding the meetings in their region.

However, despite the progressive and comprehensive approach of the WCPFC, this work has also identified some major issues where further action is needed. The organization faces seven key challenges, and how it deals with them will determine whether the RFMO will succeed or fail in the future. These challenges are to:

- cooperate effectively with all sub-regional organizations;

- overcome the dichotomy between the aims of Distant Water Fishery Nations and those of the Pacific Island Countries;

- further integrate the areas in the north and west of the Convention area;

- reconcile the conflicting interests of fisheries that use different gear types to catch the same species;

- ensure the consistent implementation of the Convention;

- avoid, or mitigate, exemptions to provisions for certain countries; and

- secure the compliance of all participants that are involved in the fishery.

All of these challenges could be addressed by either introducing new conservation and management measures or revising existing measures. The analysis has shown that current problems, such as the unsustainable management of bigeye tuna or the lack of target reference points, stem from the lack of political will of certain participants in the fishery, and not from any weakness in the legal framework. Ultimately, the critical task for the WCPFC will be to integrate the interests of all participants in the fishery and to determine how the burden of conservation is to be shared among them.

Legal documents

A Second Arrangement Implementing the Nauru Agreement Setting Forth Additional Terms and Conditions of Access to the Fisheries Zones of the Parties (1990).

A third arrangement implementing the Nauru Agreement setting forth additional terms and conditions of access to the fisheries zones of the parties (2008).

Agenda 21: Earth Summit - The United Nations Programme of Action from Rio (1992).

Agreement establishing the South Pacific Commission (SPC) (1948).

Agreement for the Establishment of the Indian Ocean Tuna Commission (1993).

Agreement for the Implementation of the Provisions of the United Nations Convention on the Law of the Sea of 10 December 1982 relating to the Conservation and Management of Straddling Fish Stocks and Highly Migratory Fish Stocks (1995).

Agreement on Port State Measures to prevent, deter and eliminate illegal, unreported and unregulated fishing (2009).

Agreement on Port State Measures to Prevent, Deter and Eliminate Illegal, Unreported and Unregulated Fishing (2009).

Agreement to Promote Compliance with International Conservation and Management Measures by Fishing Vessels on the High Seas (1993).

An Arrangement Implementing the Nauru Agreement Setting Forth Minimum Terms and Conditions of Access to the Fisheries Zones of the Parties (1982).

Assembly U. N. G., UN General Assembly Resolution 44/225: Large-Scale Pelagic Drift-Net Fishing and Its Impact on the Living Marine Resources of the World's Oceans and Seas (1989).

Assembly U. N. G., UN General Assembly Resolution 46/215: Large-Scale Pelagic Drift-Net Fishing and Its Impact on the Living Marine Resources of the World's Oceans and Seas (1991).

Case concerning the Conservation and Sustainable Exploitation of Swordfish Stocks in the South-Eastern Pacific Ocean (Chile / European Union) - Order 2000/3, 2000, ITLOS.

Case concerning the Conservation and Sustainable Exploitation of Swordfish Stocks in the South-Eastern Pacific Ocean (Chile / European Union) - Order 2001/1, 2001, ITLOS.

Case concerning the Conservation and Sustainable Exploitation of Swordfish Stocks in the South-Eastern Pacific Ocean (Chile / European Union) - Order 2003/2, 2003, ITLOS.

Case concerning the Conservation and Sustainable Exploitation of Swordfish Stocks in the South-Eastern Pacific Ocean (Chile / European Union) - Order 2005/1, 2005, ITLOS.

Case concerning the Conservation and Sustainable Exploitation of Swordfish Stocks in the South-Eastern Pacific Ocean (Chile / European Union) - Order 2007/3, 2007, ITLOS.

Case concerning the Conservation and Sustainable Exploitation of Swordfish Stocks in the South-Eastern Pacific Ocean (Chile / European Union) - Order 2008/1, 2008, ITLOS.

Case concerning the Conservation and Sustainable Exploitation of Swordfish Stocks in the South-Eastern Pacific Ocean (Chile / European Union) - Order 2009/1, 2009, ITLOS.

CCSBT, Action Plan (2000).

CCSBT, CCSBT Scientific Observer Program Standards (2003).

CCSBT, Resolution on action plans to ensure compliance with Conservation and Management Measures (2009).

CCSBT, Resolution on amendment of the Resolution on "Illegal, Unregulated and Unreported Fishing (IUU) and Establishment of a CCSBT Record of Vessels over 24 meters Authorized to Fish for Southern Bluefin Tuna" (2008).

CCSBT, Resolution on Establishing a Program for Transshipment by Large-Scale Fishing Vessels (2008).

CCSBT, Resolution on establishing the CCSBT Vessel Monitoring System (2008).

CCSBT, Resolution on the Adoption of a Management Procedure (2011).

CCSBT, Resolution on the Allocation of the Global Total Allowable Catch (2011).

CCSBT, Resolution on the Implementation of a CCSBT Catch Documentation Scheme (2006).

CCSBT, Resolution on the Implementation of a CCSBT Catch Documentation Scheme (2012).

CCSBT, Resolutions pursuant to the 2000 Action Plan (2003).

CCSBT, Strategic Plan for the Commission for the Conservation of Southern Bluefin Tuna (2011).

Chile - Measures Affecting the Transit and Importation of Swordfish - Request for Consultations by the European Communities of 26 April 2000, 2000, WTO.

Chile - Measures Affecting the Transit and Importation of Swordfish - Request for the Establishment of a Panel by the European Communities of 6 November 2000, 2000, WTO.

Chile - Measures Affecting the Transit and Importation of Swordfish - Arrangement between the European Communities and Chile - Communication by the European Communities of 23 March 2001, 2001, WTO.

Chile - Measures Affecting the Transit and Importation of Swordfish - Arrangement between the European Communities and Chile - Communication from Chile - Addendum of 28 March 2001, 2001, WTO.

Chile - Measures Affecting the Transit and Importation of Swordfish - Arrangement between the European Communities and Chile - Communication from Chile and the European Communities - Addendum of 12 November 2003, 2003, WTO.

Chile - Measures Affecting the Transit and Importation of Swordfish - Arrangement between the European Communities and Chile - Communication from the European Communities - Addendum of 21 December 2005, 2005, WTO.

Chile - Measures Affecting the Transit and Importation of Swordfish - Arrangement between the European Communities and Chile - Communication from the European Communities - Addendum of 13 December 2007, 2007, WTO.

Chile - Measures Affecting the Transit and Importation of Swordfish - Joint Communication from the European Union and Chile - Addendum of 28 May 2010, 2010, WTO.

Chilean National Fishery Law (Ley General de Pesca y Acuicultura) - Consolidated by the Supreme Decree 430 of 28 September 1991, and extended by Decreee 598 of 15 October 1999 (1991).

Code of Conduct for Responsible Fisheries (1995).

Conclusions in the General Report of the African States Regional Seminar on the Law of the Sea (Yaoundé Conclusions) (1972).

Convention for the Conservation of Southern Bluefin Tuna (1993).

Convention for the Conservation of Southern Bluefin Tuna (1993).

Convention for the Establishment of an Inter-American Tropical Tuna Commission (1949).

Convention for the Prohibition of Fishing with Long Driftnets in the South Pacific (1989).

Convention for the Strengthening of the Inter-American Tropical Tuna Commission established by the 1949 Convention between the United States of America and the Republic of Costa Rica "Antigua Convention" (2003).

Convention on Fishing and Conservation of the Living Resources of the High Seas (1958).

Convention on the Conservation and Management of Highly Migratory Fish Stocks in the Western and Central Pacific Ocean (2000).

Convention on the Conservation and Management of Highly Migratory Fish Stocks in the Western and Central Pacific Ocean (2000).

Convention on the Continental Shelf (1958).

Convention on the High Seas (1958).

Convention on the Territorial Sea and the Contiguous Zone (1958).

Data exchange agreement between the Western and Central Pacific Fisheries Commission (WCPFC) and the Secretariat of the Pacific Community (SPC) (2009).

Decision of the Commission of the European Communities - Provisions of the Council Regulation No. 3268/94, Document 300D0296, Para.17 (2000).

Declaration of Cancun (1992).

Declaration of the Organization of African Unity on the "Issues of the Law of the Sea" (Addis Ababa Declaration) (1973).

Declaration of the United Nations Conference on the Human Environment (1972).

Declaration on the Maritime Zone (Santiago Declaration) (1952).

Draft Articles for a Convention on the Conservation and Management of Highly Migratory Fish Stocks in the Western and Central Pacific Ocean (1998).

FAO Model Scheme on Port State Measures to Combat Illegal, Unreported and Unregulated Fishing (2007).

Federal Republic of Germany v. Iceland, 1974, International Court of Justice.

Final Act of the Multilateral High-Level Conference on the Conservation and Management of Highly Migratory Fish Stocks in the Western and Central Pacific (2000).

IATTC, Resolution (amended) on a Regional Vessel Register (2011).

IATTC, Resolution (amended) on establishing a list of longline fishing vessels over 24 meters (LSTLFVs) authorized to operate in the eastern pacific ocean (2011).

IATTC, Resolution (amended) on establishing a program for transipment by large scale vessels (2011).

IATTC, Resolution on a Multinational Program for the Conservation of Tuna in the Eastern Pacific Ocean in 2011-2013 (2011).

IATTC, Resolution on fleet capacity (1998).

IATTC, Resolution on fleet capacity (2001).

IATTC, Resolution on IATTC bigeye tuna statistical document program (2003).

IATTC, Resolution on Northern Albacore Tuna (2005).

IATTC, Resolution on scientific observers for longline vessels (2011).

IATTC, Resolution on the conservation of sharks caught in association with fisheries in the eastern pacific ocean (2005).

IATTC, Resolution on the establishment of a Vessel Monitoring System (VMS) (2004).

IATTC, Resolution to establish a list of vessels presumed to have carried out illegal, unreported and unregulated fishing activities in the eastern pacific ocean (2005).

ICCAT, Criteria for the allocation of fishing possibilities (2001).

ICCAT, Recommendation Amending the Recommendation by ICCAT to Establish a Multi-annual Recovery Plan for Bluefin Tuna in the Eastern Atlantic and Mediterranean (2010).

ICCAT, Recommendation by ICCAT Amending the Recommendation 08-12 on an ICCAT Bluefin Tuna Catch Documentation Program (2009).

ICCAT, Recommendation by ICCAT Concerning a Limit on Bluefin Tuna Size and Fishing Mortality (1974).

ICCAT, Recommendation by ICCAT concerning minimum standards for the establishment of a Vessel Monitoring System in the ICCAT Convention area (2003).

ICCAT, Recommendation by ICCAT concerning the amendment of the forms of the ICCAT bluefin/bigeye/swordfish statistical documents (2003).

ICCAT, Recommendation by ICCAT concerning the ban on landings and transhipments of vessels from non-Contracting Parties identified as having committed a serious infringement (1998).

ICCAT, Recommendation by ICCAT Concerning the Establishment of an ICCAT Record of Vessels 20 Meters in Length Overall or Greater Authorized to Operate in the Convention Area (2011).

ICCAT, Recommendation by ICCAT concerning the ICCAT Bigeye Tuna Statistical Document Program (2001).

ICCAT, Recommendation by ICCAT Concerning the ICCAT Bluefin Tuna Statistical Document Program (1992).

ICCAT, Recommendation by ICCAT concerning the trade sanction against St. Vincent and the Grenadines (2002).

ICCAT, Recommendation by ICCAT establishing a programme for transhipment (2006).

ICCAT, Recommendation by ICCAT establishing a Swordfish Statistical Document Program (2001).

ICCAT, Recommendation by ICCAT for a Revised ICCAT Port Inspection Scheme (1997).

ICCAT, Recommendation by ICCAT Further Amending Recommendation 09-10 Establishing a List of Fishing Vessels Presumed to be Engaged in Illegal, Unreported and Unregulated (IUU) Fishing Activities in the ICCAT Convention Area (2011).

ICCAT, Recommendation by ICCAT on a multi-year conservation and management program for bigeye tuna (2004).

ICCAT, Recommendation by ICCAT on a Yellowfin Size Limit (1972).

ICCAT, Recommendation by ICCAT on an Electronic Bluefin Tuna Catch Document Programme (eBCE) (2010).

ICCAT, Recommendation by ICCAT on an Electronic Statistical Document Pilot Program (2006).

ICCAT, Recommendation by ICCAT on an ICCAT Bluefin Tuna Catch Documentation Program (2007).

ICCAT, Recommendation by ICCAT on an ICCAT Bluefin Tuna Catch Documentation Program (2008).

ICCAT, Recommendation by ICCAT on supplemental regulatory measures for the management of Atlantic yellowfin tuna (1993).

ICCAT, Recommendation by ICCAT Regarding Belize and Honduras Pursuant to the 1994 Bluefin Tuna Action Plan Resolution (1996).

ICCAT, Recommendation by ICCAT regarding Bolivia pursuant to the 1998 Resolution concerning the unreported and unregulated catches of tuna by large-scale longline vessels in the Convention area (2000).

ICCAT, Recommendation by ICCAT Regarding Panama Pursuant to the 1994 Bluefin Tuna Action Plan Resolution (1996).

ICCAT, Rules of procedure (2012).

International Convention for the Conservation of Atlantic Tunas (1966).

International Plan of Action for Reducing Incidental Catch of Seabirds in Longline Fisheries (1999).

International Plan of Action for Reducing Incidental Catch of Seabirds in Longline Fisheries, Technical note on some optional technical and operational measures for reducing the incidental catch of seabirds in longline fisheries (1999).

International Plan of Action for the Conservation and Management of Sharks (1999).

International Plan of Action for the Management of Fishing Capacity (1999).

International Plan of Action to Prevent, Deter, and Eliminate Illegal, Unreported and Unregulated Fishing (IUU) (2001).

IOTC, Recommendation 12/14 On Interim Target And Limit Reference Points (2012).

IOTC, Resolution 01/03 Establishing a Scheme to promote compliance by Non-Contracting Party vessels with resolutions established by IOTC (2001).

IOTC, Resolution 01/06 Concerning the IOTC bigeye tuna statistical document programme (2001).

IOTC, Resolution 02/01 Relating to the establishment of an IOTC programme of inspection in port (2001).

IOTC, Resolution 03/01 On the limitation of fishing capacity of Contracting Parties and Cooperating non-Contracting Parties (2003).

IOTC, Resolution 05/01 On conservation and management measures for bigeye tuna (2005).

IOTC, Resolution 05/03 relating to the establishment of an IOTC programme of inspection in port (2005).

IOTC, Resolution 06/03 On establishing a Vessel Monitoring System (2006).

IOTC, Resolution 06/05 On the limitation of fishing capacity, in terms of number of vessels, of IOTC contracting parties and co-operating non contracting parties (superseded by Resolution 09/02) (2006).

IOTC, Resolution 07/02 Concerning the establishment of an IOTC Record of Vessels Authorised to operate in the IOTC area (2007).

IOTC, Resolution 07/05 Limitation of fishing capacity of IOTC Contracting Parties and Cooperating non-Contracting Parties in terms of number of longline vessels targeting swordfish and albacore (superseded by Resolution 09/02) (2007).

IOTC, Resolution 09/02 On the implementation of a limitation of fishing capacity of Contracting Parties and Cooperating non-Contracting Parties (2009).

IOTC, Resolution 09/03 On establishing a list of vessels presumed to have carried out illegal, unregulated and unreported fishing in the IOTC area (2009).

IOTC, Resolution 10/01 For the Conservation and Management of Tropical Tunas Stocks in the IOTC Area of Competence (2010).

IOTC, Resolution 10/01 for the conservation and management of tropicaltunas stocks in the IOTC area of competence (2010).

IOTC, Resolution 10/11 on Port State Measures to prevent, deter and eliminate illegal, unreported and unregulated fishing (2010).

IOTC, Resolution 11/03 On Establishing A List Of Vessels Presumed To Have Carried Out Illegal, Unreported And Unregulated Fishing In The IOTC Area of Competence (2011).

IOTC, Resolution 11/04 On a regional observer scheme (2011).

IOTC, Resolution 11/05 On establishing a programme for transipment by large-scale fishing vessels (2011).

IOTC, Resolution 12/01 On The Implementation Of The Precautionary Approach (2012).

Johannesburg Declaration on Sustainable Development (2002).

Majuro Declaration (1997).

Memorandum of Understanding between the CCSBT and ICCAT Secretariats for Transshipment at sea by Large-Scale Fishing Vessels (2009).

Memorandum of Understanding between the CCSBT and IOTC Secretariats for Monitoring Transshipment at Sea by Large-Scale Tuna Longline Fishing Vessels (2009).

Memorandum of Understanding between WCPFC and CCSBT (2006).

Memorandum of Understanding between WCPFC and CCSBT (2006).

Memorandum of Understanding between WCPFC and IATTC (2009).

Memorandum of Understanding between WCPFC and IOTC (2009).

Nauru Agreement Concerning Cooperation on the Management of Fisheries of Common Interest (1982).

Niue Treaty on Cooperation in Fisheries Surveillance and Law Enforcement in the Pacific Region (1992).

Palau Arrangement for the Management of the Purse Seine Fishery in the Western and Central Pacific (1992).

Palau Arrangement for the Management of the Purse Seine Fishery in the Western and Central Pacific (as amended by VDS Working Group Meeting-Honiara, 7 & 13 October 2005) (2005).

Palau Arrangement for the Management of the Purse Seine Fishery in the Western and Central Pacific (Amended 11th September 2010) (2010).

Presidential Declaration Concerning Continental Shelf (1947).

Presidential Decree No. 781 (1947).

Report of The Thirteenth Session of the Asian-African Consultative Committee (1972).

Rio Declaration on Environment and Development (1992).

Rio Declaration on Environment and Development (1992).

Rules of the Tribunal, 2009, ITLOS.

Second United Nations Conference on the Law of the Sea (1960).

South Pacific Forum Fisheries Agency Convention (1979).

Southern Bluefin Tuna Case - Australia and New Zealand v. Japan, Arbitral Award of August 4, 2000, 2000, First Arbitral Tribunal constituted under Part XV ("Settlement of Disputes"), Annex VII ("Arbitration") of the United Nations Convention of the Law of the Sea (UNCLOS).

Southern Bluefin Tuna Cases (New Zealand v. Japan; Australia v. Japan) - Annex 2 Australia's Statement of Claim dated 15 July 1999, 1999, ITLOS.

Southern Bluefin Tuna Cases (New Zealand v. Japan; Australia v. Japan) - Annex 2 New Zealand's Statement of Claim dated 15 July 1999, 1999, ITLOS.

Southern Bluefin Tuna Cases (New Zealand v. Japan; Australia v. Japan) - Order 1999/4, 1999, ITLOS.

Southern Bluefin Tuna Cases (New Zealand v. Japan; Australia v. Japan) - Order 1999/5, 1999, ITLOS.

Southern Bluefin Tuna Cases (New Zealand v. Japan; Australia v. Japan) - Request for the Prescription of Provisional Measures submitted by Australia 1999, ITLOS.

Southern Bluefin Tuna Cases (New Zealand v. Japan; Australia v. Japan) - Request for the Prescription of Provisional Measures submitted by New Zealand, 1999, ITLOS.

Southern Bluefin Tuna Cases (New Zealand v. Japan; Australia v. Japan) - Response of the Government of Japan to Request For Provisional Measures & Counter-Request For Provisional Measures, 1999, ITLOS.

Statute of the International Court of Justice (1945).

Te Vaka Moana Arrangement (2010).

Te Vaka Toa Arrangement (2010).

The Declaration of Latin American States on the Law of the Sea (The Lima Declaration) (1970).

The Declaration of Santo Domingo (1972).

The Federated States of Micronesia Arrangement for Regional Fisheries Access (1994).

The General Agreement on Tariffs and Trade (1994).

The Montevideo Declaration on the Law of the Sea (1970).

Third United Nations Conference on the Law of the Sea (1973-1982).

Truman Proclamation on Policy of the United States With Respect to the Natural Resources of the Subsoil and Sea Bed of the Continental Shelf (1945).

U.N. Document A/AC.138/SC.II/L.38, article 10, reproduced in III SBC Report 1973, at 82,84 (Canada, India, Kenya and Sri Lanka) (1973).

U.N. Document A/AC.138/SC.II/L.9, section III SBC Report 1972, at 175, 176 (USA) (1972).

Understanding concerning the conservation of swordfish stocks in the South Eastern Pacific Ocean (2010).

United Kingdom of Great Britain and Northern Ireland v. Iceland, 1974, International Court of Justice.

United Nations Conference on Environment and Development (UNCED) (1992).

United Nations Conference on the Human Environment (UNCHE) (1972).

United Nations Conference on the Law of the Sea (1958).

United Nations Convention on the Law of the Sea (1982).

Vienna Convention on the Law of Treaties with final act of the conference, declarations and resolutions (1969).

WCPFC IATTC Memorandum of Cooperation on the Exchange and Release of Data (2009).

WCPFC ISC Memorandum of Understanding (2007).

WCPFC SPC-OFP Revised Memorandum of Understanding (2009).

WCPFC, Conservation and Magement Measures for Bigeye and Yellowfin Tuna in the Western and Central Pacific Ocean (2008).

WCPFC, Conservation and Management for Sharks (2009).

WCPFC, Conservation and Management Measure - Charter Notification Scheme (2009).

WCPFC, Conservation and Management Measure - Commission Vessel Monitoring System (2006).

WCPFC, Conservation and Management Measure - Commission Vessel Monitoring System (2007).

WCPFC, Conservation and Management Measure - Commission Vessel Monitoring System (2012).

WCPFC, Conservation and Management Measure - Cooperating Non-Members (2004).

WCPFC, Conservation and Management Measure - Cooperating Non-Members (2009).

WCPFC, Conservation and Management Measure - Record of Fishing Vessels and Authorization to Fish (2004).

WCPFC, Conservation and Management Measure - Specifications for the Marking and Identification of Fishing Vessels (2004).

WCPFC, Conservation and Management Measure - WCPFC Record of Fishing Vessels and Authorization to Fish (2009).

WCPFC, Conservation and Management Measure - Western and Central Pacific Fisheries Commission Boarding and Inspection Procedures (2006).

WCPFC, Conservation and Management Measure for Bigeye, Yellowfin and Skipjack Tuna in the Western and Central Pacific Ocean (2012).

WCPFC, Conservation and Management Measure for Compliance Monitoring Scheme (2010).

WCPFC, Conservation and Management Measure for Compliance Monitoring Scheme (2012).

WCPFC, Conservation and Management Measure for North Pacific Albacore (2005).

WCPFC, Conservation and Management Measure for North Pacific Striped Marlin (2010).

WCPFC, Conservation and Management Measure for Oceanic Whitetip Shark (2012).

WCPFC, Conservation and Management Measure for Pacific Bluefin Tuna (2009).

WCPFC, Conservation and Management Measure for Pacific Bluefin Tuna (2010).

WCPFC, Conservation and Management Measure for Protection of Cetaceans from Purse Seine Fishing Operations (2012).

WCPFC, Conservation and Management Measure for Sharks (2010).

WCPFC, Conservation and Management Measure for Sharks in the Western and Central Pacific Ocean (2006).

WCPFC, Conservation and Management Measure for South Pacific Albacore (2005).

WCPFC, Conservation and Management Measure for South Pacific Albacore (2010).

WCPFC, Conservation and Management Measure for Striped Marlin in the Southwest Pacific (2006).

WCPFC, Conservation and Management Measure for Swordfish in the Southwest Pacific (2006).

WCPFC, Conservation and Management Measure for Temporary Extension of CMM 2008-01 (2012).

WCPFC, Conservation and Management Measure for the Eastern High-Seas Pocket Special Management Area (2010).

WCPFC, Conservation and Management Measure for the Regional Observer Programme (2007).

WCPFC, Conservation and Management Measure for Vessels without Nationality (2009).

WCPFC, Conservation and Management Measure on Regulation of Transshipment (2009).

WCPFC, Conservation and Management Measure on the Application of High Seas FAD Closures and Catch Retention (2009).

WCPFC, Conservation and Management Measure Prohibiting Fishing on Data Buoys (2009).

WCPFC, Conservation and Management Measure to Establish a List of Vessels presumed to Have Carried out Illegal, Unreported and Unregulated Fishing Activities in the WCPO (2007).

WCPFC, Conservation and Management Measure to establish a List of Vessels presumed to have carried out Illegal, Unreported and Unregulated fishing activities in the WCPO (2010).

WCPFC, Conservation and Management Measure to Mitigate the Impact of Fishing For Highly Migratory Fish Stocks on Seabirds (2006).

WCPFC, Conservation and Management Measure to Mitigate the Impact of Fishing For Highly Migratory Fish Stocks on Seabirds (2007).

WCPFC, Conservation and Management Measure to Monitor Landings of Purse Seiners at Ports so as to Ensure Reliable Catch Data by Species (2009).

WCPFC, Conservation and Management Measure to Prohibit the use of Large Scale Driftnets on the High Seas in the Convention Area (2008).

WCPFC, Conservation and Management Measures for Bigeye and Yellowfin Tuna in the Western and Central Pacific Ocean (2005).

WCPFC, Conservation and Management Measures for Bigeye and Yellowfin Tuna in the Western and Central Pacific Ocean (2006).

WCPFC, Conservation and Management of Sea Turtles (2008).

WCPFC, Conservation and Management of Sharks (2008).

WCPFC, Conservation and Management of Swordfish (2008).

WCPFC, Conservation and Management of Swordfish (2009).

WCPFC, Joint FFA/ EU proposal: WCPFC Catch Documentation Scheme Proposed Intercessional Working Group Operations and Terms of Reference (2012).

WCPFC, Resolution I establishing a Preparatory Conference for the Establishment of the Commission for the Conservation and Management of Highly Migratory Fish Stocks in the Western and Central Pacific Ocean (2001).

WCPFC, Scientific Data to be Provided to the Commission - refined and adopted at the Fourth Regular Session of the Commission, Tumon, Guam, USA, 2-7

WCPFC, Service Level Agreement between WCPFC and FFA (2011).

WCPFC, Standards, specifications and procedures (SSP) for the fishing vessel monitoring system (VMS) of the Western and Central Pacific Fisheries Commission (WCPFC) (2008).

WCPFC-FFA Memorandum of Understanding (2009).

World Summit on Sustainable Development (WSSD) (2002).

Reports

CCSBT, Objectives and principles for the design and implementation of an experimental fishing program, Canberra, Australia

CCSBT, Report of the Special Meeting, Canberra (16-18 November 2000).

CCSBT, Report of the Seventh Meeting of the Stock Assessment Group, Tokyo, Japan (4-11 September 2006).

CCSBT, Report of the Independent Expert Part 2, (September 2008).

CCSBT, Report of the Performance Review Working Group Part 1 - Self Assessment, (3-4 July 2008).

CCSBT, Report of the Sixteenth Meeting of the Scientific Committee Bali, Indonesia (19-28 July 2011).

CCSBT, Report of the Special Meeting, (23-27 August 2011).

FAO, Report of the Nineteenth Session of the Committee on Fisheries, FAO Fisheries Report - No. 459, Rome, Italy (8-12 April 1991).

FAO, Report of the FAO Council - 102nd Session, Rome, Italy (9-20 November 1992).

FAO, Report of the Technical Consultation on High Seas Fishing, FAO Fisheries Report - No. 484, Rome, Italy (7-15 September 1992).

FAO, Committee on Fisheries, 23rd session Rome, Italy (15-19 February 1999).

FAO, Committee on Fisheries, 24rd Session Rome, Italy (26 February - 2 March 2001).

FAO, Report of the Technical Consultation to Review Port State Measures to Combat Illegal, Unreported and Unregulated Fishing, FAO Fisheries Report - No. 759, Rome, Italy (31 August - 2 September 2004).

FAO, Report of the Twenty-seventh Session of the FAO Committee on Fisheries, Rome, Italy (5-9 March 2007).

FAO, Progress in the Implementation of the Code of Conduct for Responsible Fisheries, Related International Plans of Action and Strategy, Twenty-eighth Session of the Committee on Fisheries, Rome, Italy (2-6 March 2009).

FAO, Report of the Twenty-eighth Session of the Committee on Fisheries, Rome, Italy (2-6 March 2009).

FFA, Record of Proceedings of the Multilateral High-Level Conference on South Pacific Tuna Fisheries, Honiara, Solomon Islands (5-9 December 1994).

ICCAT, Conference of Plenipotentiaries of the States Parties to the international convention for the conservation of Atlantic tunas, Paris, France (9-10 July 1984).

ICCAT, Conference of Plenipotentiaries of the Contracting Parties to the International Convention for the Conservation of Atlantic Tunas, Madrid, Spain (4-5 June 1992).

ICCAT, Report of the ICCAT Ad Hoc Working Group Meeting on the Precautionary Approach, Miami, USA (13-14 May 1998).

ICCAT, Report of the 2010 ICCAT Working Group on Stock Assessment Methods, Madrid, Spain (21-23 April 2010).

ICCAT, Report of the 2011 Joint Meeting of the ICCAT Working Group on Stock Assessment Methods and Bluefin Tuna Species Group to Analyze Assessment Methods Developed Under the GBYP and Electronic Tagging, Madrid, Spain (27 June - 1 July 2011).

ICCAT, Report for Biennial Period, 2010-11 PART II (2011) - Vol. 2 (Standing Committee on Research & Statistics), Madrid, Spain (3-7 October 2011).

ICCAT, Report of the 3rd Meeting of the Working Group on the Future of ICCAT, Madrid, Spain (28-31 May 2012).

ICSP7, Report of the Seventh Round of Informal Consultations of States Parties to the Agreement for the Implementation of the Provisions of the United Nations Convention on the Law of the Sea of 10 December 1982 relating to the Conservation and Management of Straddling Fish Stocks and Highly Migratory Fish Stocks, New York, USA (11-12 March 2008).

IOTC, Report of the Fourteenth Session of the IOTC Scientific Committee, Mahé, Seychelles (12-17 December 2011).

IOTC, Report of the Technical Committee Meeting on Allocation Criteria, Nairobi, Kenya (16-18 February 2011).

IOTC, Report of the Sixteenth Session of the Indian Ocean Tuna Commission, Fremantle, Australia (22–26 April 2012).

ISC, ISC10 Plenary Report, Victoria, B.C., Canada (21-26 July 2010).

ISC, Report of the Pacific Bluefin Tuna Working Group Workshop, Nanaimo, Canada (6-9 July 2010).

United Nations General Assembly, Report of the Review Conference on the Agreement for the Implementation of the Provisions of the United Nations Convention on the Law of the Sea of 10 December 1982 relating to the Conservation and Management of Straddling Fish Stocks and Highly Migratory Fish Stocks, New York, USA (22-26 May 2006).

United Nations General Assembly, Report of the Resumed Review Conference on the Agreement for the Implementation of the Provisions of the United Nations Convention on the Law of the Sea of 10 December 1982 relating to the Conservation and Management of Straddling Fish Stocks and Highly Migratory Fish Stocks, New York, USA (24-28 May 2010).

WCED, Report of the World Commission on Environment and Development: Our Common Future (11 December 1987).

WCPFC, MHLC 2 - Report of the Conference, Majuro, Republic of Marshall Islands (10-13 June 1997).

WCPFC, MHLC 3 - Report of the Conference, Tokyo, Japan (22-26 June 1998).

WCPFC, Final Report of the Preparatory Conference for the Establishment of the Commission for the Conservation of Highly Migratory Fish Stocks in the Western and Central Pacific Ocean on all Matters within its Mandate pursuant to Paragraph 9 of Resolution I, Pohnpei, Federated States of Micronesia (7 December 2004).

WCPFC, WCPFC Preparatory Conference: List of Documents, Pohnpei, Federated States of Micronesia (10 December 2004).

WCPFC, Summary Report - First Regular Session of the Scientific Committee, Noumea, New Caledonia (8-19 August 2005).

WCPFC, Summary Report - Second Regular Session of the Scientific Committee, Manila, Philippines (13-24 August 2006).

WCPFC, Summary Report - Third Regular Session of the Commission, Apia, Samoa (11–15 December 2006).

WCPFC, Summary Report - Third Regular Session of the Scientific Committee, Honolulu, USA (13-24 August 2007).

WCPFC, Summary Report - Fifth Regular Session of the Commission, Busan, Korea (8-12 December 2008).

WCPFC, Summary Report - Fourth Regular Session of the Scientific Committee, Port Moresby, Papua New Guinea (11-22 August 2008).

WCPFC, Summary Report - Fifth Regular Session of the Scientific Committee, Port Vila, Vanuatu (10-21 August 2009).

WCPFC, Summary Report - Fifth Regular Session of the Technical and Compliance Committee Pohnpei, Federated States of Micronesia (1–6 October 2009).

WCPFC, Summary Report - Sixth Regular Session of the Commission, Papeete, French Polynesia (7–11 December 2009).

WCPFC, Summary Report - Seventh Regular Session of the Commission, Honolulu, USA (6-10 December 2010).

WCPFC, Summary Report - Sixth Regular Session of the Scientific Committee, Nuku'alofa, Tonga (10-19 August 2010).

WCPFC, Summary Report - Sixth Regular Session of the Technical and Compliance Committee Pohnpei, Federated States of Micronesia (30 September - 5 October 2010).

WCPFC, Charter Notifications for 2011, Pohnpei, Federated States of Micronesia (28 September - 4 October 2011).

WCPFC, Report of the third session of the WPEA OFP Project Steering Committee, Pohnpei, Federated States of Micronesia (13 August 2011).

WCPFC, Summary Report - Seventh Regular Session of the Scientific Committee, Pohnpei, Federated States of Micronesia (9-17 August 2011).

WCPFC, Summary Report - Seventh Regular Session of the Technical Compliance Committee, Koror, Palau (5-9 December 2011).

WCPFC, Tunas and billfishes in the Eastern Pacific Ocean in 2010, Pohnpei, Federated States of Micronesia (9-17 August 2011).

WCPFC, Revised template for the 2012 Annual Report (Part 2), Pohnpei, Federated States of Micronesia (24 May 2012).

WCPFC, Summary Report - Eighth Regular Session of the Commission, Tumon, USA (26-30 March 2012).

WCPFC, Summary Report - Ninth Regular Session of the Commission, Manila, Philippines (2-6 December 2012).

Bibliography

Agnew D. J., Aldous D., Lodge M., Miyake P. and Parkes G., Allocation issues for WCPFC tuna resources (London: Marine Resources Assessment Group Ltd 2006).

Agnew D. J., Pearce J., Pramod G., Peatman T. and Watson R., 'Estimating the Worldwide Extent of Illegal Fishing', PLoS ONE 4(2): e4570 (2009), 1-8.

Aires-da-Silva A. and Maunder M. N., Status of Bigeye Tuna in the Eastern Pacific Ocean in 2010 and Outlook for the Future (La Jolla, USA: IATTC 2011).

Aires-da-Silva A. and Maunder M. N., Status of Yellowfin Tuna in the Eastern Pacific Ocean in 2010 and Outlook for the Future (La Jolla, USA: IATTC 2011).

Allen R. L., 'International management of tuna fisheries - Arrangements, challenges and a way forward', FAO Fisheries and Aquaculture Technical Paper - Volume 536 (Rome: FAO 2010).

Allen R. L., Joseph J., Squires D. and Stryjewski E., Conservation and Management of Transnational Tuna Fisheries (Ames: Wiley-Blackwell 2010).

Allen R. L., 'The Inter American Tropical Tuna Commission', Symposium on World Tuna Fisheries Commemorating the 50th Anniversary of the Establishment of the Inter American Tropical Tuna Commission (La Jolla: IATTC 2001).

Amanda H., Elizabeth H. and Liam C., 'WCPFC8 disappoints on enhanced management of bigeye and yellowfin stocks', FFA Fisheries Trade News - Volume 5 (Honiara: Forum Fisheries Agency 2012).

Amoa F., 'Introductory Note - The Niue Treaty on Cooperation in Fisheries Surveillance and Law Enforcement in the Pacific Region', International legal materials - Vol. 32 (1993), 136-137.

Aqorau T. and Bergin A., 'Ocean governance in the Western Pacific purse seine fishery - The Palau Arrangement', Marine Policy - Vol. 21(2) (1997), 173-186.

Aqorau T. and Bergin A., 'The Federated States Of Micronesia Arrangement for Regional Fisheries Access', The International Journal of Marine and Coastal Law - Vol. 12 (1997), 37-80.

Aqorau T., Analysis of the responses of the Pacific Island States to the fisheries provisions of the Law of the Sea Convention (Wollongong: University of Wollongong 1998).

Aqorau T., 'Cooperative Management of Shared Fish Stocks in the South Pacific', FAO Fisheries Report No. 695, Supplement - Papers presented at the Norway- FAO expert consultation on the management of shared fish stocks (Rome: FAO 2002).

Aqorau T., 'Illegal Fishing and Fisheries Law Enforcement in Small Island Developing States: The Pacific Islands Experience', The International Journal of Marine and Coastal Law - Vol.15(1) (2000), 37-63.

Aqorau T., 'Moving towards a rights-based fisheries management regime for the tuna fisheries in the Western and central pacific ocean', International Journal of Marine and Coastal Law - Vol. 22(1) (2007), 125-142.

Aqorau T., 'Recent Developments in Pacific Tuna Fisheries: The Palau Arrangement and the Vessel Day Scheme', International Journal of Marine & Coastal Law - Vol. 24(3) (2009), 557-581.

Aqorau T., 'Tuna Fisheries Management in the Western and Central Pacific Ocean: A Critical Analysis of the Convention for the Conservation and Management of Highly Migratory Fish Stocks in the Western and Central Pacific Ocean and Its Implications for the Pacific', International Journal of Marine & Coastal Law - Vol.16(3) (2001), 379-431.

Aranda M., de Bruyn P. and Murua H., 'A report review of the tuna RFMOs: CCSBT, IATTC, IOTC, ICCAT and WCPFC', EU FP7 project n°212188 TXOTX (Pasaia: AZTI Tecnalia 2010).

Attard D. J., The exclusive economic zone in international law (New York: Oxford University Press 1987).

Bailey M., Flores J., Pokajam S. and Sumaila U. R., 'Towards better management of Coral Triangle tuna', Ocean & Coastal Management - Vol. 63 (2012), 30-42.

Balton D. A., 'Strengthening the Law of the Sea: The new agreement on straddling fish stocks and highly migratory fish stocks', Ocean Development and International Law - Vol. 27(1-2) (1996), 125-151.

Balton D. A., The Compliance Agreement, In: E. Hey, 'Developments in international fisheries law' (London: Kluwer Law International 1999).

Barnes R., The Convention on the Law of the Sea: An Effective Framework for Domestic Fisheries Conservation?, In: D. Freestone, R. Barnes and D. Ong, 'The Law of the Sea Progress and Prospects ' (New York: Oxford University Press 2006).

Begg G. A., Friedland K. D. and Pearce J. B., 'Stock identification and its role in stock assessment and fisheries management: an overview', Fisheries Research - Vol. 43(1-3) (1999), 1-8.

Bergin A., 'Political and Legal Control over Marine Living Resources- Recent Developments in South Pacific Distant Water Fishing', The international journal of marine and coastal law - Vol. 9(3) (1994), 298-309.

Bialek D., 'Australia and New Zealand v Japan: Southern Bluefin Tuna Case', Melbourne Journal of International Law - Vol. 1 (2000), 1-9.

Birnie P. W. and Boyle A. E., International law and the environment (New York: Oxford University Press 1999).

Birnie P. W., Boyle A. E. and Redgwell C., International law and the environment (New York: Oxford University Press 2009).

Boyle A. and Evans M. D., 'The Southern Bluefin Tuna Arbitration', The International and Comparative Law Quarterly (2001), 447-452.

Brownlie I., Principles of Public International Law (New York: Oxford University Press 2008).

Buonaccorsi V. P., Reece K. S., Morgan L. W. and Graves J., 'Geographic distribution of molecular variance within the blue marlin (Makaira nigricans): A hierarchical analysis of allozyme, single-copy nuclear DNA, and mitochondrial DNA markers', Evolution - Vol. 53(2) (1999), 568-579.

Burke W. T., 'Annex 1 - 1982 Convention on the Law of the Sea provisions on conditions of access to fisheries subject to national jurisdiction', Report of the expert consultation on the conditions of access to the fish resources of the exclusive economic zones (Rome: FAO 1983).

Burke W. T., 'Impacts of the UN Convention of the Law of the Sea on tuna regulation', FAO legislative study (Rome: FAO 1982).

Burke W. T., The new international law of fisheries: UNCLOS 1982 and beyond (Oxford: Clarendon Press 1994).

Caddy J. F. and Mahon R., 'Reference points for fisheries management ', FAO Fisheries Technical Paper - Volume 347 (Rome: FAO 1995).

Campbell R., Identifying possible Limit Reference Points for the key target species in the WCPFC (Tonga: WCPFC 2010).

Churchill R. and Lowe A. V., The law of the sea (Manchester: Manchester University Press 1988).

Churchill R. and Lowe A. V., The law of the sea (Manchester: Manchester University Press 1999).

Churchill R. and Owen D., The EC common fisheries policy (New York: Oxford University Press 2010).

Clarke S., Best Practice Study of Fish Catch Documentation Schemes (Brisbane: Marine Resources Assessment Group Ltd - Asia Pacific 2010).

Collette B. B. and Nauen C. E., 'Scombrids of the world : an annotated and illustrated catalogue of tunas, mackerels, bonitos, and related species known to date', FAO species catalogue v. 2 (Rome: FAO 1983).

Collette B. B., Carpenter K. E., Polidoro B. A., Juan-Jordá M. J., Boustany A., Die D. J., Elfes C., Fox W., Graves J., Harrison L. R., McManus R., Minte-Vera C. V., Nelson R., Restrepo V., Schratwieser J., Sun C.-L., Amorim A., Brick Peres M., Canales C., Cardenas G., Chang S.-K., Chiang W.-C., de Oliveira Leite N., Harwell H., Lessa R., Fredou F. L., Oxenford H. A., Serra R., Shao K.-T., Sumaila R., Wang S.-P., Watson R. and Yáñez E., 'High Value and Long Life—Double Jeopardy for Tunas and Billfishes', Science - Vol. 333 (2011), 291-292.

Cordonnery L., 'A note on the 2000 Convention for the Conservation and Management of Tuna in the Western and Central Pacific Ocean', Ocean Development and International Law - Vol. 33(1) (2002), 1-15.

Cox A., 'Quota Allocation in International Fisheries', OECD Food, Agriculture and Fisheries Papers No. 22 (Paris: OECD Publishing 2009).

Cox A., Renwrantz L. and Kelling I., Strengthening regional fisheries management organisations (Paris: OECD 2009).

Dahmani M., 'The fisheries regime of the exclusive economic zone', Publications on ocean development - No. 11 (Dordrecht: Martinus Nijhoff 1987).

Davies C. and Basson M., Approaches for identification of appropriate reference points and implementation of MSE within the WCPO (Manila: WCPFC 2009).

Davies N., Hoyle S., Harley S., Langley A., Kleiber P. and Hampton J., Stock assessment of bigeye tuna in the western and central Pacific Ocean (Pohnpei: WCPFC 2011).

Davies P. and Redgewell C., The International Legal Regulation of Straddling Fish Stocks, In: I. Brownlie and J. Crawford, 'British Yearbook of International Law - Vol.67' (Oxford: Oxford University Press 1996).

de Bruyn P., Murua H. and Aranda M., 'The Precautionary Approach to fisheries management: How this is taken into account by tuna regional fisheries management organisations (RFMOs)', Marine Policy - Vol. 38 (2013), 397-406.

de Yturriaga J. A., The International Regime of Fisheries: from UNCLOS 1982 to the Presidential Sea (The Hague: Kluwer Law International 1997).

Di Salvo C. J. P. and Raymond L., 'Defining the precautionary principle: an empirical analysis of elite discourse', Environmental Politics - Vol. 19(1) (2010), 86-106.

Doulman D. J. and Swan J., 'A guide to the background and implementation of the 2009 FAO Agreement on Port State Measures to Prevent, Deter and Eliminate Illegal, Unreported and Unregulated Fishing', FAO Fisheries and Aquaculture Circular No. 1074 (Rome: FAO 2012).

Doulman D. J., Development and expansion of the tuna purse seine fishery, In: D. J. Doulman, 'Tuna issues and perspectives in the Pacific islands region' (Honolulu: East-West Center 1987).

Doulman D. J., Fisheries co-operation: the case of the Nauru Group, In: D. J. Doulman, 'Tuna Issues and Perspectives in the Pacific Islands Region' (Honolulu: East-West Center 1987).

Ebert K., Rechtsvergleichung - Einführung in die Grundlagen (Bern: Stämpfli & CIE AG 1978).

Edeson W. R., Freestone D. and Gudmundsdottir E., Legislating for sustainable fisheries: a guide to implementing the 1993 FAO Compliance Agreement and 1995 UN Fish Stocks Agreement (Washington: World Bank 2001).

Edeson W. R., 'International Plan of Action on Illegal Unreported and Unregulated Fishing: The Legal Context of a Non-Legally Binding Instrument', The International Journal of Marine and Coastal Law - Vol. 16(4) (2001), 603-624.

Edeson W. R., 'The Code of Conduct for Responsible Fisheries: An Introduction', The International Journal of Marine and Coastal Law- Vol. 11(2) (1996), 233-238.

Edwards H., 'When Predators Become Prey: The Need for International Shark Conservation', Ocean and Coastal Law Journal - Vol. 12(2) (2007), 305-354.

Eiriksson G., The International Tribunal for the Law of the Sea (The Hague: Martinus Nijhoff Publishers 2000).

Engler Palma M. C., Allocation of Fishing Opportunities in Regional Fisheries Management Organizations: A Legal Analysis in the Light of Equity (Halifax: Dalhousie University 2010).

Erickson A. L., 'Out of Stock: Strengthening International Fishery Regulations to Achieve a Healthier Ocean', North Carolina Journal of International Law and Commercial Regulation - Vol. 34(1) (2008), 281 to 324.

Fabra A., Gascón V., Marrero M., Lieberman S. and Sack S., Closing the gap: Comparing tuna RFMO port State measures with the FAO Agreement on Port State Measures (Philadelphia: Pew Environment Group 2011).

FAO Fishery Resources Division and Fishery Policy and Planning Division, 'Fisheries management', FAO Technical Guidelines for Responsible Fisheries - No. 4 (Rome: FAO 1997).

FAO, 'FAO yearbook', Fishery and Aquaculture Statistics 2007 (Rome: FAO 2007).

FAO, 'Fishing operations: Vessel monitoring systems', FAO Technical Guidelines for Responsible Fisheries - No. 1 (Rome: FAO 1998).

FAO, 'Guidelines for developing an at-sea fishery observer programme', FAO Technical Paper - No. 414 (Rome: FAO 2002).

FAO, Model Scheme on Port State Measures to Combat Illegal, Unreported and Unregulated Fishing (Rome: FAO 2007).

FAO, 'Precautionary approach to capture fisheries and species introductions', FAO Technical Guidelines for Responsible Fisheries - Vol.2 (Lysekil: FAO 1996).

FAO, 'The Exclusive Economic Zone: A historical perspective', Essays in memory of Jean Carroz (Sri Lanka: FAO 1987).

FAO, The Standard Specifications for the Marking and Identification of Fishing Vessels (Rome: FAO 1989).

FAO, The State of World Fisheries and Aquaculture 2012 (Rome: FAO 2012).

FAO, TUNA - A global perspective (Rome: FAO 2013).

Fenichel E. P., Tsao J. I., Jones M. L. and Hickling G. J., 'Real options for precautionary fisheries management', Fish and Fisheries - Vol. 9(2) (2008), 121-137.

Fisheries Forum Agency, Strategic Plan 2005-2020 (Honiara: FFA 2005).

Flothmann S., von Kistowski K., Dolan E., Lee E., Meere F. and Album G., 'Closing Loopholes: Getting Illegal Fishing Under Control', Scienceexpress - Vol. 328 (2010), 1235-1236.

Foundation I. S. S., 'Status of the world fisheries for tuna: Management of tuna stocks and fisheries', Technical Report 2012-07 (Washington: ISSF 2012).

Franckx E., 'Pacta Tertiis and the Agreement for the Implementation of the Provisions of the United Nations Convention on the Law of the Sea of 10 December 1982 relating to the Conservation and Management of Straddling Fish Stocks and Highly Migratory Fish Stocks ', FAO Legal Papers - Vol. 8 (Rome: FAO 2000).

Franckx E., 'Pacta Tertiis and the Agreement for the Implementation of the Provisions of the United Nations Convention on the Law of the Sea of 10 December 1982 relating to the Conservation and Management of Straddling Fish Stocks and Highly Migratory Fish Stocks', FAO Legal Papers - Vol. 8 (Rome: FAO 2000).

Freestone D. and Makuch Z., The New International Environmental Law of Fisheries: The 1995 UN Straddling Stocks Agreement, In: G. Handl, J. Brunnée and P. Sands, 'Yearbook of International Environmental Law - Vol. 7' (Oxford: Clarendon Press 1998).

Freestone D., International Fisheries Law Since Rio: The Continued Rise of the Precautionary Principle, In: A. Boyle and D. Freestone, 'International Law and Sustainable Development - Past Achievements and Future Challenges' (New York: Oxford University Press 1999).

Gaertner M., 'The Dispute Settlement Provisions of the Convention on the Law of the Sea: Critique and Alternatives to the International Tribunal for the Law of the Sea', San Diego Law Review - Vol. 19 (1982), 577-597.

Gamboa M. J., A Dictionary of International Law and Diplomacy (Quezon City: Phoenix Press 1973).

Garcia S. M., Caddy J. F., Csirke J., Grainger R. and Majkowski J., 'World review of highly migratory species and straddling stocks', FAO Fisheries Technical Paper - Vol. 337 (Rome: FAO 1994).

Garcia S. M., 'The Precautionary Approach to Fisheries and its Implications for Fishery Research, Technology and Management: An Updated Review', FAO Fisheries Technical Paper - No. 350 (Rome: FAO 1996).

Garcia S. M., Zerbi A., Aliaume C., Do Chi T. and Lasserre G., 'The ecosystem approach to fisheries. Issues, terminology, principles, institutional foundations, implementation and outlook', FAO Fisheries Technical Paper- No. 443 (Rome: FAO 2003).

Gianni M. and Simpson W., Flags of Convenience, Transhipment, Re-Supply and At-Sea Infrastructure in Relation to IUU Fishing, In: OECD, 'Fish Piracy: Combating Illegal, Unreported and Unregulated Fishing' (Paris: OECD Publishing 2004).

Gillett R. and Cartwright I., The future of Pacific Island fisheries (Noumea: SPC 2010).

Gillett R., 'A short history of industrial fishing in the Pacific islands', RAP PUBLICATION 2007/22 (Bangkok: FAO Regional Office for Asia and the Pacific 2007).

Gillett R., 'Catch Attribution in the Western and Central Pacific Fisheries Commission', Paper prepared by Gillett, Preston and Associates for the Seventh Regular Session of the Technical and Compliance Committee Meeting (Pohnpei: WCPFC 2011).

Gillett R., 'Marine fishery resources of the Pacific Islands', FAO Fisheries and Aquaculture Technical Paper - Vol. 537 (Rome: FAO 2010).

Gillett R., 'Review of the state of world marine fishery resources', FAO Fisheries Technical Paper - Vol. 457 (Rome: FAO 2005).

Gilman E. L., 'Bycatch governance and best practice mitigation technology in global tuna fisheries', Marine Policy - Vol. 35(5) (2011),

Gordon H. S., 'The Economic Theory of a Common-Property Resource: The Fishery', Journal of Political Economy - Vol. 62(2) (1954), 124-142.

Grafton Q., Hannesson R., Shallard B., Sykes D. and Terry J., The Economics of Allocation in Tuna Regional Fisheries Management Organizations, In: R. Allen, J. Joseph and D. Squires, 'Conservation and Management of Transnational Tuna Fisheries' (Iowa: Blackwell Publishing 2010).

Grafton R. Q., Hannesson R., Shallard B., Sykes D. and Terry J., 'The Economics of Allocation in Tuna Regional Fisheries Management Organizations (RFMOs)', Economics and Environment Network Working Paper, EEN0612, (Canberra: Australian National University 2006).

Grotius H., Mare Liberum, sive de jure quod Batavis competit ad Indicana commercia dissertatio (Lugduni Batavorum: Ludovici Elzevirij 1609).

Guillotreau P., Salladarré F., Dewals P. and Dagorn L., 'Fishing tuna around Fish Aggregating Devices (FADs) vs free swimming schools: Skipper decision and other determining factors', Fisheries Research - Vol. 109(2-3) (2011), 234-242.

Hampton J., 'Working Paper MHLC-2: The Convention Area, presented at the 12th Meeting of the Standing Committee on Tuna and Billfish', (Papeete: Oceanic Fisheries Program, Secretariat of the Pacific Community, Noumea, New Caledonia 1999).

Hanich Q. A., Parris H. and B.M. T., 'Sovereignty and cooperation in regional Pacific tuna fisheries management: Politics, economics, conservation and the vessel day scheme', Australian Journal of Maritime and Ocean Affairs - Vol. 2(1) (2010), 2-15.

Hanich Q., Control, Cooperation and 'Paricipatory Rights' in the Western and Central Pacific Ocean Tuna Fisheries, In: Q. Hanich and M. Tsamenyi, 'Navigating Pacific fisheries : legal and policy trends in the implementation of international fisheries instruments in the Western and Central Pacific region' (Wollongong: Australian National Centre for Ocean Resources and Security 2009).

Hanich Q., 'Distributing the bigeye conservation burden in the western and central pacific fisheries', Marine Policy - Vol. 36(2) (2012), 327-332.

Hanich Q., Schonfield C. and Cozens P., Oceans of Opportunity: The Limits of Maritime Claims in the Western and Central Pacific Region, In: Q. Hanich and M. Tsamenyi, 'Navigating Pacific fisheries : legal and policy trends in the implementation of international fisheries instruments in the Western and Central Pacific region' (Wollongong: Australian National Centre for Ocean Resources and Security 2009).

Hanich Q., Teo F. and Tsamenyi M., 'A collective approach to Pacific islands fisheries management: Moving beyond regional agreements', Marine Policy - Vol. 34(1) (2010), 85-91.

Hannesson R., 'The exclusive economic zone and economic development in the Pacific island countries', Marine Policy - Vol. 32(6) (2008), 886-897.

Harley S. and Hampton J., Status of Tuna Stocks in the Western and Central Pacific Ocean and Scientific Challenges, In: Q. Hanich and M. Tsamenyi, 'Navigating Pacific fisheries : legal and policy trends in the implementation of international fisheries instruments in the Western and Central Pacific region' (Wollongong: Australian National Centre for Ocean Resources and Security 2009).

Havice E., 'The structure of tuna access agreements in the Western and Central Pacific Ocean: Lessons for Vessel Day Scheme planning', Marine Policy - Vol. 34(5) (2010), 979-987.

Haward M. and Bergin A., 'The political economy of Japanese distant water tuna fisheries', Marine Policy - Vol. 25(2) (2001), 91-101.

Hayashi M., The Straddling and Highly Migratory Fish Stocks Agreement, In: E. Hey, 'Developments in international fisheries law' (The Hague: Kluwer Law International 1999).

Hedley C., Molenaar E. J. and Elferink A. G., 'The Implications of the UN Fish Stocks Agreement (New York, 1995) for Regional Fisheries Organisations and International Fisheries Management', Working paper prepared for the European Parlament (Luxembourg: Directorate-General for Research 2003).

Henriksen T., Hønneland G. and Sydnes A. K., 'Law and politics in ocean governance: the UN Fish Stocks Agreement and regional fisheries management regimes', Publications on ocean development (Leiden: Martinus Nijhoff Publishers 2006).

Herrick S. F., Rader B. and Squires D., 'Access fees and economic benefits in the Western Pacific United States purse seine tuna fishery', Marine Policy - Vol. 21(1) (1997), 83-96.

Hey E., The Fisheries Provisions of the LOS Convention, In: E. Hey, 'Developments in international fisheries law' (The Hague: Kluwer Law International 1999).

Hilborn R., Maguire J.-J., Parma A. M. and Rosenberg A. A., 'The Precautionary Approach and risk management: can they increase the probability of successes in fishery management?', Canadian Journal of Fisheries & Aquatic Sciences - Vol. 58 (2001), 99-107.

Ho P. S. C., 'The impact of the UN Fish Stocks Agreement on Taiwan's participation in international fisheries fora', Ocean Development and International Law - Vol. 37(2) (2006), 133-148.

Holder J. and Lee M., Environmental Protection, Law and Policy (New York: Cambridge University Press 2007).

Hollick A. L., 'The Origins of 200-Mile Offshore Zones', The American Journal of International Law - Vol. 71(3) (1977), 494-500.

Honer A., Das explorative Interview, In: A. Honer, 'Kleine Leiblichkeiten - Erkundungen in Lebenswelten' (Wiesbaden: Verlag für Sozialwissenschaften 2011).

Horn N., Einführung in die Rechtswissenschaft und Rechtsphilosophie (Heidelberg: C.F. Müller Verlag 2004).

Hunt C., 'Management of the South Pacific tuna fishery', Marine Policy - Vol. 21(2) (1997), 155-171.

Hurry G. D., Hayashi M. and Maguire J. J., Report of the independent Review - ICCAT (Madrid: ICCAT 2008).

IATTC, Fishery Status Report No. 1 (Tunas and Billfishes in the Eastern Pacific Ocean in 2002) (La Jolla: IATTC 2003).

IATTC, The fishery for tunas and billfishes in the Eastern Pacific Ocean in 2010 (La Jolla: IATTC 2011).

ICCAT, 'Expert Panel on CITES Listing Proposals - Regional Fisheries Bodies', CoP14 Doc. 68 Annex - Annex 1 - Summary of measures taken historically by ICCAT for bluefin tuna (Madrid: ICCAT 2010).

International Seafood Sustainability Foundation, 'KOBE III Bycatch Joint Technical Working Group: Harmonisation of Purse-seine Data Collected by Tuna-RFMO Observer Programmes', ISSF Technical Report 2012-12 (Washington: ISSF 2012).

IOTC, Approaches to Allocation Criteria in Other Tuna Regional Fishery Management Organizations (Nairobi, Kenya: IOTC 2011).

IOTC, Report of the IOTC Performance Review Panel (Victoria: IOTC 2009).

Itano D. G. and Holland K. N., 'Movement and vulnerability of bigeye (Thunnus obesus) and yellowfin tuna (Thunnus albacares) in relation to FADs and natural aggregation points', Aquatic Living Resources - Vol. 13(4) (2000), 213-223.

Joseph J., 'Managing Fishing Capacity of the World Tuna Fleet', FAO Fisheries Circular - No. 982 (Rome: FAO 2003).

Joseph J., Squires D., Bayliff W. and Groves T., Addressing the Problem of Excess Fishing Capacity in Tuna Fisheries, In: R. Allen, J. Joseph and D. Squires, 'Conservation and Management of Transnational Tuna Fisheries' (Iowa: Wiley-Blackwell 2010).

Juda L., The United Nations Fish Stocks Agreement, In: O. S. Stokke and Ø. B. Thommessen, 'Yearbook of International Co-operation on Environment and Development ' (London: Earthscan Publications 2001).

Kaye S. B., International fisheries management (The Hague: Kluwer Law International 2001).

Kingsbury B., 'The Tuna-Dolphin Controversy, The World Trade Organization, and the Liberal Project to Reconceptualize International Law', (1994), 1-40.

Koh T., A Constitution for the Oceans (Montego Bay: United Nations 1983).

Kolody D., Polacheck T., Basson M. and Davies C., 'Salvaged pearls: lessons learned from a floundering attempt to develop a management procedure for Southern Bluefin Tuna', Fisheries Research - Vol. 94(3) (2008), 339-350.

Kuemlangan B., 'National Legislative Options to Combat IUU Fishing', FAO Fisheries Report - No. 666 (Rome: FAO 2000).

Lack M., Catching On? Trade-related Measures as a Fisheries Management Tool (Cambridge: TRAFFIC International 2007).

Langley A., Hoyle S. and Hampton J., Stock assessment of yellowfin tuna in the western and central Pacific Ocean (Rev.1 - 03 August 2011) (Pohnpei: WCPFC 2011).

Larenz K. and Canaris C. W., Methodenlehre der Rechtswissenschaft (Berlin: Springer Verlag 1995).

Le Gallic B., 'The use of trade measures against illicit fishing: Economic and legal considerations', Ecological Economics - Vol. 64(4) (2008), 858-866.

Leggett K., 'The Southern Bluefin Tuna Cases: ITLOS Order on Provisional Measures', Review of European Community & International Environmental Law - Vol. 9(1) (2000), 76-79.

Lewis A. D., The South Pacific Commission, In: K. Hinman, 'Getting ahead of the curve: conserving the Pacific Ocean's tunas, swordfish, billfishes and sharks' (Leesburg: National Coalition for Marine Conservation 2000).

Lodge M. W. and Nandan S. N., 'Some Suggestions towards Better Implementation of the United Nations Agreement on Straddling Fish Stocks and Highly Migration Fish Stocks of 1995', International Journal of Marine & Coastal Law - Vol. 20 (2005), 345-379.

Lodge M. W., Anderson D., Løbach T., Munro G., Sainsbury K. and Willock A., Recommended Best Practices for Regional Fisheries Management Organizations (London: Chatham House 2007).

Lodge M. W., 'Minimum terms and conditions of access: Responsible fisheries management measures in the South Pacific region', Marine Policy - Vol. 16(4) (1992), 277-305.

Lodge M. W., 'The development of the Palau Arrangement for the management of the western Pacific purse seine fishery', Marine Policy - Vol. 22(1) (1998), 1-28.

Lodge M., 'Recommended best practices for regional fisheries management organizations', Report of an independent panel to develop a model for improved governance by Regional Fisheries Management Organizations (London: Chatham House 2007).

Losada S., Lieberman S., Drews C. and Hirshfield M., 'The Status of Atlantic Bluefin Tuna', Science - Vol. 328 (2010), 1353-1355.

Lugten G., The impact of Extra Legal Factors in the Historical Development of International Fisheries Law (Hobart: University of Tasmania 1996).

Lugten G., 'The role of international fishery organizations and other bodies in the conservation and management of living aquatic resources', FAO Fisheries Circular - No. 1054 (Rome: FAO 2010).

Maguire J.-J., 'The state of world highly migratory, straddling and other high seas fishery resources and associated species', FAO Fisheries and Aquaculture Technical Paper - Vol. 495 (Rome: FAO 2006).

Majkowski J., 'Global fishery resources of tuna and tuna-like species', FAO Fisheries and Aquaculture Technical Paper - Vol. 483 (Rome: FAO 2007).

Mansfield B., Compulsory dispute settlement after the Southern Bluefin Tuna award, In: A. G. Oude Elferink and D. R. Rothwell, 'Oceans Management in the 21st Century: Institutional Frameworks and Responses' (Leiden: Martinus Nijhoff 2004).

Marashi S. H., 'Summary Information on the Role of International Fishery and other Bodies with Regard to the Conservation and Management of Living Resources of the High Seas', FAO Fisheries Circular - No. 908 (Rome: FAO 1996).

Markus T., European Fisheries Law - From Promotion to Management (Groningen: Europa Law Publishing 2009).

Markus T., 'Wege zu einer nachhaltigen EU-Fischereiaußenhandelspolitik', Europarecht (EuR) Vol. 48 (2013),

Marr S., 'The Southern Bluefin Tuna cases: the precautionary approach and conservation and management of fish resources', European Journal of International Law - Vol. 11(4) (2000), 815-831.

Matsuda Y. and Ouchi Y., 'Legal, political and economic constraints on Japanese strategies for distant water tuna and skipjack fisheries in Southeast Asian seas and the Western Central Pacific', Memoirs of the Kagoshima University Research Centre for the South Pacific (Honolulu: East-West Environmental and Policy Institute 1984).

Maunder M. N., Reference points, decision rules, and management strategy evaluation for tunas and associated species in the eastern Pacific Ocean - IATTC Stock Assessment Report 2013 (La Jolla: IATTC 2013).

Maunder M. N., Updated Indicators of Stock Status for Skipjack Tuna in the Eastern Pacific Ocean (La Jolla: IATTC 2011).

McDonald J. M., 'Appreciating the precautionary principles as an ethical evolution in ocean management', Ocean Development and International Law - Vol. 26(3) (1995), 255-286.

McDorman T. L., 'Implementing Existing Tools: Turning Words Into Actions Decision-Making Processes of Regional Fisheries Management Organisations (RFMOs)', The International Journal of Marine and Coastal Law - Vol. 20 (2005), 423-457.

Meltzer E. and Fuller S., The quest for sustainable international fisheries : regional efforts to implement the 1995 United Nations Fish Stocks Agreement : an overview for the May 2006 review conference (Ottawa: NRC Research Press 2009).

Meltzer E., 'Global Overview of Straddling and Highly Migratory Fish Stocks; Maps and Charts Detailing RFMO Coverage and Implementation', The International Journal of Marine and Coastal Law - Vol. 20 (2005), 571-604.

Michaels R., The Functional Method of Comparative Law, In: M. Reimann and R. Zimmermann, 'The Oxford Handbook of Comparative Law' (New York: Oxford University Press 2006).

Miyake M. P., Miyabe N. and Nakano H., 'Historical trends of tuna catches in the world', FAO Fisheries and Aquaculture Technical Paper - Vol. 467 (Rome: FAO 2004).

Miyake M., Guillotreau P., Sun C.-H. and Ishimura G., 'Recent developments in the tuna industry', FAO Fisheries and Aquaculture Technical Paper - Vol. 543 (Rome: FAO 2010).

Molenaar E. J., 'Non-Participation in the Fish Stocks Agreement: Status and Reasons', The International Journal of Marine and Coastal Law - Vol. 26(2) (2011), 195-234.

Molenaar E. J., 'Participation, Allocation and Unregulated Fishing: The Practice of Regional Fisheries Management Organisations', International Journal of Marine and Coastal Law - Vol.18 (2003), 457-480.

Molenaar E. J., Regional Fisheries Management Organizations: Issues of Participation, Allocation and Unregulated Fishing, In: A. G. Oude Elferink and D. R. Rothwell, 'Oceans Management in the 21st Century: Institutional Frameworks and Responses' (Leiden: Martinus Nijhoff 2004).

Molenaar E. J., 'The Concept of "Real Interest" and Other Aspects of Co-operation through Regional Fisheries Management Mechanisms', The International Journal of Marine and Coastal Law - Vol.15(4) (2000), 475-531.

Mooney-Seus M. L. and Rosenberg A. A., Regional fisheries management organizations: progress in adopting the precautionary approach and ecosystem-based management (London: Chatham House 2007).

Moore G., The Code of Conduct for Responsible Fisheries, In: E. Hey, 'Developments in international fisheries law' (The Hague: Kluwer Law International 1999).

Munro G. R., 'Internationally shared fish stocks, the high seas, and property rights in fisheries', Marine Resource Economics - Vol. 22(4) (2007), 425-443.

Munro G. R., van Houtte A. and R. W., 'The conservation and management of shared fish stocks- legal and economic aspects', FAO Fisheries Technical Paper - Vol. 465 (Rome: FAO 2004).

Murphy D. D., 'The tuna-dolphin wars', Journal of World Trade - Vol. 40(4) (2006),

Nakamura I., 'Billfishes of the world: An annotated and illustrated catalogue of marlins, sailfishes, spearfishes and swordfishes known to date', FAO species catalogue - Vol. 5 (Rome: FAO 1985).

Nandan S. N., 'Statement of the chairman, ambassador Satya N. Nandan, on 4 August 1995, upon the adoption of the agreement for the implementation of the provisions of the United Nations Convention on the Law of the Sea of 10 December 1982 relating to the conservation and management of straddling fish stocks and highly migratory fish stocks', Sixth session of the United Nations Conference on straddling fish stocks and highly migratory fish stocks (New York, USA: United Nations 1995).

Nédélec C. and Prado J., 'Definition and classification of fishing gear categories', FAO Fisheries Technical Paper - No. 222 (Rome: FAO 1990).

Nordquist M. H., United Nations Convention on the Law of the Sea, 1982: A Commentary Volume II Article 1 to 85 Annexes I & II Final Act, Annex II (Dordrecht: Martinus Nijhoff 1993).

O'Connell D. A., 'Tuna, Dolphins, and Purse Seine Fishing in the Eastern Tropical Pacific: The Controversy Continues', UCLA Journal of Environmental Law and Policy - Vol. 23(1) (2005), 77-100.

OECD, Fishing for Coherence - Proceedings of the Workshop on Policy Coherence for Development in Fisheries (Paris: OECD Publishing 2006).

OECD, 'Strengthening the International Commission for the Conservation of Atlantic Tunas (ICCAT)', Strengthening Regional Fisheries Management Organisations (Paris: OECD Publishing 2009).

OECD, 'Transition to Responsible Fisheries: Economic and Policy Implications', (Paris: OECD Publishing 2000).

Örebech P., Sigurjonsson K. and McDorman T. L., 'The 1995 United Nations Straddling and Highly Migratory Fish Stocks Agreement: Management, Enforcement and Dispute Settlement', The International Journal of Marine and Coastal Law - Vol. 13 (1998), 119-141.

Orellana M. A., 'The Swordfish Dispute between the EU and Chile at the ITLOS and the WTO', Nordic Journal of International Law - Vol. 71 (2002), 55-81.

Orrego Vicuña F., 'The exclusive economic zone: regime and legal nature under international law', Cambridge studies in international and comparative law. New series (Cambridge: Cambridge University Press 1989).

Pacific Islands Forum Fisheries Agency, 'The Western Pacific Purse Seine Fishery: A Summary of Concerns', Forum Fisheries Agency Report - Number 90/27 (Honiara: Forum Fisheries Agency 1990).

Palma M. A., Tsamenyi M. and Edeson W. R., Promoting Sustainable Fisheries: The International Legal and Policy Framework to Combat Illegal, Unreported and Unregulated Fishing (Leiden: Martinus Nijhoff Publishers 2010).

Parris H. and Lee A., Allocation Models in the Western and Central Pacific Fisheries Commission, In: Q. Hanich and M. Tsamenyi, 'Navigating Pacific Fisheries Legal and Policy Trends in the Implementation of International Fisheries Instruments in the Western and Central Pacific Ocean' (Wollongong: Australian National Centre for Ocean Resources and Security 2009).

Pauly D., Christensen V., Dalsgaard J., Froese R. and Torres F., 'Fishing down marine food webs', Science - Vol. 279(860) (1998), 860-863.

Pitcher T., Kalikoski D., Pramod G. and Short K., 'Not honouring the code', Nature - Vol. 457(5) (2009), 658-659.

Polacheck T., 'Experimental catches and the precautionary approach: the Southern Bluefin Tuna dispute', Marine Policy - Vol. 26(4) (2002), 283-294.

Polacheck T., 'Politics and independent scientific advice in RFMO processes: A case study of crossing boundaries', Marine Policy, Vol. 36(1) (2012), 132-141.

Proelß A., Meeresschutz im Völker- und Europarecht – Das Beispiel des Nordostatlantiks (Berlin: Duncker & Humblot 2004).

Rayfuse R., 'The Future of Compulsory Dispute Settlement Under the Law of the Sea Convention', Victoria University of Wellington Law Review - Vol. 36 (2005), 683-711.

Reid C., Squires D., Jeon Y., Rodwell L. and Clarke R., 'An analysis of fishing capacity in the western and central Pacific Ocean tuna fishery and management implications', Marine Policy - Vol. 27(6) (2003), 449-469.

Ress G., 'Die Bedeutung der Rechtsvergleichung für das Recht internationaler Organisationen', Zeitschrift für ausländisches öffentliches Recht und Völkerrecht - Vol. 36 (1976), 227-279.

Richards A. H., 'Problems of drift-net fisheries in the South Pacific', Marine Pollution Bulletin - Vol. 29(1-3) (1994), 106-111.

Röben V., 'The Southern Bluefin Tuna Cases: Re-Regionalization of the Settlement of Law of the Sea Disputes?', Zeitschrift für ausländisches öffentliches Recht und Völkerrecht - Vol. 62 (2002), 61-72.

Roheim C. A. and Sutinen J., 'Trade and Marketplace Measures to Promote Sustainable Fishing Practices, ICTSD Natural Resources, International Centre for Trade and Sustainable Development', Series Issue Paper No. 3 (Geneva: 2006).

Romano C., 'The Southern Bluefin Tuna dispute: Hints of a world to come ... like it or not', Ocean Development and International Law - Vol. 32(4) (2001), 313-348.

Russell D. A. and Vander Zwaag D. L., Recasting Transboundary Fisheries Management Arrangements in Light of Sustainability Principles (Leiden: Martinus Nijhoff 2010).

Safina C., Tuna Conservation, In: B. A. Block and E. D. Stevens, 'Tuna: Physiology, ecology, and evolution' (San Diego: Academic Press 2001).

Schaefer K. M., Reproductive biology of tunas, In: B. A. Block and E. D. Stevens, 'Tuna: Physiology, ecology and evolution' (San Diego: Academic Press 2001).

Schneider G., Current Debate Over Potential Use of Trade Measures (Washington: U.S. Department of Commerce - National Marine Fisheries Service 2000).

Secretariat of the Pacific Community, Strategic plan 2010-2013 (Noumea: SPC 2009).

Secretariat of the Pacific Community, WCPFC Tuna Fishery Yearbook (Noumea: Western and Central Pacific Fisheries Commission 2010).

Secretariat of the Pacific Community, WCPFC Tuna Fishery Yearbook (Noumea: WCPFC 2010).

Serdy A., International Fisheries Law and the Transferability of Quota: Principles and Precedents, In: R. Allen, J. Joseph and D. Squires, 'Conservation and Management of Transnational Tuna Fisheries' (Iowa: Wiley-Blackwell 2010).

Serdy A., 'One fin, two fins, red fins, bluefins: some problems of nomenclature and taxonomy affecting legal instruments governing tuna and other highly migratory species', Marine Policy - Vol. 28(3) (2004), 235-247.

Shanks S., 'Introducing a transferable fishing day management regime for Pacific Island countries', Marine Policy - Vol. 34(5) (2010), 988-994.

Sharp A., The Effectiveness or Not of the New Port State Measures in the Battle to Control Illegal, Unregulated and Unreported Fishing (Auckland: University of Auckland - Department of Commercial Law 2010).

Shelton D. and Kiss A. C., Guide to international environmental law (Leiden: Martinus Nijhoff 2007).

Sibert J. and Hampton J., 'Mobility of tropical tunas and the implications for fisheries management', Marine Policy - Vol. 27(1) (2003), 87-95.

Sodik D. M., 'Non-Legally Binding International Fisheries Instruments and Measures to Combat Illegal, Unreported and Unregulated Fishing', Australian International Law Journal - Vol. 15(1) (2008), 129-164.

Stoll P.-T. and Vöneky S., 'The Swordfish Case: Law of the Sea v. Trade', Zeitschrift für ausländisches öffentliches Recht und Völkerrecht - Vol. 62 (2002), 21-35.

Swan J., 'Decision-making in Regional Fishery Bodies or Arrangements- the evolving role of RFBs and international agreement on decision-making processes.', FAO Fisheries Circular - No. 995 (Rome: FAO 2004).

Swan J., 'Fishing Vessels operating under open registers and the exercise of flag State responsibilities - Information and options', FAO Fisheries Circular - No. 980 (Rome, Italy: FAO 2002).

Sydnes A. K., 'Establishing a regional fisheries management organisation for the Western and Central Pacific tuna fisheries', Ocean & Coastal Management - Vol. 44(11-12) (2001), 787-811.

Tahindro A., 'Conservation and management of transboundary fish stocks: Comments in light of the adoption of the 1995 agreement for the conservation and management of straddling fish stocks and highly migratory fish stocks', Ocean Development and International Law - Vol. 28(1) (1997), 1-58.

Tarasofsky R. G., 'Regional Fisheries Organizations and the World Trade Organization: Compatibility or Conflict?', 2003).

Tarasofsky R., Enhancing the Effectiveness of Regional Fisheries Management Organizations through Trade and Market Measures (London: Chatham House 2007).

Tarte S., 'Negotiating a Tuna Management Regime for the western and central Pacific: The MHLC process 1994', Journal of Pacific History - Vol. 34(3) (1999), 273-280.

Tarte S., The Convention for the Conservation and Management of Highly Migratory Fish Stocks in the Western and Central Pacific Ocean: Implementation Challenges from a Historical Perspective, In: Q. Hanich and M. Tsamenyi, 'Navigating Pacific fisheries: legal and policy trends in the implementation of international fisheries instruments in the Western and Central Pacific region' (Wollongong: Australian National Centre for Ocean Resources and Security 2009).

Techera E. J., 'Good Environmental Governance: Overcoming Fragmentation in International Law for Shark Conservation and Management', American Society of International Law Proceedings - Vol. 105 (2011), 103-107.

The International Consortium of Investigative Journalists, Looting the Seas (Washington: The Center for Public Integrity 2012).

TRAFFIC W., WWF & Traffic Statement to WCPFC – 8th Regular Session (Guam: WWF & TRAFFIC 2012).

Treves T., 'Dispute-Settlement Clauses in the law of the Sea Convention and their Impact on the Protection of the Marine Environment: Some Observations', Review of European Community & International Environmental Law - Vol. 8(1) (1999), 6-9.

Trujillo E., 'The WTO Appellate Body Knocks Down U.S. "Dolphin-Safe" Tuna Labels But Leaves a Crack for PPMs', The American Society of International Law ASIL - Vol. 16(25) (2012), 1-5.

Tsamenyi B. M., Co-operation in Fisheries Law Enforcement in the South Pacific Region: An Analysis of the Niue Treaty (Bangkok: SEAPOL Tri-Regional Conference on Current Issues in Ocean Law Policy and Management: Southeast Asia, North Pacific, and Southwest Pacific 1994).

Tsamenyi B. M., 'Treaty on Fisheries between the Governments of Certain Pacific Island States and the Government of the United States of America: The Final Chapter in United States Tuna Policy', The Brooklyn Journal of International Law - Vol. 15(2) (1989), 183-222.

Valencia M. J., 'Domestic politics fuels Northeast Asian maritime disputes ', Asia Pacific Issues - No. 43 (2000), 1-8.

Van Dyke J. M. and Haftel S., Tuna management in the Pacific: an analysis of the South Pacific Forum Fisheries Agency (Honolulu: 1981).

Van Dyke J. M., Allocating Fish Across Jurisdictions, In: R. Allen, J. Joseph and D. Squires, 'Conservation and Management of Transnational Tuna Fisheries' (Iowa: Wiley-Blackwell 2010).

Vitzthum W. G., Handbuch des Seerechts (München: C. H. Beck 2006).

Warbrick C., McGoldrick D. and Anderson D. H., 'The Straddling Stocks Agreement of 1995 - an Initial Assessment', International & Comparative Law Quarterly - Vol. 45(2) (1996), 463-475.

WCPFC, 'WCPFC-IATTC Overlap Area', Discussion Paper for WCPFC (Koror: WCPFC 2011).

WCPFC, 'West Pacific East Asia Oceanic Fisheries Management Project Steering Committee', Summary Report (Port Vila: WCPFC 2009).

Willock A. and Lack M., Follow the leader - Learning from experience and best practice in regional fisheries management organizations (Sydney: WWF International and TRAFFIC International 2006).

Wolfrum R., The Protection of the Marine Environment after the Rio Conference: Progress or Stalemate?, In: U. Beyerlin, 'Recht zwischen Umbruch und Bewahrung: Völkerrecht, Europarecht, Staatsrecht (Beiträge zum ausländischen öffentlichen Recht und Völkerrecht)' (Berlin: Springer Verlag 1995).

Worm B., Davis B., Kettemer L., Ward-Paige C. A., Chapman D., Heithaus M. R., Kessel S. T. and Gruber S. H., 'Global catches, exploitation rates, and rebuilding options for sharks', Marine Policy - Vol. 40 (2013), 194-204.

Wright A., Stacey N. and Holland P., 'The cooperative framework for ocean and coastal management in the Pacific Islands: Effectiveness, constraints and future direction', Ocean & Coastal Management - Vol. 49(9-10) (2006), 739-763.

WWF, WWF Position – 9th Regular Session of the Western and Central Pacific Fisheries Commission (WCPFC) (Manila: WWF 2012).

Yesaki M., 'Observations on the biology of yellowfin (Thunnus albacares) and skipjack (Katsuwonus pelamis) tunas in Philippine waters', FAO./UN D P. Indo-Pac. Tuna Dev. Mgt.Programme (Rome: FAO 1983).

Zhu J., Chen Y., Dai X., Harley S. J., Hoyle S. D., Maunder M. N. and Aires-da-Silva A. M., 'Implications of uncertainty in the spawner–recruitment relationship for fisheries management: An illustration using bigeye tuna (Thunnus obesus) in the eastern Pacific Ocean', Fisheries Research - Vol. 119-120 (2012), 89-93.